Springer Series in
SOLID-STATE SCIENCES 125

Springer
*Berlin
Heidelberg
New York
Barcelona
Hong Kong
London
Milan
Paris
Singapore
Tokyo*

Springer Series in
SOLID-STATE SCIENCES

Series Editors:
M. Cardona   P. Fulde   K. von Klitzing   R. Merlin   H.-J. Queisser   H. Störmer

126 **Physical Properties of Quasicrystals**
    Editor: Z.M. Stadnik
127 **Positron Annihilation in Semiconductors**
    Defect Studies
    By R. Krause-Rehberg and H.S. Leipner

Volumes 1–125 are listed at the end of the book.

H. Fukuyama   N. Nagaosa   (Eds.)

Physics and Chemistry of Transition Metal Oxides

Proceedings of the
20th Taniguchi Symposium
Kashikojima, Japan, May 25–29, 1998

With 146 Figures and 8 Tables

Springer

Professor Hidetoshi Fukuyama
The University of Tokyo, Department of Physics
Tokyo 113, Japan
e-mail: fukuyama@phys.s.u-tokyo.ac.jp

Professor Naoto Nagaosa
The University of Tokyo, Department of Applied Physics
Tokyo 113, Japan
e-mail: nagaosa@appi.t.u-tokyo.ac.jp

Series Editors:

Professor Dr., Dres. h. c. Manuel Cardona
Professor Dr., Dres. h. c. Peter Fulde*
Professor Dr., Dres. h. c. Klaus von Klitzing
Professor Dr., Dres. h. c. Hans-Joachim Queisser
Max-Planck-Institut für Festkörperforschung, Heisenbergstrasse 1, D-70569 Stuttgart, Germany
* Max-Planck-Institut für Physik komplexer Systeme, Nöthnitzer Strasse 38
  D-01187 Dresden, Germany

Professor Dr. Roberto Merlin
Department of Physics, 5000 East University, University of Michigan
Ann Arbor, MI 48109-1120, USA

Professor Dr. Horst Störmer
Dept. Phys. and Dept. Appl. Physics, Columbia University, New York, NY 10023 and
Bell Labs., Lucent Technologies, Murray Hill, NJ 07974, USA

ISSN 0171-1873

ISBN 3-540-65187-X Springer-Verlag Berlin Heidelberg New York

Library of Congress Cataloging-in-Publication Data applied for.

Die Deutsche Bibliothek – CIP-Einheitsaufnahme
Physics and chemistry of transition metal oxides: proceedings of the 20th Taniguchi Symposium, Kashikojima, Japan, May 25–29, 1998 / H. Fukuyama; N. Nagaosa. –
Berlin; Heidelberg; New York; Barcelona; Budapest; Hong Kong; London; Milan; Paris; Singapore; Tokyo: Springer, 1999
(Springer series in solid-state sciences; 125)
ISBN 3-540-65187-X

This work is subject to copyright. All rights are reserved, whether the whole or part of the material is concerned, specifically the rights of translation, reprinting, reuse of illustrations, recitation, broadcasting, reproduction on microfilm or in any other way, and storage in data banks. Duplication of this publication or parts thereof is permitted only under the provisions of the German Copyright Law of September 9, 1965, in its current version, and permission for use must always be obtained from Springer-Verlag. Violations are liable for prosecution under the German Copyright Law.

© Springer-Verlag Berlin Heidelberg 1999
Printed in Germany

The use of general descriptive names, registered names, trademarks, etc. in this publication does not imply, even in the absence of a specific statement, that such names are exempt from the relevant protective laws and regulations and therefore free for general use.

Typesetting: camera-ready copies by the authors
Cover concept: eStudio Calamar Steinen
Cover production: *design & production* GmbH, Heidelberg

SPIN: 10685399   57/3142 - 5 4 3 2 1 0 – Printed on acid-free paper

Preface

This volume contains papers presented at the Twentieth Taniguchi Symposium on the Theory of Condensed Matter, which was held for May 25–29, 1998, at Kashikojima, Ise, Japan. This Symposium had focused on " Physics and Chemistry of Transition Metal Oxides ". Experimental and theoretical studies on transition metal oxides have a long history both in physics and chemistry. However studies on this family of materials have undergone revolutionary developments after the discovery in 1986 of high-temperature superconductivity in copper oxides, which had disclosed in a very clear way the importance of the fine control of carrier density. Through the high T_c cuprates a new cocept of "Doped Mott Insulators" has emerged. This is in contrast to the conventional "Doped Band Insulators".

In the former the transport properties are essentially entangled with the magnetism, which is not the case in doped semiconductors. For many years such an entanglement between magnetism and transport properties had been noted experimentally in magnetic metals and other transition metal oxides but the lack of precise determination of the carrier density have prohibited the sound and microscopic understanding of the interesting phenomena observed experimentally. In the wake of high T_c cuprates the realization of the importance of the carrier density control has revived extensive and intensive studies on these transition metal oxides and the horizon of the understanding have expanded including the colossal magnetoresistance and the charge and orbital orderings. The results of recent studies in this direction are reported in this volume.

The present Symposium was the Grand Finale of the series, all of which had been financially supported by Taniguchi Foudations. It is to be remembered that the activities of this Foundations have been based on the unique philosophy of Mr. Toyosaburo Taniguchi. For the memories of us all, the Address of Welcome he gave at the Reception of the First Symposium held in 1978 is contained here. At the same time the Histories of the Taniguchi Symposium in the Division of Theory of Condensed Matter are also given by Professor Junjiro Kanamori, who has been steering this Division since the beginning.

On behalf of all the participants we deeply thank late Mr. Taniguchi and the Taniguchi Foundations for their generous support of many years, which contributed greatly to the mutual understandings among colleagues throughout the world.

Tokyo, September 1998
Hide Fukuyama
Naoto Nagaosa

Address of Welcome

Toyosaburo Taniguchi

I am very pleased and honored to be here this evening with so many distinguished guests and scholars from so many countries. I wish to extend a warm and sincere welcome to the participants in the First International Symposium, Division of Theory of Condensed Matter of the Taniguchi Foundation.

I sincerely feel that the need is greater than ever today for Japan to seriously consider how best we can strengthen and foster international understanding and friendship between nations, peoples and societies, and how best we can contribute towards the establishment of peace and prosperity in the world.

For more than twenty years, I have been supporting a seminar conference in the field of mathematics in which distinguished scholars from all over the world such as the gentlemen gathered here today, engage in free discussions which resulted, so I hear, in fruitful and satisfactory developments and growth on a global basis. The conception of this symposium is mostly unique, in that the participants, both Japanese and foreign, live together in community-living, albeit for a limited period of time.

I have heard from members of some of these study groups that this type of set-up, backgrounding the symposium, has helped to strengthen their ties of friendship and understanding with their colleagues on a more personal basis. With this knowledge in the background, I strongly determined to reorganize the Taniguchi Foundation only a few years ago. I decided to put in additional funds and strengthen the Foundation, and it is no exaggeration to say that what developed in the mathematics groups led me to make the determination and put the plan into execution. Fortunately, the Japanese Government has granted us permission to reorganize the Foundation.

In order to translate more effectively and intensively the objectives of the Foundation into action with the funds available, it becomes necessary, in my belief, to select those fields which are not necessarily in the limelight of popular interest, which means those fields which are low in funds. I would rather choose a project from modest unimpressive academic fields for the Foundation, than pick something that stands out in gaudy, gorgeous popular acclaim.

In particular, I would desire to see promising younger groups of scholars throughout the world assume an even more active role in our symposia, for, they have the greater potential to develop their abilities in adapting themselves to the changing international scene. With world-shaking events occurring daily, deep-rooted changes are taking place in every structure. It is really the young generation who will lead us into a tomorrow of promise. I am convinced that their experiences in the symposia could be instrumental in forging the bonds of mutual understanding between peoples, between societies and between nations.

I share your hopes that those present in the symposia will enjoy themselves and come to exchange their views openly, frankly and thoroughly in a spirit of mutual trust and understanding leading to the establishment of a lasting friendship for the benefits of all participants. I wish to add that the number of the participants is being preferably limited to about 15 or 20 persons rather than more people, because of my desire to see a situation of living and studying matters together in a close-knit community set-up. The more the participants increase in number, the more the substance of the symposium becomes vague and the perfume is gone, and I would beg your best understanding of our basic philosophies and thinking on the number of the participants.

The steady progress that the Foundation has made during such a short period of time, has been possible only thanks to the continuing support and cooperation of all concerned. Today, the Foundation extends its activities to nearly ten divisions, in addition to Theory of Condensed Matter, such as Mathematics, Business History, Medical History, Biophysics, Ethnology, Neurobiology in Vision, Brain Sciences as well. I am convinced that were my late father, the founder of this Foundation in 1929, to know of this, he would be rendered unbelievably happy.

It is my fervent hope and prayer that these symposia shall not be temporary in nature and shall continue on for ten years or more under the guidance of Mr.H.Shima, President of the Foundation, who is the former President of National Space Development Agency of Japan.

This reception should be a welcome relief from the tight symposium schedule, and I hope that you can relax and enjoy yourself at the table with all of us. This is probably the best season of the year to visit Kyoto, the 1200-year-old capital of Japan which is widely known as a veritable treasure house of cultural and historical interest in Japan. While I realize that your time in Kyoto-area is limited, I hope you will be enjoying yourself.

In conclusion, I would like to end these words of welcome with our earnest prayer for the great success of the symposium throughout all sessions, and also sincerely hope that you will have a very pleasant return flight to your country bearing heartwarming memories of your stay in Japan and with us.

Thank you.

Toyosaburo Taniguchi

It was a pleasure for me to hold the closing address at this, the 20th and final Taniguchi Symposium of Theoretical Solid-State Physics. I would like, here too, to express my heartfelt thanks to Mr. T. Taniguchi. Two decades ago, he had the inspiration to inaugurate this pioneering series of symposia, which have contributed so much to progress in solid-state theory and thereby helped to pave the way to one of the few Nobel Prizes awarded in this field.

Robert B. Laughlin

Taniguchi Symposium: History and Activities of the Division of Theory of Condensed Matter

Junjiro Kanamori
Chairman of the Steering Committee
Division of the Theory of Condensed Matter of the Taniguchi Symposium

The history and purpose of the Taniguchi symposium is well explained in the preceding address given by Mr. Toyosaburo Taniguchi on the occasion of the dinner party at the First Symposium of the Division of Theory of Condensed Matter in 1978. After supplementing it by adding a brief biography of Mr. Taniguchi, I would like to summarize the twenty years history of our division, since the present symposium is the last one of the series.

Mr. Toyosaburo Taniguchi passed away at the age of 94 in 1994. He established the Taniguchi Foundation with his inheritance in 1929 just after the demise of his father who was a successful businessman in the textile industry. His father's company and several other companies have been unified as the Toyobo Company since 1931, which keeps the top position in the Japanese textile industry to the present. Mr. Taniguchi who worked in Toyobo for his life became the president and subsequently the chairman of the board of directors in his own right. He was also the president of the Textile Industry Society of Japan when the textile problem was the main issue of the Japan-US. trade negotiation between the Prime Minister Sato of Japan and the President Nixon of US. Mr. Taniguchi was a rare character who spent his personal money unsparingly to promote basic sciences in Japan. In 1930s the Foundation donated the first cyclotron in Japan to Osaka University and later shared the construction budget for another cyclotron at Kyoto University with the Government evenly. Incidentally only three cyclotrons had existed in Japan before the Second World War, two at Osaka and Kyoto Universities and one at the Research Institute for Physics and Chemistry known as Riken. After the war, his foundation had stopped its activity because of inflation. However, he started helping mathematicians by his pocket money in 1956 for organizing a series of small size international symposia. He had two eminent mathematicians among his classmates at a gymnasium. As was mentioned in his address, he was deeply impressed by the free atmosphere of the meetings and reorganized the Taniguchi Foundation in 1976 to start similar symposia in various fields of sciences which counted 18 finally. He donated a land to the Foundation, which was equivalent to about 20 million dollars at that time. He adopted the unusual policy that the Foundation should spend the whole money in about ten years and dissolve. Actually the Foundation which is going to dissolve next year has lasted for about twenty years since then because the price of the land went up.

Mr. Taniguchi wanted to help people in those fields of science which were not in the limelight of popular interest, as was mentioned in his address. He did not like at all to seek publicity. Though he seldom mentioned his opinion on the management of symposia except for the limit of the number of participants, he once opposed explicitly to a press release of a symposium.

I met Mr. Taniguchi for the first time in 1977 to propose the Theory of Condensed Matter as candidate for an addition to the Taniguchi Symposia. Before this interview I had discussed about the proposal with Professors Ryogo Kubo and Takeo Matsubara. The interview itself was, however, my responsibility. After listening to my explanation for about an hour, he said at the end that he did not understand any detail but he could say OK. Then he added that your field seemed to be not so popular as particle physics. At the subsequent creation of the Division of the Theory of Condensed Matter, we organized the steering committee whose members were R.Kubo (Chairman), T.Matsubara, Kei Yosida, Eiichi Hanamura and J.Kanamori (Secretary and Chairman after Kubo's demise). Hidetoshi Fukuyama and Ayao Okiji joined the committee since 1993. Each year we discussed the scope of the symposium for next year and selected a person to take up the responsibility for organizing it. The past symposia are listed below. The list reflects the developments in our field in the past twenty years. However, we note that we have selected those topics which were suitable for intensive discussions in a symposium of a small size. For example we deferred the high temperature superconductivity to conferences of other styles when it was in a fever. All symposia were quite successful at least in promoting mutual understanding and friendship particularly among young people. I can mention also that many cases of collaboration were born after the symposia. Many participants expressed his appreciation afterwards and I may say that the Taniguchi Symposium achieved international fame.

I would like to mention the fact that the late Professor Kubo had not only led the discussion of the steering committee, but also participated actively in most of the symposia until 1992. Since then his health declined gradually and he passed away in 1995. I believe that his profound and penetrating understanding of physics and his warm personality impressed most of participants of the symposia.

Finally I would like to express our sincere thanks to Mr. Toyosaburo Taniguchi and the Taniguchi Foundation for giving us these rare opportunities of fostering friendship and collaboration on behalf of the steering committee and all the participants of the symposia. I hope that the symposia will make long-standing contributions to the progress of our field in the coming century.

The Taniguchi Symposia in the Division of Theory of Condensed Matter

The year and title of each symposium are listed together with the name of the person who represented the organizing body. Springer Verlag published the Proceedings of the symposia except for the first one, which was not published, being a compilation of preprints.

1978	The d states in transition metals and alloys (J.Kanamori).
1979	Relaxation of elementary excitations (E.Hanamura).
1980	Electron correlation and magnetism in narrow-band systems (T.Moriya).
1981	Anderson Localization (Y.Nagaoka).
1982	Topological disorder in condensed matter (T.Ninomiya).
1983	Superconductivity in magnetic and exotic materials (T.Matsubara).
1984	Dynamical processes and ordering on solid surfaces (A.Yoshimori).
1985	Theory of heavy fermions and valence fluctuations (T.Kasuya).
1986	Quantum Monte Carlo methods in equilibrium and nonequilibrium systems (M.Suzuki).
1987	Core-level spectroscopy in condensed systems (J.Kanamori).
1988	Space-time organization in macromolecular fluids (F.Tanaka).
1989	Quasicrystals (T.Fujiwara).
1990	Molecular dynamics simulations (F.Yonezawa).
1991	Transport phenomena in mesoscopic systems (H.Fukuyama).
1992	Interatomic potential and structural stability (K.Terakura).
1993	Correlation effects in low-dimensional electron systems (A.Okiji).
1994	Spectroscopy of Mott insulators and correlated metals (A.Fujimori).
1995	Elementary processes in excitations and reactions on solid surfaces (A.Okiji).
1996	Relaxations of excited states and photo-induced structural phase transitions (K.Nasu).
1997	no symposium
1998	Physics and chemistry of transition metal oxides (H.Fukuyama).

The 20th Taniguchi International Symposium, Shima, Kanko Hotel, May 25-29, 1998

Fourth row (left to right): B. Keimer, A. Fujimori, K. Miyake, S.-C. Zhang, N. Nagaosa
Third row: S. Ishihara, Y. Maeno, Y. Tokura, Y. Kitaoka, K. Torakura, M. Sato, M. Takano, K. Yamada, H. Shiba
Second row: S. Uchida, Z.-X. Shen, M. Ogata, M. Sigrist, E. Hanamura, N.P. Ong, J. Akimitsu, M. Imada, Y. Endoh, S. Maekawa
First row: J.C. Campuzano, T.M. Rice, P.A. Lee, J.B. Goodenough, J. Kanamori, H. Fukuyama, R.B. Laughlin, G. Sawatzky, H. Lotsch

Contents

Part I	Introduction

Introduction
By J. Kanamori, H. Fukuyama, and N. Nagaosa 3

Part II	Overviews

Localized to Itinerant Electronic Transitions in Perovskite-Related Structures
By J.B. Goodenough and J.-S. Zhou 9

Crystal Distortions in Magnetic Compounds
By J. Kanamori (With 2 Figures) 19

Present Status of the First-Principles Electronic Structure Calculations
for the Strongly Correlated Transition-Metal Oxides
By K. Terakura, Z. Fang, and I.V. Solovyev (With 6 Figures) 34

Aspects of Coupled Spin-Orbital Degrees of Freedom
in d- and f-Electron Systems
By H. Shiba and R. Shiina (With 1 Figure) 45

Part III	Manganites

Anisotropic Charge Dynamics in Layered Manganite Crystals
By Y. Tokura, T. Kimura, and T. Ishikawa (With 5 Figures) 55

Roles of Orbitals in Transition Metal Oxides
By Y. Endoh, K. Hirota, Y. Murakami, H. Nojiri, S. Ishihara, S. Maekawa,
H. Kimura, T. Fukuda, and N. Okamoto (With 7 Figures) 69

Theory of Anomalous X-ray Scattering
in a Variety of Orbital Ordered Manganites
By S. Ishihara and S. Maekawa (With 5 Figures) 84

Nonlinear Optical Responses of Strongly Correlated Electronic System
By E. Hanamura, Y. Tanabe, and M. Fiebig (With 6 Figures) 95

Part IV High T_c Cuprates

Fermi-Liquid Versus Pseudogap Behaviors
in Filling-Control Transition-Metal Oxides
By A. Fujimori, T. Yoshida, A. Ino, T. Mizokawa, C. Kim, Z.-X. Shen, Y. Taguchi,
T. Katsufuji, Y. Tokura, H. Eisaki, S. Uchida, and K. Kishio (With 6 Figures) ... 111

Metal-Insulator and Superconductor-Insulator Transitions
in Correlated Electron Systems
By M. Imada, F.F. Assaad, H. Tsunetsugu, and Y. Motome (With 2 Figures) 120

High Temperature Superconductors as a Member of Transition Metal Oxides
By S. Maekawa (With 7 Figures) 136

Electronic Structure of Underdoped Superconductors
and Related Mott Insulators
By Z.-X. Shen (With 9 Figures) 144

Destruction of the Fermi Surface in Underdoped Cuprates
By J.C. Campuzano, H. Ding, M.R. Norman, and M. Randeria
(With 12 Figures) 152

Effects of Stripe Order on Charge Dynamics in La-Cuprates
By S. Uchida, N. Ichikawa, T. Noda, and H. Eisaki (With 9 Figures) 163

Spin Excitations in Pure and Zn-Substituted $YBa_2Cu_3O_{6+x}$
By B. Keimer (With 9 Figures) 173

Static and Dynamical Incommensurate Spin Correlations in High-T_c Cuprates
of $La_{2-x}Sr_xCuO_{4+y}$
By K. Yamada, R.J. Birgeneau, Y. Endoh, T. Fukase, M. Greven, M. Fujita,
K. Hirota, S. Hosoya, M.A. Kastner, Y.M. Kim, H. Kimura, K. Kurahashi,
C.H. Lee, S.H. Lee, Y.S. Lee, H. Matsushita, G. Shirane, T. Suzuki, S. Ueki,
and S. Wakimoto (With 3 Figures) 182

Experimental Studies on Singlet Formation in High-T_c Oxides
and Quantum Spin Systems
By M. Sato (With 7 Figures) 192

Thermal Conductivity as a Probe of Quasi-Particles in the Cuprates
By N.P. Ong, K. Krishana, Y. Zhang, and Z.A. Xu (With 4 Figures) 202

A New-Type Superconducting State Locally Induced Around a Vortex
in the Two-Dimensional t-J Model
By M. Ogata (With 4 Figures) 212

Instability of a 2-Dimensional Landau-Fermi Liquid Due to Umklapp Scattering
By N. Furukawa and T.M. Rice (With 5 Figures) 221

Antiferromagnetism, Singlet and Disorder
By H. Fukuyama and H. Kohno (With 6 Figures) 231

The Spin Gap and Superconducting States of Underdoped Cuprates
By P.A. Lee (With 1 Figure) 241

Gauge Field, Confinement, and Superconductivity
in Underdoped High-T_c Cuprates
By N. Nagaosa (With 1 Figure) 250

A Progress Report on the SO(5) Theory of High T_c Superconductivity
By S.-C. Zhang (With 1 Figure) 260

Nested Spin-Fluctuation Scenario for Anomalous Metallic
Phase of High-T_c Cuprates
By K. Miyake and O. Narikiyo (With 6 Figures) 267

Part V New Materials: Ladders, Ruthenates and Others

Novel Transition Metal Oxides Prepared at High Pressure
and Their Electronic Properties
By M. Takano, Z. Hiroi, M. Azuma, S. Kawasaki, R. Kanno, and T. Takeda
(With 7 Figures) 279

Spin and Charge Dynamics in the Hole-Doped Two-Leg Ladder Compound
$Sr_{14-x}Ca_xCu_{24}O_{41}$
By J. Akimitsu, T. Nagata, H. Fujino, N. Motoyama, H. Eisaki, S. Uchida,
H. Takahashi, T. Nakanishi, N. Môri, M. Nishi, K. Kakurai, S. Katano, M. Hiroi,
M. Sera, and N. Kobayashi (With 7 Figures) 289

NMR Probe of Magnetic Order and Spin Correlation
in Hole-Doped Ladder Cuprates
By Y. Kitaoka, T. Mito, S. Ohsugi, K. Magishi, S. Matsumoto, T. Nagata,
J. Akimitsu, N. Motoyama, H. Eisaki, and S. Uchida (With 10 Figures) 299

Metal-Insulator and Magnetic Transitions in Layered Ruthenates
By Y. Maeno, S. Nakatsuji, and S. Ikeda (With 5 Figures) 313

Spin Triplet Superconductivity in Sr_2RuO_4 – A New Territory
for Experimentalists and Theorists
By M. Sigrist, C. Honerkamp, D. Agterberg, T.M. Rice, M.E. Zhitomirsky,
and A. Furusaki (With 1 Figure) 323

Concluding Remarks
By R.B. Laughlin (With 1 Figure) 332

List of Contributors 335

Part I

Introduction

Introduction

J. Kanamori[1], H. Fukuyama[2], and N. Nagaosa[2]

[1] 3-1-1 Hyakurakuen, Nara, Japan
[2] Department of Applied Physics, University of Tokyo, 7-3-1 Hongo, Bunkyo, Tokyo, Japan

Transition-metal oxides show rich variety of phenomena, e.g., Mott transition, high-T_c superconductivity, ferromagnetism, antiferromagnetism, low-spin/high-spin transitions, ferroelectricity, antiferroelectricity, colossal magnetoresistance, charge ordering, and bipolaron formation. These cover almost all the interesting phenomena in condensed matter physics/chemistry. The main actors in these phenomena are the d-electrons of the transition-metal ions surrounded by oxygen ions. Nominally the oxygen valency is O^{2-} attracting two electrons, and the s-electrons of transition-metal atom are removed. This is the reason why the d-electrons play the major role in transition-metal oxides compared with the transition metals.

The d-electrons have internal degrees of freedom, i.e., the spin and the orbital, in addition to the charge. These degrees of freedom give rise to the rich behavior listed above. The d-orbitals extends to reach the oxygen ions and are subject to the crystal fields. This gives rise to the splitting of the d-orbitals. In the octahedral symmetry, which corresponds to the three-dimensional perovskite structure, the five orbitals are divided into two e_g-orbitals ($x^2 - y^2, 3z^2 - r^2$) and three t_{2g}-orbitals (xy, yz, zx). When the symmetry is reduced, there occurs further splitting. For example, in the two-dimensional perovskite structure, the degeneracy of the e_g-orbitals is lifted and the $x^2 - y^2$ orbital is highest in energy. In high-T_c cuprates, only this $x^2 - y^2$ orbital is believed to be relevant for low-energy phenomena. However, in the general cases, the mutual interactions among the spin, orbital and charge determine the physical properties of the system.

These subjects associated with transition-metal oxides are interesting from both points of view of physics and chemistry. Actually the studies on this family of matrerials have a long history in both disciplines. However, the discovery of high-T_c cuprates in 1986 had not only demonstrated the diverse possibilities hidden in materials around us but also opened new ways to revisit the transition-metal oxides in general, which have led to various new findings. The results of these recent findings have been discussed in this Taniguchi Symposium.

The d-electrons are ambivalent having both localized and itinerant features (Pt. II.1). Therefore two approaches, i.e., from the atomic localized electron picture and from the band calculation, are possible. Here the first-principle band calculations give important and useful information, and hence can be the starting point of the discussion (Pt. II.2). Based on this, the interaction between electrons and electron-phonon can be taken into account. Especially when the degeneracy of the orbitals is present, the lattice distortion which reduces the symmetry and lifts the degeneracy stabilizes the electronic system. This is called Jahn-Teller instability, which was first studied for molecules. In solids, this instability leads to the collective structural phase transition called the collective Jahn-Teller effect (Pt. II.3). On the other hand, the electron-electron interaction is the origin of the Mott insulator, an insulator with odd integer electrons per transition-metal ion due to the repulsive force between electrons. This insulator is distinct from the band insulator because the single-particle band picture would give metallic behavior in such cases. Therefore, we need the many-body treatment of this insulator, which is one of the main issues in the present volume. One important feature here is that local spin moments, and sometimes local orbital polarization, are induced, which are coupled with the exchange interactions and lead to various magnetic and orbital orderings (Pts. II.4 and III.1–3). Furthermore, carrier doping into this Mott insulator has become possible, and the concept of a doped Mott insulator is now playing the central role in the physics and chemistry of transtion-metal oxides. These doped carriers are many-body objects different from those in the band insulators, e.g., realized in semiconductors.

One of the most interesting materials today is the manganese oxides. These materials are characterized by a large Hund coupling between the e_g- and t_{2g}-electrons, which gives rise to the double-exchange interaction. This interaction comes from the fact that the effective transfer integrals of electrons between the two sites is multiplied by the overlap integral of the spin wavefunctions of these two sites. To gain the kinetic energy, the parallel spin alignment is preferable, and this interaction is ferromagnetic. Therefore, the strong interplay between magnetism and transport is expected, as actuallly seen in (Pt. III.1,2), for example, the colossal magnetoresistance, and also the optical properties (Pt. III.4).

Of course, the most remarkable phenomenon in transition-metal oxides is the high-temperature superconductivity realized in cuprates with d^9 configurations without orbital degeneracy. Because the high-T_c superconductivity occurs near the Mott insulator, i.e., with a rather small doping concentration $x \approx 0.2$, the understanding of the Mott transition and the anomalous metallic state next to it are considered to be essential. In order to avoid the strong Coulomb interaction, the pairing symmetry of the Cooper pair is $d_{x^2-y^2}$. There occurs also the competition/interplay between the antiferromagnetism and the d-wave superconductivity.

Many experimental means, i.e., transport (Pt. IV.6, 9 and 10), optical spectra (Pt. IV.6), photoemission spectroscopy (Pt. IV.1, 4 and 5), neutron scattering (Pt. IV.7,8), and NMR, have been employed to clarify the properties of the high-T_c cuprates, most of which are reported in this volume. Typically there are two pictures for the normal state, one is the hole carrier doped into the Mott insulator and the other is the metallic electrons with a large Fermi surface. These two aspects appear in different ways depending on the physical quantity one is looking at. Another important feature is that the physical properties change as functions of the doping concentration x and temperatures. The superconducting transition temperature T_c has a maximum at $x = x_c \cong 0.2$, and $x < x_c$ is called the underdoped region, $x \cong x_c$ the optimally doped region, and $x > x_c$ the overdoped region. In the normal state of the underdoped region, the spin gap or pseudogap phenomenon has been observed in neutron scattering, NMR, optical spectra, photoemission spectroscopy, specific heat, etc.; it can be understood as due to the spin-singlet formation.

Several theoretical concepts and methods have been proposed to describe this strongly correlated state. One sound approach is the numerical one using computers. The quantum Monte Carlo (Pt. IV.2) and exact diagonalization (Pt. IV.3) are the two representative ones. The analytical approaches are closely connected to physical intuition and require some approximations. One is to take into account the AF spin fluctuation based on the Fermi-liquid theory (Pt. IV.17). On the other hand, the spin-charge separation, first proposed by P.W. Anderson, has been the subject of intensive studies for many years (Pt. IV.13–15). In this idea, the two kinds of the quasi-particles are introduced, i.e., spinons and holons, to describe the magnetic and transport properties, respectively. Another approach is to divide the Fermi surface into several parts, and the renormalization-group equation is constructed (Pt. IV.12). A symmetry-based theory has been also proposed based on the SO(5) group, which tries to unify the d-wave superconductivithy and antiferromagnetism (Pt. IV.16). The proximity of these two states becomes apparent also in the nonuniform situation as in vortices (Pt. IV.11).

There are several other interesting compounds discussed in the present volume (Pt. V.1). One is the ladder compounds of cuprates, where the superconductivity has recently been discovered (Pt. V.2,3). In the two-leg spin ladder system, it is now known that the finite gap occurs in the triplet excitation in contrast to the $S = 1/2$ 1D Heisenberg antiferromagnet. Then, it is expected that the doped spin ladder has some relevance to the underdoped high-T_c cuprates. Another remarkable recent finding is the p-wave superconductivity in $SrRuO_4$ (Pt. V.4,5), where the $4d$ electrons in t_{2g}-orbitals contribute to the Fermi surface and also superconductivity. This is the first example of the 2D p-wave superconductivity, where the time-reversal symmetry is broken.

We believe these interesting findings are just the beginning of even more surprises oxides will show us in years to come.

Part II

Overviews

Localized to Itinerant Electronic Transitions in Perovskite-Related Structures

J.B. Goodenough and J.-S. Zhou

Texas Materials Institute, ETC 9.102
University of Texas at Austin, Austin, TX 78712-1063

Abstract. Experiments on several perovskite-related transition-metal oxides at the cross-over from localized to itinerant electronic behavior reveal strong electron coupling not only to static, but also to dynamic oxygen displacements. A discontinuous change in the mean kinetic energy of the electrons results in a first-order transition at the cross-over and a breakdown of the Brinkman-Rice strong-correlation model before long-range magnetic order is stabilized. The superconductive copper oxides appear to stabilize a distinguishable thermodynamic state below a $T_\ell \approx 300$ K where a multicenter polaron gas condenses into a polaron liquid that progressively orders into mobile stripes as the temperature is lowered to T_c. It is suggested that itinerant vibronic states are formed as a result of coupling of itinerant electrons to dynamic oxygen displacements along the Cu-O bond axes in the mobile stripes. Such vibronic states would be responsible for a remarkable transfer of spectral weight to states propagating parallel to the bond axes. The transfer of spectral weight is responsible for the anisotropy of the superconductive gap and an enhancement of the thermoelectric power in the interval $T_c < T < T_\ell$ that is uniquely associated with the high-T_c superconductive phenomenon.

Introduction

Transition-metal oxides with perovskite-related structures provide an opportunity to study the transition from localized to itinerant electronic behavior in a three-dimensional (3D) MO_3 array or a 2D MO_2 sheet of corner-shared octahedra in which the dominant interactions of interest are $(180° - \phi)$ M-O-M interactions between $3d^n$ configurations on the transition-metal atoms M. Transitions in single-valent systems need to be distinguished from those occurring in mixed-valent MO_3 arrays or MO_2 sheets. In this brief paper, attention is focused on evidence for dynamic electron-lattice interactions that give rise to isotropic ferromagnetic superexchange spin-spin coupling, "correlation fluctuations", or the formation of "itinerant vibronic states".

Perovskite Structure

The ideal structures of AMO_3 perovskites and related intergrowth oxides such as $AO \cdot AMO_3 = A_2MO_4$ are, respectively, cubic and tetragonal. In this paper, the larger A cation is a lanthanide, yttrium, or an alkaline earth and M is a first-long-period transition-metal atom. In each structure, the deviation from unity of a tolerance factor

$$t \equiv (A-O)/\sqrt{2}(M-O) \tag{1}$$

is a measure of the mismatch between equilibrium bond lengths (A - O) and (M - O) in the ideal structures. Equilibrium (A - O) and (M - O) bond lengths at ambient temperature and pressure may be calculated from the sums of tabulated empirical ionic radii obtained from x-ray-diffraction data [1, 2]. A larger thermal-expansion coefficient of the equilibrium (A - O) bond makes dt/dT > 0. Normally the equilibrium (A - O) bond is more compressible, which makes dt/dP < 0 [3]; but, as discussed below, a dt/dP > 0 is found at a cross-over from localized to itinerant electronic behavior.

At < 1 places the M - O bonds under a compressive stress, the A - O bonds under a tensile stress. These stresses are relieved by a cooperative rotation of the $MO_{6/2}$ octahedra that bends the M-O-M bond angle from 180° to (180° - ϕ). A cooperative rotation about the cubic [110] axis reduces the symmetry to orthorhombic (Pbnm) with $c/a > \sqrt{2}$. A cooperative rotation about the [111] axis would give rhombohedral $\bar{R}3c$ symmetry.

Cooperative oxygen displacements may be superimposed on the cooperative rotations. In $LaMnO_3$, for example, oxygen displacements reflect Jahn-Teller ordering of the occupied electron orbitals that changes the orthorhombic axial ratio to $c/a < \sqrt{2}$. Static, cooperative oxygen displacements are directly measurable with x-ray and neutron diffraction and have been identified in ferroelectric, Jahn-Teller, and disproportionation transitions as well as in charge-density-wave "stripes". On the other hand, direct measurement of dynamic, cooperative oxygen displacements requires a fast experimental probe; dynamic displacements are not revealed by a conventional diffraction experiment.

Electronic Considerations

Localized $3d^n$ configurations are described by crystal-field theory; itinerant 3d electrons by tight-binding band theory. Localized d-orbital wave functions of an M cation in an octahedral site may be formulated as

$$\psi_t = N_\pi(f_t - \lambda_\pi \phi_\pi) \quad \text{and} \quad \psi_e = N_\sigma(f_e - \lambda_s \phi_s - \lambda_\sigma \phi_\sigma) \tag{2}$$

where f_t and f_e are, respectively, the ionic threefold-degenerate t orbitals that only π-bond with the oxygen ligands and the twofold-degenerate e orbitals that only σ-bond with the oxygen ligands. The ϕ_π, ϕ_s, and ϕ_σ are appropriately symmetrized O -$2p_\pi$, 2s, and $2p_\sigma$ orbitals, and the covalent-mixing parameters $\lambda \equiv b^{ca}/\Delta E$ contain the cation-anion resonance integral $b^{ca} \equiv (f, H'\phi) \approx \varepsilon(f,\phi)$ and the energy ΔE corresponding to an O -$2p_\pi$, 2s, or $2p_\sigma$ electron transfer. The cubic-field splitting

$$\Delta_c = \Delta_M + (\lambda_\sigma^2 - \lambda_\pi^2)\Delta E_p + \lambda_s^2 \Delta E_s \tag{3}$$

contains a small electrostatic contribution Δ_M; the primary contribution comes from $\lambda_\sigma > \lambda_\pi > \lambda_s$.

The interatomic M-O-M interactions involve spin-dependent resonance integrals $b_{ij} \equiv (\psi_i, H'\psi_j) \approx \varepsilon(\psi_i, \psi_j)$. If the $3d^n$ configurations retain a localized spin, spin-dependent integrals

$$t_{ij}^{\uparrow\uparrow} = b_{ij}\cos(\theta_{ij}/2) \quad \text{or} \quad t_{ij}^{\uparrow\downarrow} = b_{ij}\sin(\theta_{ij}/2) \tag{4}$$

must be used, where θ_{ij} is the angle between localized spins on neighboring atoms M_i and M_j. The tight-binding bandwidth is $W \approx 2zt_{ij}$ where z is the number of like nearest-neighbor M atoms. A $W < U_{eff}$, where U_{eff} is the electron on-site electrostatic energy between $3d^n$ and $3d^{n+1}$ configurations, leaves localized $3d^n$ configurations; a $W > U_{eff}$ transforms the localized electrons to itinerant electrons occupying narrow π^* or σ^* bands.

Zaanen, Sawatzky, and Allen [4] have emphasized that a distinction should be made between systems where the energy gap is U_{eff} between $3d^n$ and $3d^{n+1}$ configurations and those where the gap is Δ between the top of the O-$2p^6$ bands and the empty $3d^{n+1}$ redox energy in an ionic model. At the cross-over from a U_{eff} to a Δ gap in the ionic model, the perturbation expansion defining the covalent-mixing parameters is no longer applicable to the basis wave functions associated with the holes; the holes need to be treated in molecular-orbital theory either within a polaronic complex or within a second phase. Nevertheless, they occupy antibonding states of d-orbital symmetry consisting of strongly hybridized O - 2p and M - 3d states. The system $La_{1-x}Sr_xMnO_3$ appears to represent a transition from localized to itinerant electronic behavior where the parent $LaMnO_3$ compound has a U_σ gap. The parent compounds of the superconductive copper oxides, on the other hand, have CuO_2 sheets with a Δ gap in the ionic model. Oxidation introduces holes into $x^2 - y^2$ molecular orbitals that, in the overdoped compositions, become transformed into itinerant-electron states of an antibonding $x^2 - y^2$ band.

Since $U_\pi > U_\sigma$ and $W_\pi < W_\sigma$, it is possible to have localized t^3 configurations in the presence of itinerant σ^* electrons. This situation arises in mixed-valent, ferromagnetic $La_{1-x}Sr_xMnO_3$ perovskites where the Hund intraatomic exchange field couples the itinerant σ^*- electron spins parallel to the localized t^3 configurations to give a tight-binding bandwidth

$$W_\sigma \approx 12 t_\sigma^{\uparrow\uparrow} \sim \varepsilon_\sigma \lambda_\sigma^2 \cos\phi <\cos(\theta_{ij}/2)> \tag{5}$$

We have discussed elsewhere [5] the origin of the colossal magnetoresistance (CMR) found in the manganese oxides; here we draw attention to the consequences of strong, dynamic electron-lattice interactions in some other oxides with perovskite-related structures.

Single-valent Systems

1. **$LaMn_{1-x}Ga_xO_3$**

Stoichiometric $LaMnO_3$ undergoes a static, cooperative Jahn-Teller deformation to the O'-orthohombic structure below a $T_t \approx 600°C$; below $T_N \approx 130$ K, it becomes a

Type A antiferromagnet with ferromagnetic superexchange in the *a-b* planes and antiferromagnetic coupling between planes [6]. An antisymmetric superexchange term cants the spins from collinear to give a weak ferromagnetic component. Substitution of Ga^{3+} for Mn^{3+} suppresses the static, cooperative Jahn-Teller deformation; and at the O' to O transition, the Mn - O - Mn interactions become isotropically ferromagnetic [7,8]. According to the rules for the sign of the superexchange interactions, this finding demonstrates that a dynamic Jahn-Teller coupling of the σ-bonding e electrons to the two optical-mode lattice vibrations of E_g symmetry creates local "vibronic" states that correlate a dominant electron transfer from half-filled e orbitals on one atom to an empty e orbital on the neighboring atoms. Mn^{3+}-ion clustering to minimize the elastic energy associated with dynamic Jahn-Teller interactions was also appreciated in the 1960's [9].

2. Strongly correlated $\pi^{*1}\sigma^{*0}$ systems

The Ti^{3+} and V^{4+} ions each contain a single $3d^1$ configuration, and $La_{1-x}Y_xTiO_3$ and $Sr_{1-x}Ca_xVO_3$ would represent spin-1/2 systems with a U_π gap were the $3d^1$ configurations localized. In fact, each system has strongly correlated, itinerant π* electrons at the threshold of a transition from localized to itinerant electronic behavior [10]. In these single-valent systems, the π* bands are narrowed by substitution of a smaller, more acidic A cation.

Inoue *et al* [11] have reported photoemission spectroscopy (PES) data showing the coexistence of coherent and incoherent states with a continuous shift of spectral weight from coherent to incoherent states as the π* band narrows with increasing x in $Sr_{1-x}Ca_xVO_3$. Nevertheless, the system remains an enhanced Pauli paramagnet for all x. A maximum in the π*-electron effective mass m* at an intermediate value of x was also found. In order to determine whether the maximum in m* is due to a perturbation of the periodic potential by dissimilar A cations or to correlation fluctuations implicit in the PES data, we measured the temperature dependence of the thermoelectric power α(T) for $CaVO_3$ under different hydrostatic pressures [12]. Our data showed an increase in α(300 K) with pressure whereas Pt showed the expected decrease in α(300 K). Moreover, pressure increased the low-temperature phonon-drag component of $CaVO_3$ but left that of Pt unchanged. These results confirmed that m* increases with W_π as a result of the coexistence of incoherent localized-state fluctuations coexisting with coherent itinerant-electron states. We concluded that the Brinkman-Rice relationship

$$m*/m = [1-(U/U_c)^2]^{-1} \qquad (6)$$

for strongly correlated itinerant electrons breaks down with the onset of incoherent-state fluctuations. As the on-site electrostatic energy U approaches the critical value U_c, the evolution of the electronic state in a perovskite-related structure is not continuous, as envisaged by Hubbard, but undergoes a first-order transition from itinerant to localized electronic behavior.

Stoichiometric $LaTiO_3$ is a Pauli paramagnetic metal above a $T_N \approx 140$ K and becomes a Type G antiferromagnet below T_N [13] without exhibiting the cooperative

Jahn-Teller deformation to be expected for a localized-electron collinear-spin antiferromagnet [14]. PES data [15] have shown for this system also the coexistence of incoherent and coherent electronic states, which has suggested to us [16] that the measured [17] pressure dependence $dT_N/dP > 0$ does not signal a localized-spin configuration [18], as was originally inferred [17], but a transfer of spectral weight from incoherent to coherent states and a T_N that increases with the density of states $N(\varepsilon_F)$ at the Fermi energy. This interpretation is consistent with an observed decrease in T_N with decreasing width of the π^* band as the smaller Y^{3+} ion is substituted for La^{3+} in $La_{1-x}Y_xTiO_3$. Significantly, as T_N decreases to zero with increasing x, a ferromagnetic Curie temperature T_C increases with x, reaching a $T_C = 30$ K in the ferromagnetic insulator $YTiO_3$ having a U_π gap between the $3d^1$ and $3d^2$ redox energies [19].

3. A $t^6\sigma^{*2}$ system $La_{1-x}Nd_xCuO_3$

The metallic perovskite system $La_{1-x}Nd_xCuO_3$ contains a half-filled σ^* band. In the absence of localized spins, $t_\sigma = b_\sigma$ and the factor $< \cos(\theta_{ij}/2) >$ becomes unity in equation (8). Introduction of a smaller Nd^{3+} ion increases ϕ, which decreases $\cos\phi$ and narrows W_σ, whereas pressure increases W_σ without introducing any additional perturbation of the periodic potential. We have measured $\alpha(T)$ under different hydrostatic pressures for x = 0.0, 0.25, and 0.50 [20]. In $LaCuO_3$, $\alpha(T)$ exhibited a negative phonon-drag component with a maximum at a $T_{max} = 70$ K, typical of a conventional metal. However, the phonon-drag component was enhanced by the application of pressure as in $CaVO_3$. The x = 0.25 sample showed not only enhancement with pressure of the phonon-drag component, but also an increase with pressure of $\alpha(300\text{ K})$ as in $CaVO_3$ and opposite to Pt, which is indicative of an increase in m* with pressure. A similar situation was found for the x = 0.50 sample, but the phonon-drag component was more strongly suppressed. These results indicate that the $La_{1-x}Nd_xCuO_3$ perovskite system has a narrow σ^* band with an electronic heterogeneity similar to that observed in $CaVO_3$ and with m* approaching a maximum value in $LaCuO_3$.

4. The $LnNiO_3$ Perovskites

The $LnNiO_3$ perovskite family has been prepared from Ln = La to Lu and Y. Whereas $LaNiO_3$ can be prepared easily at ambient pressure, the other members of the $LnNiO_3$ family were first obtained only with a high-pressure preparation. Moreover, $LaNiO_3$ has a low-spin configuration $t^6\sigma^{*1}$ and is a "bad metal" exhibiting an enhanced Pauli paramagnetism down to lowest temperatures [21]. $PrNiO_3$, on the other hand, undergoes a first-order transition to an antiferromagnetic insulator below a $T_{IM} = 130$ K; and as the size of the lanthanide ion Ln^{3+} decreases, T_{IM} increases to over 450 K in $EuTiO_3$ [22]. The antiferromagnetic phase has the larger volume, and pressure decreases sharply T_{IM}, i.e. $dt/dP > 0$, a $|dT_{IM}/dP| > 5$ K/kbar being reported [23] for $NdNiO_3$. The antiferromagnetic order is unusual; ferromagnetic (111) Ni planes having a nickel-atom moment $\mu_{Ni} \approx 0.8 \mu_B$ are alternately coupled ferromagnetically and antiferromagnetically [24]. This order can be rationalized if a charge-density wave is also stabilized in which ferromagnetic itinerant-electron Ni-O-Ni (111) blocks are coupled antiferromagnetically via Ni-O-Ni superexchange [25]. In this model, the oxygen within a ferromagnetic block transfer strongly O-$2p_\sigma$ electrons back to the low-spin Ni^{III} ions whereas the oxygen between blocks are more ionically bonded to their nickel near neighbors.

Medarde et al [26] have reported a decrease of 10 K in the T_{IM} of $PrNiO_3$ on the exchange of ^{18}O for ^{16}O. Pressure reduces the Ni-O bond length and hence the bond angle ϕ whereas $^{18}O/^{16}O$ exchange reduces the frequency $\omega_o \sim M_o^{-1/2}$ of the axial vibration of an oxygen atom of mass M_o, so we conclude that T_{IM} varies sensitively with an $\omega_o(\phi)$.

The stabilization of an antiferromagnetic phase below a T_{IM} in $PrNiO_3$ indicates that $LaNiO_3$ may, like $CaVO_3$ and $La_{1-x}Nd_xCuO_3$, be in the transitional region with strong-correlation fluctuations giving rise to an electronic heterogeneity. Preliminary measurements show that the phonon-drag component of $\alpha(T)$ for $LaNiO_3$ is suppressed, which supports this conjecture. Within this picture, stabilization of the charge-density wave in $PrNiO_3$ occurs where the concentration of strong-correlation fluctuations becomes large enough for them to condense into an ordered array.

The low-spin Ni^{III} ions would be strong Jahn-Teller ions like the high-spin Mn^{3+} ions of $LaMnO_3$ were the single e electron per Ni^{III} ion localized. Therefore, it is interesting to compare the $^{18}O/^{16}O$ isotope shift found in $PrNiO_3$ with that found for the Curie temperature T_c in the mixed-valent, ferromagnetic manganese oxides exhibiting a CMR [27, 28]. In our experiments [28], we chose an O'-orthorhombic sample $(La_{0.25}Nd_{0.75})_{0.7}Ca_{0.3}MnO_3$ that was at the threshold of the O' to O-orthorhombic transition and applied pressure to see how the $^{18}O/^{16}O$ isomer shift changes on going from the static to the dynamic Jahn-Teller regime. We found at $P=P_c$ a change from a second-order magnetic transition with no isotope shift of T_c in the O' phase to a first-order transition in the O phase. In the ^{16}O sample the O' to O transition occurs at a $P_c \approx 1.5$ kbar whereas in the ^{18}O sample it is shifted to a $P_c \approx 10.5$ k bar. Moreover, the isotope-shift parameter $\alpha_0 = d \ln T_c / d \ln M_o$, where M_o is the oxygen mass, has a maximum value of about 5 at 10.5 kbar and decreases with further increase in pressure. Zhao et al [27] also noted a decrease in α_0 with increasing T_c in the O-orthorhombic phase. Here also is evidence for strong electron coupling to dynamic, cooperative oxygen displacements associated with a first-order magnetic transition.

Equilibrium M-O Bond Lengths

Thus far, we have shown by both magnetic and isotope measurements that orbital degeneracies at strong Jahn-Teller ions result in electron coupling to dynamic, cooperative oxygen displacements. Moreover, the existence of an electronically heterogeneous region in which there is a transfer of spectral weight from incoherent to coherent states with increasing bandwidth implies a segregation into two phases via strong electron coupling to dynamic, cooperative oxygen displacements. In order to demonstrate that a first-order phase change, and hence phase segregation, may occur at the cross-over from localized to itinerant electronic behavior, we turn to the virial theorem for central force fields, which states

$$2<T> + <V> = 0 \tag{7}$$

If the mean kinetic energy <T> of the electronic system increases discontinuously at a cross-over from localized to itinerant electronic behavior, then the mean potential energy <V> must decrease discontinuously. For antibonding electrons, a discontinuous decrease in <V> means a discontinuous shortening of the equilibrium M - O bond length, which leads to a double-well potential at cross-over [29]. A double-well potential would be manifest by the following observations: (1) a first-order phase change at cross-over that results in a dynamic phase segregation into coherent and incoherent electronic states or the stabilization of a static charge-density wave and (2) an anomalously high compressibility of the localized-electron M - O bond at cross-over where pressure transfers spectral weight from incoherent to coherent states.

In the closing section, we apply to mixed-valent systems the consequences of the virial theorem and electron coupling to dynamic oxygen displacements as illustrated by the superconductive copper oxides.

Copper-Oxide Superconductors

The copper oxides have intergrowth structures in which CuO_2 sheets have a partially occupied, antibonding $x^2 - y^2$ band. The simplest superconductive system is $La_{2-x}Sr_xCuO_4$. The parent compound La_2CuO_4 is an antiferromagnetic insulator, but T_N drops precipitously to zero with increasing x, and superconductivity appears for x > 0.05. Three compositional ranges are generally recognized: an underdoped range $0.05 < x \leq 0.10$ in which the superconductive critical temperature T_c increases with x and the hole-rich superconductive phase segregates from the antiferromagnetic parent phase below about 240 K, an optimally doped range $0.10 < x \leq 0.20$ in which T_c reaches a maximum near x=0.15, and an overdoped range $0.20 < x < 0.30$ in which T_c drops, probably step-wise, to zero with increasing x and the charge carriers change from holes to electrons [30]. The phase segregation in the underdoped region occurs via clustering of oxygen interstitials in $La_2CuO_{4+\delta}$ where the interstitial oxygen atoms are mobile to well below room temperature [31], but in $La_{2-x}Sr_xCuO_4$ it can only occur by cooperative oxygen displacements. Investigation of stoichiometric $La_{2-x}Sr_xCuO_4$ has proven instructive since only the CuO_2 sheets are superconductive, the occupancy of the x^2-y^2 band is unambiguously determined by the value of x, and x can be varied over the entire range of superconductive compositions.

1. **Underdoped CuO_2 Sheets**

At low carrier concentrations, holes introduced into the (x^2-y^2) band are polaronic despite the fact that they occupy x^2-y^2 orbitals that can no longer be described by crystal-field theory. Normally, non-adiabatic polarons are confined to a single metal site even where the lowest unoccupied states of the $MO_{6/2}$ cluster must be described by molecular-orbital rather than crystal-field theory. However, in La_2CuO_4 a pseudo Jahn-Teller deformation may create at low-spin Cu^{III} ions two shorter and two longer Cu-O bonds within a CuO_2 sheet. The elastic energy of the local deformation is reduced by enlarging the polaron to include, for example, the four nearest-neighbor copper centers. Calculation [32] has predicted a single-hole polaron containing 6± 1 pseudo Jahn-Teller copper centers with a small motional enthalpy $\Delta H_m \approx 0.01$ eV because the multicenter polaron moves one center at a time via dynamic, cooperative oxygen-atom displacements. Experimentally, we have reported [33] a temperature-independent thermoelectric power above 240 K, which signals the presence of a non-adiabatic polaron gas described by the statistical term

$$\alpha(300 \text{ K}) = -\frac{k}{e} \ln[2(1-k_1 x)/k_1 x] \tag{8}$$

that includes the spin degeneracy. The parameter k_1 corresponds to the mean number of copper centers in a polaron; and in the range $0 < x \leq 0.10$, a good fit was obtained with $k_1 = 5.3$. Below 240 K we anticipate a dynamic segregation into antiferromagnetic and superconductive phases in the absence of a magnetic field. However, Baebinger et al [34] have shown an increase in the resistivity $\rho(T)$ with decreasing temperature below 50 K in single crystals of $La_{2-x}Sr_xCuO_4$ with $0.08 < x \leq 0.15$ in which superconductivity was suppressed by a magnetic field H = 60 T. It is tempting to interpret this result as a manifestation of polaronic motion in these high fields with a $\Delta H_m < kT$ only at lowest temperatures. The multicenter polarons can be considered a segregated hole-rich phase in which the holes occupy molecular orbitals within a matrix of localized electrons at Cu^{2+} ions.

2. Optimum carrier concentration at x_c

Phase segregation in the underdoped compositions indicates that the superconductive phase is a thermodynamically distinguishable phase [33]. Moreover, the pressure dependence of the tolerance factor, $dt/dP > 0$, which is manifest by an orthorhombic to tetragonal transition at a critical pressure P_c in the superconductive compositional range [30], indicates that the superconductive phase is stabilized at the cross-over from localized to itinerant electronic behavior. A $dT_c/dP > 0$ in the orthorhombic phase becomes $dT_c/dP = 0$ in the tetragonal phase, which shows that T_c is lowered by bending the Cu-O-Cu bond angle from 180°.

The ambient pressure thermoelectric power $\alpha(T)$ obtained [35] for the $x = 0.15$ sample shows two features: at temperatures $T > T_t \approx 300$ K, $\alpha(T)$ is nearly temperature-independent, which indicates the charge carriers form a polaron gas; in the interval $T_c < T < T_t$, $\alpha(T)$ exhibits an enhancement with a $T_{max} \approx 140$ K, which is too high for an acoustic-mode phonon drag. We [35] have interpreted this curve to signal the condensation of the polaron gas into a polaron liquid below T_t, but we did not specify the nature of the polaron liquid. However, it was pointed out [32] that the elastic energy of a polaronic liquid would be minimized by the formation of stripes at the time Bianconi et al [36] were first presenting evidence for stripes in their XAFS data.

Tranquada et al [37] have identified static stripes parallel to [100] and [010] axes in alternate CuO_2 sheets of a non-superconductive, low-temperature-tetragonal (LTT) phase $La_{0.88}Ba_{0.12}CuO_4$. In this phase, the stripes are pinned by cooperative rotations of the $CuO_{6/2}$ octahedra alternately about [100] and [010] axes of successive CuO_2 sheets. In the orthorhombic phase, rotations about [110] axes at 45° to the stripes does not pin the stripes. Pinning the stripes suppresses superconductivity; superconductivity is regained even in the LTT phase if the stripes become mobile [38].

We have shown [39] that the enhancement of $\alpha(T)$ below T_l with a $T_{max} \approx 140$ K is associated exclusively with the superconductive phase, which by inference means it is associated exclusively with the existence of mobile stripes ordered parallel to [100] and [010] axes in alternate CuO_2 sheets. Why this should be so is clarified by recent angle-resolved PES measurements by Norman *et al* [40] of the temperature dependence of the Fermi surface. As Coleman [41] has illustrated, the data show a dramatic transfer of spectral weight into directions of the Cu-O bonds in the CuO_2 sheets. Such a transfer of spectral eight requires a stabilization of itinerant-electron states propagating along the [100] and [010] axes. A strong coupling of itinerant-electron states to the optical phonon modes traveling parallel to the propagation vector $\mathbf{Q}$ of a mobile stripe would increase with greater ordering of the stripes on lowering the temperature below T_l. The electron-phonon coupling would create itinerant vibronic states having a stabilization energy proportional to $(\mathbf{k} \cdot \mathbf{Q})^2$ or, taking account of perpendicular stripes in alternate planes, to $(\cos\theta + \sin\theta)(\cos\theta - \sin\theta) \sim \cos 2\theta$, where θ is the angle between $\mathbf{k}$ and $\mathbf{Q}$ in a particular plane. An exceptional flatness of $\varepsilon_k(\mathbf{k})$ near the M point of the Brillouin zone was already noted by Dessau *et al* [42] in 1993, and we have emphasized that the enhancement in the $\alpha(T)$ below T_l indicates an increase in the asymmetry of the $\varepsilon_k(\mathbf{k})$ curve about ε_F according to the relation:

$$\alpha(T) = -\frac{k}{e} \int \frac{(\varepsilon - \varepsilon_F)}{kT} \frac{\sigma(\varepsilon)}{\sigma} d\varepsilon \qquad (9)$$

The increasing transfer of spectral weight with decreasing temperature accounts well for the increase in $\alpha(T)$ with decreasing temperature in the range $T_{max} < T < T_l$. We therefore conjecture that the decrease in $\alpha(T)$ with decreasing temperature in the range $T_c < T < T_{max}$ is due to the onset of vibronic Cooper-pairs from itinerant vibronic states.

We wish to thank the NSF and the Robert A. Welch Foundation, Houston, TX, for financial assistance.

References

1. R.D. Shannon and C.T. Prewitt, Acta Crystallogr. **B25**, 725 (1969)
2. R.D. Shannon and C.T. Prewitt, Acta Crystallogr. **B26**, 1046 (1970)
3. J.B. Goodenough, J.A. Kafalas, and J.M. Longo, in *Preparative Methods in Solid State Chemistry*, P. Hagenmuller, ed. (Academic Press, New York, 1972) Chap. 1
4. J. Zaanen. G.A. Sawatsky, and J.W. Allen, Phys. Rev. Lett. **55**, 418 (1985)
5. J.B. Goodenough and J.S. Zhou in MRS Symp. Proc. **474**, (1998)

6. E.O. Wollan and W.C. Koehler, Phys. Rev. **100**, 545 (1955)
7. J.B. Goodenough, A. Wald, R.J. Arnott, and N. Menyuk, Phys. Rev. **124**, 373 (1961)
8. J. Töpfer and J.B. Goodenough, Eu. J. Solid State & Inorg. Chem. **34**, 467 (1997)
9. J.B. Goodenough, J. Appl. Phys. **36**, 2342 (1965)
10. J.B. Goodenough, Appl. Phys. **39**, 403 (1968)
11. I.H,. Inoue et al, Phys. Rev. Lett. **74**, 2539 (1995)
12. J.-S. Zhou and J.B. Goodenough, Phys. Rev. B**54**, 13393 (1996)
13. Y. Okimoto et al, Phys. Rev. B**51**, 9581 (1995)
14. J.B. Goodenough, Phys. Rev. **171**. 466 (1968)
15. A. Fujimori et al, Phys. Rev. Lett. **69**, 1796 (1992)
16. J.B. Goodenough and J.-S. Zhou, Chem. Mater. (in press)
17. Y. Okada et al, Phys. Rev. B**48**, 9677 (1993)
18. J.B. Goodenough , Prog. Solid State Chem. **5**, 145 (1971)
19. K. Kumagai et al, Phys. Rev. B**48**, 7636 (1993)
20. J.-S. Zhou, W.B. Archibald, and J.B. Goodenough, Phys. Rev. B**57**, R2017 (1998)
21. J.B. Goodenough and P.M. Raccah, J. Appl. Phys. **36**, 1031 (1965)
22. J.B. Torrance et al, Phys. Rev. B**45**. 8209 (1992)
23. X. Obradors et al, Phys. Rev. B**47**, 12353 (1993); P.C. Cranfield, J.D. Thompson, S.W. Cheong, and L.W. Rupp, Phys. Rev. B**47**, 12357 (1997)
24. J.L. García-Muñoz, P. Lacorre, and R. Cywinski, Phys. Rev. B**51**, 15197 (1995); J. Rodríguez-Carvajal et al, Phys. Rev. B**57,** 456 (1998)
25. J.B. Goodenough, J. Solid State Chem. **127**, 126 (1996)
26. M. Medarde et al, Phys. Rev. Lett, **80**, 2397 (1998)
27. G.M. Zhao, K. Konder, H. Keller, and K.A. Müller, Nature **381**, 676 (1996)
28. J.-S. Zhou and J.B. Goodenough, Phys. Rev. Lett. **80**, 2665 (1998)
29. J.B. Goodenough, Ferroelectrics **130**, 77 (1992)
30. J.-S. Zhou, H. Chen, and J.B. Goodenough, Phys. Rev. B **49**, 9084 (1994)
31. J.C. Grenier et al, Physica C **202**, 209 (1992)
32. G.I. Bersuker and J.B. Goodenough, Physica C **274**, 267 (1997)
33. J.B. Goodenough, J.-S. Zhou, and J. Chan, Phys. Rev. B **47**, 5275 (1993)
34. G.S. Baebinger et al, Phys. Rev. Lett. **77**, 5427 (1996)
35. J.B. Goodenough and J.-S. Zhou, Phys. Rev. B **49**, 4251 (1994)
36. A. Bianconi et al, Phys. Rev. Lett. **76**, 3412 (1996); Phys. Rev. B **54,** 4310, 12018 (1996)
37. J.M. Tranquada et al, Nature **375**, 561 (1995); Phys. Rev. B**54**. 7489 (1996)
38. J.-S. Zhou and J.B. Goodenough, Phys. Rev. B**56**, 6288 (1997)
39. J.-S. Zhou and J.B. Goodenough, Phys. Rev. B**51**, 3104 (1995); Phys. Rev. Lett. **77**, 151 1190 (1996)
40. M.R. Norman et al, Nature **392**, 157 (1998)
41. P. Coleman, Nature **392**, 134 (1998)
42. D.S. Dessau et al, Phys. Rev. Lett **71**, 2781 (1993)

Crystal Distortions in Magnetic Compounds

Junjiro Kanamori

Professor Emeritus, Osaka University
3-1-1 Hyakurakuen, Nara 631-0024, Japan

Abstract. A theoretical overview of the cooperative Jahn-Teller effect in the insulating phase is given. We obtain an effective Hamiltonian of an interaction between the orbital states of the Jahn-Teller ions through a canonical transformation, which associates each electronic state with a local lattice distortion, and by use of the mean field approximation. The effective Hamiltonian yields a simple unified picture of cooperative distortions of various types. The competing effect of the spin-orbit coupling is discussed also. Electron itinerancy is briefly discussed at the end.

1. Introduction

I present a review of theoretical discussions about the cooperative effect of the Jahn-Teller coupling in the insulating phase of magnetic compounds of transition elements. We assume that 3d electrons are localized and subject to ligand field acting on the orbital states. We take into account the exchange interaction between neighboring magnetic ions and the spin-orbit coupling within an atom in the cases where these interactions are relevant to crystal distortions. We concentrate on the cases where magnetic ions have a doubly or triply degenerate orbital degeneracy in an undistorted environment.

I would like to lay an emphasis on the distinction between a cooperative lattice distortion and local distortions around a given ion, which will persist even at high temperatures above the transition. The discussion is based on our papers [1-4] published many years ago. Preceding our work, Wojtowicz [5] had treated the cooperative lattice distortion as an alignment of the tetragonal axes of distorted octahedrons or tetrahedrons of cation-anion complexes. The theory prefixed the magnitude of the local distortion and introduced an interaction between neighboring complexes ad hoc. We started, on the other hand, with a microscopic Hamiltonian of interactions between the electronic states and ion displacements. Our theory derives the local and bulk distortions as consequences of the interaction. In the discussion of the cooperative phenomena, we derive an effective Hamiltonian in which the electronic orbital states interact directly with bulk distortions. This effective Hamiltonian is obtained in the mean field approximation applied to an interaction between the electronic states of the Jahn-Teller ions. The interaction is obtained through a canonical transformation, which associates a local distortion with each electronic state of a given Jahn-Teller ion. This local distortion will persist even in the absence of a long range order. However, our theory has been misunderstood occasionally as neglecting the local distortions at all, probably because the initial argument was not quite transparent. In this paper we shall discuss the capability and limit of our old treatment in a more systematic way. A reviving interest in cooperative lattice distortions brought by recent exploitation of the remarkable properties of Mn-perovskites [6] would justify such an attempt.

When an orbital degeneracy is present, various types of cooperative phenomena appear, depending upon a relative magnitude of the Jahn-Teller coupling to competing and/or modifying interactions such as the spin-orbit coupling, the interatomic exchange interaction, the interatomic itinerancy, etc. After summarizing the Jahn-Teller effect of a single ion in section 2, we discuss in sections 3 and 4 the cooperative lattice distortion that takes place independently of the onset of a cooperative magnetism. We discuss in section 3 the case where all the ions are in the same electronic states. In section 4 we discuss more complicated cases bearing in mind Mn-perovskites. In section 5 we discuss the role of the spin-orbit coupling; the magnetostriction which is a lattice distortion accompanying a cooperative magnetism and dependent on the direction of magnetization vectors is expected in the case where the spin-orbit coupling dominates. A possible intermediate case where the magnetostriction might appear at a temperature lower than the Curie temperature of magnetic ordering will be discussed also. In the final section(§ 6) we discuss briefly the effect of itinerancy of d electrons.

2. Orbital degeneracy and the Jahn-Teller effect

The following is a summary of the Jahn-Teller effect on the degenerate states of a single transition element cation based on the theoretical discussions given by Van Vleck and many other people [7-9]. In a ligand field of cubic symmetry, the d levels are split into the doubly degenerate dγ orbitals, $d(x^2-y^2)$ and $d(2z^2-x^2-y^2)$ and the triply degenerate dε ones, $d(yz)$, $d(zx)$ and $d(xy)$. In the tetrahedral coordination of anions, the dγ level is lower in energy than the dε level, while the situation is reversed in the octahedral coordination. For more than one d electrons we can find easily the lowest energy configuration having an orbital degeneracy by filling the dε and dγ levels from the bottom with the high or low spin assumption. Table 1 shows the examples in the high spin case.

It is convenient to introduce the pseudo-angular momentum operators of magnitude 1/2, σ, for the subspace spanned by the doubly degenerate states and those of magnitude 1, ℓ, for that spanned by the triply degenerate states. Operators acting in the subspace can be expressed in terms of these operators. We express the Jahn-Teller coupling of the degenerate orbital states with the vibration modes corresponding to tetragonal and orthorhombic distortions Q_2 and Q_3 as

$$H_{JT} = gc^{1/2}(Q_2\Omega_2 + Q_3\Omega_3), \qquad (1)$$

where g is a coupling constant; c is the elastic constant defined by the expression of the potential energy for Q's given by $H_e = (c/2)(Q_2^2 + Q_3^2)$; and the operators Ω_2 and Ω_3 are defined by

$\Omega_2 = \sigma_x$, $\Omega_3 = \sigma_z$ and

$$\Omega_2 = (1/2^{1/2})(\ell_x^2 - \ell_y^2), \quad \Omega_3 = (1/6^{1/2})(2\ell_z^2 - \ell_x^2 - \ell_y^2) \qquad (2)$$

in the doubly and triply degenerate cases, respectively. The normal coordinates Q_2 and Q_3 are illustrated in Fig. 1.

H_{JT} can be diagonalized easily. In the triply degenerate case we find that the three wavefunctions ψ_i (i = x, y and z) satisfying $\ell_i\psi_i = 0$, respectively, are the eigenfunctions. The eigenvalues of Ω_2 for ψ_x, ψ_y, and ψ_z are $-1/2^{1/2}$, $1/2^{1/2}$, 0, respectively and those of Ω_3 for are $1/6^{1/2}$, $1/6^{1/2}$, $-2/6^{1/2}$. If the ground state energy of $H_e + H_{JT}$ obtained by use of these eigenvalues is minimized with respect to Q_2 and Q_3, we find that the lowest energy is given by $-g^2/3$ and the lowest energy state corresponds to a tetragonal volume conserving distortion with the tetragonal axis in the either x or y or z direction. The sign of the tetragonal distortion, either elongation or contraction along the principal axis depends on the sign of the coupling constant g, $g>0$ corresponding to elongation; g for Ni^{2+} at the tetrahedral site is of positive sign and that for Cu^{2+} at the same site of negative sign.

In the doubly degenerate case, we obtain as $\pm gc^{1/2}(Q_2^2 + Q_3^2)^{1/2}$ as the eigenvalues with eigenfunctions dependent on the ratio of Q_2 and Q_3. In this case any distortion corresponding to a point on the circle of the radius = $|g|/c^{1/2}$ in the $Q_3 - Q_2$ plane is of lowest energy. We can show, however, that higher order terms such as the anharmonic term cubic in Q's in the potential energy or a higher order coupling between σ's and quadratic terms of Q's favor distortions with an elongated tetragonal axis. [1] As is shown in Fig.2 it is convenient to express the distortion by a two-dimensional vector in the $Q_3 - Q_2$ plane in which the cubic symmetry manifests itself as a three-fold symmetry. The higher order terms mentioned above produce an anisotropy energy proportional to $\cos 3\theta$ in this plane.

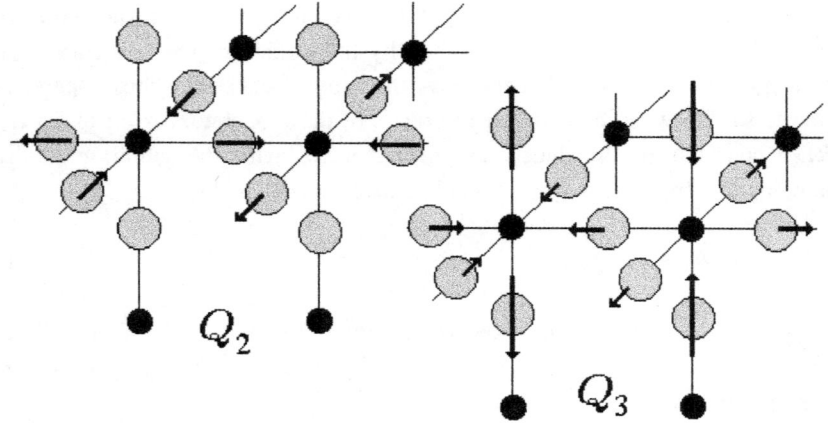

Fig. 1. The Q_2 and Q_3 modes with opposite phase at neighboring octahedrons on an idealized perovskite lattice. The small solid circles correspond to Mn atoms and the gray circles to O atoms in the case of $LaMnO_3$, La atoms being omitted.

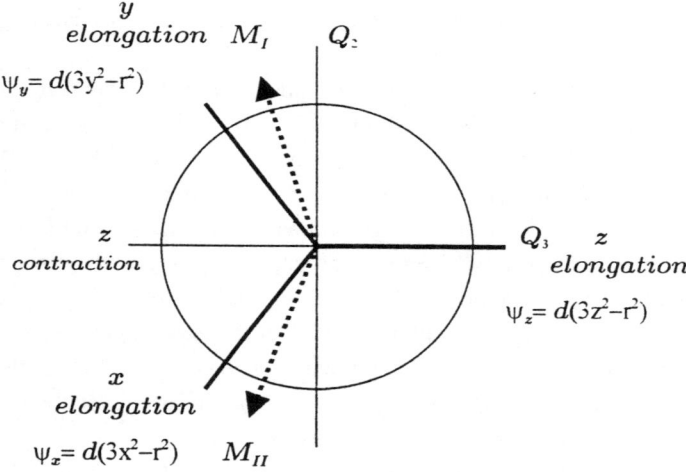

Fig. 2. Two-dimensional representation of Q_2, Q_3 and wave functions of the doubly degenerate level. The positive Q_3 direction corresponds to an elongation in the z direction of the real space (accompanied by contraction in the x and y directions to conserve the volume); the directions in the Q_3-Q_2 plane corresponding to elongation in the x and y directions are shown also in the figure.. These directions correspond to $\theta = 0$, $2\pi/3$, $4\pi/3$ of the angle defined by Eq.(13) given in section 3 which yield the wave functions Ψx, Ψy and Ψz favoring the elongation in the x,y and z directions. M_I and M_{II} represent the sublattice magnetization vectors discussed in section 4. The canting results in a contraction in the z direction.

3. The cooperative Jahn-Teller effect — The renormalized Hamiltonian —

$Mn^{2+}\underline{Mn^{3+}}_2O_4, \underline{Cu^{2+}}Fe^{3+}_2O_4$, $\underline{Ni^{2+}}Cr^{3+}_2O_4$ and $\underline{Cu^{2+}}Cr^{3+}_2O_4$ listed in Table 1 are typical examples which show a spontaneous crystal distortion from cubic to tetragonal through a first order phase transition. c/a is larger than 1 for the first two examples of the doubly degenerate case, while the tetragonal phase of $\underline{Ni^{2+}}Cr^{3+}_2O_4$ has c/a>1 and that of $\underline{Cu^{2+}}Cr^{3+}_2O_4$ c/a <1 in the triply degenerate case. In the latter case the opposite signs of 1-c/a in the Ni and Cu chromites is in accord with the signs of the single ion Jahn-Teller coupling constant g, suggesting a picture that tetragonally distorted cation-anion complexes align their principal axes below the transition temperature. We emphasize, however, that the local distortion around a Jahn-Teller ion should not be kept constant through the phase transition and the relation of the order parameter to the bulk crystal distortion must be clarified.

Our theory given in [1,2] answers these questions. We express Q's of each Jahn-Teller ion in terms of bulk strains and normal coordinates of lattice vibrations. Bulk strains are necessary if we assume the periodic boundary condition for the latter normal coordinates. In other words, the lattice constants appearing in the definition of the periodic boundary conditions should be regarded as variables expressed in terms of bulk strains. We define the normal coordinates for a fictitious crystal in which the Jahn-Teller effect is absent and uniformly distorted as prescribed by bulk strains. Then strictly speaking, the quantities associated with the normal modes are

Table 1 Configurations of orbital degeneracy in the high spin states.

Tetrahedral Coordination

Degeneracy	2	3*
Configurations	$d^1(d\gamma^1)$, $d^6(d\gamma_\uparrow^2 d\varepsilon_\uparrow^3 d\gamma_\downarrow^1)$	$d^3(d\gamma_\uparrow^2 d\varepsilon_\uparrow^1), d^4(d\gamma_\uparrow^2 d\varepsilon_\uparrow^2)$ $d^8(d\gamma_\uparrow^2 d\varepsilon_\uparrow^3 d\gamma_\uparrow^2 d\varepsilon_\uparrow^1), d^9$
Examples**	$\underline{Fe^{2+}}Cr^{3+}_2O_4$(spinel A site)	$\underline{Ni^{2+}}Cr^{3+}_2O_4, \underline{Cu^{2+}}Cr^{3+}_2O_4$

Octahedral Coordination

Degeneracy	2	3*
Configurations	$d^4(d\varepsilon_\uparrow^3 d\gamma_\uparrow^1)$, $d^9(d\gamma_\uparrow^2 d\varepsilon_\uparrow^3 d\gamma_\downarrow^1)$	$d^1(d\varepsilon^1), d^2(d\varepsilon_\uparrow^2)$ d^6, d^7
Examples**	La $\underline{Mn^{3+}}O_3$(perovskite) $Mn^{2+}\underline{Mn^{3+}}_2O_4, \underline{Cu^{2+}}Fe^{3+}_2O_4$(spinel B site)	$\underline{Fe^{2+}}O, \underline{Co^{2+}}O$(NaCl)

*For d^3, d^8 in the tetrahedral case and d^2, d^7 in the octahedral one, the indicated configurations are mixed with other states of the same symmetry to some extent by the coulomb interaction; for example, $d\gamma^2 d\varepsilon^1$ in the case of d^3 mixes with $d\gamma^1 d\varepsilon^2$.
** The ions with orbital degeneracy are underlined.

dependent on bulk strains. We retain, however, the lowest order terms only in the expansion in powers of bulk strains in the following discussion. If we try to calculate higher order terms such as the anharmonic energy and higher order couplings with the electronic states explicitly, we need to be careful about the definition of the normal coordinates.

For simplicity we assume the case where a unit cell contains one Jahn-Teller ion. The argument can be extended easily to the cases of more-than-one Jahn-Teller ions in a unit cell. [2] H_{JT} for the assembly of the Jahn-Teller ions is expressed as

$$H_{JT} = g_0 [V(c_{11}-c_{12})/N]^{1/2} (u_2 \Sigma_j \Omega_{2j} + u_3 \Sigma_j \Omega_{3j})$$
$$+\Sigma_{k,s}(h\nu_{ks}/N)^{1/2} q_{ks} \Sigma_j exp(i k \cdot R_j)(g_{ks2}\Omega_{2j}+g_{ks3}\Omega_{3j}), \qquad (3)$$

where V is the volume, N the number of unit cells, u_2 and u_3 are the bulk strains defined below, R_j denotes the positions of the Jahn-Teller ions, q_{ks} the normal coordinates with ks specifying the wave number and mode, $h\nu_{ks}$ the energy quantum of the mode. u_2 and u_3 are defined in terms of bulk strains e_{xx}, etc. as

$$u_2=(1/2^{1/2})(e_{xx}-e_{yy}) \quad \text{and} \quad u_3=(1/6^{1/2})(2e_{zz}-e_{xx}-e_{yy}). \qquad (4)$$

The coupling constants for the acoustic mode will show singular dependence on the direction of k, while g_{os2} and g_{os3} for the acoustic modes with $k=0$ vanish, since the modes corresponds to a uniform translation. Other features of the ks dependence of the coupling constants g_{ks2} and g_{ks3} will be discussed below. The potential energy associated with u's and q's is expressed as

$$H_e = V [(c_{11}-c_{12})/2](u_2^2+u_3^2) + \Sigma_{ks}(h\nu_{ks}/2) q_{ks}^* q_{ks}. \qquad (5)$$

3.1 Derivation of the effective Hamiltonian in the triply degenerate case

We discuss the triply degenerate case first, where the operators Ω_2 and Ω_3 are commutable with each other. In this case we can carry out a contact transformation to eliminate the Jahn-Teller coupling of the normal coordinates by shifting the origin of q's. The physical meaning of the transformation is to associate a local lattice distortion with each electronic state specified by ψ_i (i = x, y and z) satisfying $\ell_i \psi_i = 0$. We obtain a new expression of $H_e + H_{JT}$ as

$$H_e + H_{JT} = V(c_{11}-c_{12})(u_2^2+u_3^2)/2 + \Sigma_{ks}(h\nu_{ks}/2) q_{ks}^* q_{ks}$$

$$+ g_0 [V(c_{11}-c_{12})/N]^{1/2}(u_2 \Sigma_j \Omega_{2j} + u_3 \Sigma_j \Omega_{3j})$$

$$-(1/2N)\Sigma_{k,s}\Sigma_{jm} exp[i k \cdot (R_j-R_m)](g_{ks2}\Omega_{2j}+g_{ks3}\Omega_{3j})(g_{ks2}^*\Omega_{2m}+g_{ks3}^*\Omega_{3m}). \qquad (6)$$

The last term of Eq.(6) can be divided into the self-energy with $j = m$ and the interaction terms with $j \neq m$. The self energy is a constant, because of the symmetry relations $\Sigma_{k,s}g_{ks2}g_{ks3}{}^* = 0$, $\Sigma_{k,s}g_{ks2}g_{ks2}{}^* = \Sigma_{k,s}g_{ks3}g_{ks3}{}^*$ and $\Omega_2^2 + \Omega_3^2 = 2/3$. We drop the subscripts 2 and 3 in the sum for the self energy hereafter. The self energy represents an energy lowering arising from the local distortion associated with each electronic state. On the other hand, the interaction terms $j \neq m$ can produce various types of the ordering through the ks dependence of the coupling constants, g_{ks2} and g_{ks3}. In general, the interaction will be of a relatively short range, corresponding to a quadrupole or higher multipole interaction. However, the interaction with bulk strains gives rise to an effective interaction of infinite range, which justifies the mean field approximation. [10] In the following, we discuss first the case where the cooperative lattice distortion is 'ferromagnetic', i.e. the case where all the Jahn-Teller ions are in the same electronic state on the average in the distorted phase. In the mean field approximation, we assume

$$\Omega_j \Omega_m = \Omega_j \langle \Omega \rangle + \langle \Omega \rangle \Omega_m - \langle \Omega \rangle \langle \Omega \rangle, \tag{7}$$

with Ω standing for either Ω_2 or Ω_3 and $\langle \Omega \rangle$ representing a parameter independent of j; we shall conclude later that $\langle \Omega \rangle$ should coincide with the thermal average by minimizing the free energy with respect to it. Inserting Eq.(7) into the terms $j \neq m$ of the last line of r.h.s. of Eq.(6), we obtain the expression of the interaction in the mean field approximation given by

$$H_{int} =$$

$$-(1/2N)\Sigma_{k,s}\Sigma_m\Sigma_j' exp[i\boldsymbol{k}\cdot(\boldsymbol{R}_j-\boldsymbol{R}_m)](g_{ks2}\langle\Omega_2\rangle + g_{ks3}\langle\Omega_3\rangle)(g_{ks2}{}^*\Omega_{2m} + g_{ks3}{}^*\Omega_{3m})$$

$$- c.c. + (1/2N)\Sigma_{k,s}\Sigma_m\Sigma_j'$$

$$exp[i\boldsymbol{k}\cdot(\boldsymbol{R}_j-\boldsymbol{R}_m)](g_{ks2}\langle\Omega_2\rangle + g_{ks3}\langle\Omega_3\rangle)(g_{ks2}{}^*\langle\Omega_2\rangle + g_{ks3}{}^*\langle\Omega_3\rangle), \tag{8}$$

where Σ_j' represents the sum omitting the term $j = m$ and $- c.c.$ in the second line means the complex conjugate of the first term. We add and subtract the term $j = m$ in the sums, and note $\Sigma_j exp(i\boldsymbol{k}\cdot\boldsymbol{R}_j) = 0$ unless $k=0$ to obtain

$$H_{int} = -\Sigma_s g_{os}g_{os}{}^*(\langle\Omega_2\rangle\Sigma_j\Omega_{2j} + \langle\Omega_3\rangle\Sigma_j\Omega_{3j}) + (N/2)\Sigma_s g_{os}g_{os}{}^*(\langle\Omega_2\rangle^2 + \langle\Omega_3\rangle^2)$$

$$+ (1/N)\Sigma_{k,s}g_{ks}g_{ks}{}^*(\langle\Omega_2\rangle\Sigma_j\Omega_{2j} + \langle\Omega_3\rangle\Sigma_j\Omega_{3j}) - (1/2)\Sigma_{k,s}g_{ks}g_{ks}{}^*(\langle\Omega_2\rangle^2 + \langle\Omega_3\rangle^2),$$

where we drop the subscripts 2 and 3 for the sums over s making use of the symmetry relation. Replacing the last term of the r.h.s. of Eq.(6) by the above expression of H_{int} and omitting the self energy of the local distortions and the energy of lattice vibrations, we obtain the total Hamiltonian as

$$H_{tot} = V(c_{11}-c_{12})(u_2^2 + u_3^2)/2 + g_o [V(c_{11}-c_{12})/N]^{1/2}(u_2\Sigma_j\Omega_{2j} + u_3\Sigma_j\Omega_{3j})$$

$$-[\Sigma_s g_{os}g_{os}* - (1/N)\Sigma_{k,s}g_{ks}g_{ks}*](<\Omega_2>\Sigma_j\Omega_{2j} + <\Omega_3>\Sigma_j\Omega_{3j})$$

$$+ (N/2)[\Sigma_s g_{os}g_{os}* - (1/N)\Sigma_{k,s}g_{ks}g_{ks}*](<\Omega_2>^2 + <\Omega_3>^2) \qquad (9)$$

We can calculate the free energy straightforwardly with the expression (9). By minimizing it with respect to $<\Omega_2>$ and $<\Omega_3>$, we conclude that $<\Omega_2>$ and $<\Omega_3>$ are the thermal averages of the operators $(1/N)\Sigma_j\Omega_{2j}$ and $(1/N)\Sigma_j\Omega_{3j}$, respectively. Since all the ions are assumed to be in the same state on the average in the ordered state, $<\Omega_2>$ and $<\Omega_3>$ are the thermal average of the single ion operators. Furthermore by minimizing with respect to u's, we derive

$$u_2 = -g_o[N/(c_{11}-c_{12})V]^{1/2}<\Omega_2> \text{ and } u_3 = -g_o[N/(c_{11}-c_{12})V]^{1/2}<\Omega_3>. \qquad (10)$$

We replace $<\Omega_2>$ and $<\Omega_3>$ by u_2 and u_3 in (9) to derive

$$H_{tot} = V(c_{11}-c_{12})R(u_2^2 + u_3^2)/2$$

$$+ g_o R^{1/2}[V(c_{11}-c_{12})R/N]^{1/2}(u_2\Sigma_j\Omega_{2j} + u_3\Sigma_j\Omega_{3j}) \qquad (11)$$

with the renormalizing factor R defined by

$$R = [g_o^2 + \Sigma_s g_{os}g_{os}* - (1/N)\Sigma_{k,s}g_{ks}g_{ks}*]/g_o^2. \qquad (12)$$

Equation (11) gives an effective Hamiltonian in which the electronic states couples with bulk strains. The physical meaning of the renormalizing factor will be clear if we note that $\Sigma_s g_{os}g_{os}*$ represents an additional contribution of the optical modes with k = 0 to the energy and $-(1/N)\Sigma_{k,s}g_{ks}g_{ks}*$ subtracts that of local distortions which should not contribute to the cooperative distortion. We have given so far a somewhat lengthy derivation of this effective Hamiltonian to clear up a misunderstanding that our theory does not take into account the local distortion.

The theory was successfully applied to the cooperative distortion of Ni-chromite, Cu-chromite and their mixed crystals. Equation (11) contains only two parameters, the renormalized Jahn-Teller coupling constant and renormalized elastic constant. For mixed crystals we introduce three renormalized parameters, coupling constants for Cu and Ni and the bulk elastic constant, neglecting the concentration dependence of these parameters. For the phase diagram, we need actually only two parameters, which are fixed by fitting the calculation to the transition temperatures of pure chromites. Then the calculation can reproduce well semi-quantitatively the experimental phase diagram, in particular, the boundaries of an orthorhombic phase appearing in-between the tetragonal phases with c>a and c<a extending from the pure chromites at the both ends of the concentration range. Our theory can reproduce other experimental results including the concentration and temperature dependence of distortions as well. [2] Moreover, it was successfully applied to the softening of elastic constants and the diffuse scattering of X-rays and neutrons near the phase transition in the cubic phase. [3,4]

3.2 The doubly degenerate case

The discussion for this case is complicated because Ω_2 and Ω_3 do not commute with each other. I proposed an approximation in which treats $\Omega_2 = \sigma_x$ and $\Omega_3 = \sigma_z$ as components of a classical vector. In the following I describe the gist of the approximation. We define a new mutually orthogonal pair of the wave functions of the orbital doublet as

$$\psi_\theta = cos(\theta/2)\, \psi_{1/2} + sin(\theta/2)\, \psi_{-1/2} \text{ and } \psi_{\theta+\pi} = -sin(\theta/2)\, \psi_{1/2} + cos(\theta/2)\, \psi_{-1/2}. \tag{13}$$

The diagonal elements of $\Omega_3 = \sigma_z$ with respect to this new basis are given by $cos\theta$ and $-cos\theta$; those of $\Omega_2 = \sigma_x$ are $sin\theta$ and $-sin\theta$. We define new operators Ω_2' and Ω_3' which are diagonal parts of Ω_3 and Ω_2, respectively, by

$$\Omega_2' = \begin{pmatrix} sin\theta & 0 \\ 0 & -sin\theta \end{pmatrix} \text{ and } \Omega_3' = \begin{pmatrix} cos\theta & 0 \\ 0 & -cos\theta \end{pmatrix}. \tag{14}$$

The angle θ in Eq.(14) may depend on j generally. We adopt the approximation that replaces Ω_3 and Ω_2 in the second term of r.h.s. of Eq.(3) by the commutable diagonal matrices Ω_3' and Ω_2'. Then the argument goes in parallel to that for the triply degenerate case; note $\Omega_2'^2 + \Omega_3'^2 = 1$ which yields a constant self energy of the local distortion associated with each orbital state. In the mean field approximation we assume the same angle θ for all j in the case of the 'ferromagnetic' ordering. Then we reach the same effective Hamiltonian as that given by Eq.(11) except for the fact that we retained only the diagonal parts of Ω_3 and Ω_2. The angle θ is not determined if we confine ourselves to the Hamiltonian given by Eq.(3). As was discussed explicitly in [2], we determine it by introducing the anharmonic energy and the higher order terms in the coupling between the orbital states and lattice distortions. We discuss also this problem in the next section.

Our approximation associates a static local distortion with the electronic state and neglects the vibronic effect caused by the nondiagonal parts of the electronic operators. We believe, however, that the effect is small for a large local distortion because the nondiagonal element connects between the states of different local distortions. Moreover, the vibronic effect will pertain mostly to the single ion energy and our discussion of the cooperative phenomena is essentially valid even in the cases of relatively small local distortions.

4. The orderings with $k=(\pi/a, \pi/a, \pi/a)$ and $(\pi/a, \pi/a, 0)$

In the previous paper [1] we explained the crystal structure of MnF_3 as the cooperative Jahn-Teller distortion of a crystal of the ReO_3 type. The basic lattice of the crystal is similar to that of a perovskite with the difference that the site occupied by La in $LaMnO_3$ is vacant. Mn^{3+} has a doubly degenerate orbital state in an octahedral cubic environment. In the MnF_3 case, the cooperative Jahn-Teller distortion is characterized by the wave vector $k=(\pi/a, \pi/a, \pi/a)$ with a representing the lattice constant of the cubic unit cell; it is accompanied by a bulk distortion through a mechanism similar to a canted antiferromagnetism. On the other hand, $LaMnO_3$ seems to exhibit a cooperative Jahn-Teller distortion with the wave vector, $k=(\pi/a, \pi/a, 0)$, accompanied also by a bulk distortion [11]; the structure was not known definitely at the time when the paper [1] was published. We discuss both the cases in this section.

4.1 The case $k=(\pi/a, \pi/a, \pi/a)$

We assume that Mn ions occupy the sites $R_j = (n_1 a, n_2 a, n_3 a)$ with integers n_1, n_2, n_3 in the undistorted lattice. The lattice is divided into two sublattices I and II according to the sign of $exp(ik \cdot R_j)$ for $k=(\pi/a, \pi/a, \pi/a)$. Replacing Ω_{2j} and Ω_{3j} by the corresponding diagonal matrices Ω_{3j}' and Ω_{2j}' in Eq.(6), we apply the mean field approximation. We assume that the average values of Ω_{3j}' and Ω_{2j}' satisfy the condition,

$<\Omega_{3j}'>=<\Omega_{3I}'>$ & $<\Omega_{2j}'>=<\Omega_{2I}'>$ if $j \in$ I and

$<\Omega_{3j}'>=<\Omega_{3II}'>$ & $<\Omega_{2j}'>=<\Omega_{2II}'>$ if $j \in$ II. (15)

Then $\Sigma_j exp(ik \cdot R_j)<\Omega_j'>$ will vanish except for $k=(0,0,0)$ and $(\pi/a, \pi/a, \pi/a)$, where $<\Omega_j'>$ stands for both $<\Omega_{3j}'>$ and $<\Omega_{2j}'>$. At $k=(\pi/a, \pi/a, \pi/a)$(and also for $k=(0,0,0)$) the group of symmetry operations keeping the wave vector invariant is of cubic symmetry. Then Ω_{3j}' and Ω_{2j}' interact with different normal modes with a common coupling constant denoted by g_{ns}. We define 'sublattice magnetization' operators and their average values by

$M_{2\pm} = \Sigma_{j \in I}\Omega_{2j}' \pm \Sigma_{j \in II}\Omega_{2j}'$ and $M_{3\pm} = \Sigma_{j \in I}\Omega_{3j}' \pm \Sigma_{j \in II}\Omega_{3j}'$

$<M_{2\pm}> = (N/2)(<\Omega_{2I}'> \pm <\Omega_{2II}'>)$ and $<M_{3\pm}> = (N/2)(<\Omega_{3I}'> \pm <\Omega_{3II}'>)$.

We rewrite Eq.(6) in the mean field approximation in terms of M's as before. By minimizing the free energy with respect to $<M_{2+}>$ and $<M_{3+}>$, we conclude that these quantities should be the thermal averages of the corresponding operators. Also the minimization of the free energy with respect to bulk strains u_2 and u_3 yields the conclusion that $<M_{2+}>$ and $<M_{3+}>$ should be proportional to u_2 and u_3, respectively. We can derive the renormalized Hamiltonian corresponding to Eq.(11) as

$$H_{tot} = V(c_{11}-c_{12})R(u_2^2 + u_3^2)/2 + g_o R^{1/2} [V(c_{11}-c_{12})R/N]^{1/2}(u_2 M_{2+} + u_3 M_{3+})$$

$$-(1/N)[\Sigma_s g_{\pi s} g_{\pi s}^* - (1/N)\Sigma_{k,s} g_{ks} g_{ks}^*](<M_2>M_2 + <M_3>M_3)$$

$$+(1/2N)[\Sigma_s g_{\pi s} g_{\pi s}^* - (1/N)\Sigma_{k,s} g_{ks} g_{ks}^*](<M_2>^2 + <M_3>^2), \qquad (16)$$

where the renormalizing factor R is given by Eq.(12) as before. Alternatively we may write more symmetrically

$$H_{tot} = -(G_o/N)(<M_{2+}>M_{2+} + <M_{3+}>M_{3+}) -(G_\pi/N)(<M_2>M_2 + <M_3>M_3)$$

$$+(G_o/2N)(<M_{2+}>^2 + <M_{3+}>^2) + (G_\pi/2N)(<M_2>^2 + <M_3>^2) \qquad (17)$$

or

$$H_{tot} = -[(G_o+G_\pi)/N] (<M_{2I}>M_{2I} + <M_{2II}>M_{2II} + <M_{3I}>M_{3I} + <M_{3II}>M_{3II})$$

$$-[(G_o-G_\pi)/N] (<M_{2II}>M_{2I} + <M_{2I}>M_{2II} + <M_{3II}>M_{3I} + <M_{3I}>M_{3II})$$

$$+[(G_o+G_\pi)/2N] (<M_{2I}>^2 + <M_{2II}>^2 + <M_{3I}>^2 + <M_{3II}>^2)$$

$$+[(G_o-G_\pi)/N] (<M_{2I}><M_{2II}> + <M_{3I}><M_{3II}>), \qquad (18)$$

where G_o and G_π are the renormalized coupling energies, $g_o^2 R$ and $[\Sigma_s g_{\pi s} g_{\pi s}^* - (1/N)\Sigma_{k,s} g_{ks} g_{ks}^*]$, respectively. Equation (18) is nothing but the mean field expression of the energy for a two-sublattice system. The angles θ_I and θ_{II} of Ω_I' and Ω_{II}' operators correspond to the angles which $<M_I>$ and $<M_{II}>$ make with the z-axis of the two-dimensional representation. If (G_o-G_π) is negative, the intersublattice interaction is antiferromagnetic, favoring $<M_{2I}> = -<M_{2II}>$ and $<M_{3I}> = -<M_{3II}>$.

The anisotropy energy arising from the higher order terms plays an important role in determining the configuration of lowest (free) energy; as is shown in Fig.2, a canted antiferromagnetism will be realized. The case MnF$_3$ was discussed in [1]. We determine the angles θ_I and θ_{II} by minimizing the free energy. As far as the phenomenology of the ordered state is concerned, our theory can give a reasonable picture. However, we have to admit that the theory has an ambiguity in the treatment of temperature effects. First of all, in the original paper [1], I mentioned that the canting angle would decrease with temperature to vanish at the transition. Moreover the transition was predicted to be of the second order. These conclusions were based on an expression of the anisotropy energy proportional to the cube of the order parameters. Actually the anisotropy which will persist in the disordered phase will dominate for a vanishing mean field, making the canting angle even larger. Thus the transition to the undistorted phase will be of the first order. We note also that our calculation of the partition function doubly degenerate case given in previous papers [1,2] assumes implicitly a small anisotropy in the high temperature phase, assuming that a local distortion can take any direction in the two dimensional representation. If the anisotropy is strong, the calculation of the partition function in the doubly degenerate case will be the same as that in the triply degenerate case. The electronic states will be such that the associated local distortions are fixed to be tetragonal with the principal axis directing in the x- or y- or z-direction in the real space.

4.2 The case of $k=(\pi/a, \pi/a, 0)$

For this wave vector the Q_2-like normal modes which interact with the electronic states via $\Omega_2 = \sigma_x$ belong to a different representation of the k group from the modes of the Q_3 symmetry. Thus distinguishing the coupling energies between $<M_2>M_2$ and $<M_3>M_3$, and defining the sublattices I and II according to the sign of $exp(i\boldsymbol{k}\cdot\boldsymbol{R}_j)$, we reach

$$H_{tot} = -[(G_o+G_{n2})/N] (<M_{2I}>M_{2I} + <M_{2II}>M_{2II})$$

$$-[(G_o+G_{n3})/N] (<M_{3I}>M_{3I} + <M_{3II}>M_{3II})$$

$$-[(G_o-G_{n2})/N] (<M_{2II}>M_{2I} + <M_{2I}>M_{2II})$$

$$-[(G_o-G_{n3})/N] (<M_{3II}>M_{3I} + <M_{3I}>M_{3II})$$

$$+[(G_o+G_{n2})/2N] (<M_{2I}>^2 + <M_{2II}>^2) + [(G_o+G_{n3})/2N] (<M_{3I}>^2 + <M_{3II}>^2)$$

$$+[(G_o-G_{n2})/N]<M_{2I}><M_{2II}> + [(G_o-G_{n3})/N]<M_{3I}><M_{3II}>, \qquad (19)$$

which indicates an anisotropic interaction. If we inspect the idealized perovskite lattice shown in Fig. 1, we can conclude that $G_{\pi 2}$ will be larger than $G_{\pi 3}$, making the antiferromagnetic configuration of M vectors along the x axis for the case $G_o < G_{\pi 2}$. Again the canted antiferromagnetism will be of lowest energy if the anisotropy energy favoring the directions $\theta = \pm 2\pi/3$ is taken into account.

5. The role of the spin-orbit coupling

The spin-orbit coupling can be important in the case of the triply degenerate case, where the submatrix of the orbital angular momentum is proportional to the pseudo angular momentum ℓ with a nonvanishing coefficient. Fist we discuss a paramagnetic phase with no cooperative magnetism. A competition with the local distortion energy of the Jahn-Teller effect determines whether the spin-orbit coupling will dominate or not, since the local distortion will tend to quench the orbital angular momentum. If the spin-orbit coupling should be diagonalized first, we may have still a multiplet of the total angular momentum $S+\ell$ in which the Jahn-Teller effect might be operative. In such a case, however, the transition temperature of the cooperative Jahn-Teller distortion will be much lower than that expected in the case where the spin orbit coupling is of minor importance. However, we can expect a large magnetostriction in the magnetically ordered state, if the interatomic exchange interaction is comparable with or larger than the spin-orbit coupling. At the onset of a cooperative magnetism we can expect that the state with a maximum z component of spin be favored by the exchange interaction, where the z-axis denotes the axis of easy magnetization.. Then the spin-orbit coupling favors either $\ell_z = 1$ or $\ell_z = -1$ state for the orbital angular momentum, depending upon the sign of the effective spin orbit coupling. The accompanying magnetostriction is of the sign opposite to the Jahn-Teller effect which favors the state $\ell_z = 0$. The observed distortion in CoO, FeO, $KCoF_3$, and $KFeF_3$ whose transition temperature coincides with the critical temperature of the cooperative magnetism can be explained by this mechanism quantitatively. [1, 12]

If the local distortion dominates to favor the states $\ell_i = 0$ (i = x, y and z), the cooperative Jahn-Teller effect can take place independently of the magnetism with the critical temperature T_{JT} either above or below T_c of the magnetic transition. This situation is realized in $NiCr_2O_4$, $CuCr_2O_4$ and their mixed crystals where $T_{JT} > T_c$.

Mixed crystals $NiCr_tFe_{2-t}O_4$ might present an example of intermediate cases. The transition temperature T_{JT} of the cooperative Jahn-Teller distortion of $NiCr_2O_4$ decreases rapidly by substituting Cr by Fe, i.e. decreasing t from t = 2. The substitution moves Ni atoms from A sites to B ones to make room for Fe^{3+} atoms in A sites. On the other hand, T_c of the magnetic transition increases rapidly with Fe concentration to become higher than T_{JT}. The cooperative Jahn-Teller distortion is of tetragonal symmetry with c/a>1. A phase with c/a<1 appears in the region of higher Fe concentration, where still some Ni atoms occupy A sites. Ni atoms at B sites and Fe atoms at A sites have nothing to do with crystal distortion, having no orbital degeneracy. Goodenough [13] pointed out the possibility that the c/a<1 distortion is a magnetostriction caused by the spin-orbit coupling operative at A site Ni atoms. However, the transition from cubic to the c/a<1 phase does not start at the onset of the ferromagnetism, taking place at a temperature lower than the magnetic Curie temperature. As was mentioned by Goodenough himself, this fact cast some doubt on the spin orbit coupling mechanism. I would like to propose a possible explanation for this discrepancy. Suppose that the local Jahn-Teller effect dominates to realize the state $\ell_z = 0$ (z is one of the cubic axes) around the magnetic transition temperature which is already higher than T_{JT}. The crystal will keep the cubic symmetry except for an ordinary small magnetostriction. With the development of magnetic order, the increasing molecular field can reverse the situation to realize the state $\ell_z = 1$ or $\ell_z = -1$. We can expect that this transition will take place at a temperature somewhat below the magnetic T_c. Of course we need a detailed calculation to confirm this possibility. We need to explain also the appearance of an orthorhombic phase in-between the c>a phase and the c<a one.

In the case of $CuCr_xFe_{2-x}O_4$ [13] there is no indication of the phase of a large magnetostriction. This suggests that the quenching of the orbital angular momentum persists in this mixed crystal system.

6. Electron itinerancy and lattice distortion

Experimental results on Mn perovskites indicate that the electron itinerancy competes with the Jahn-Teller distortion. I have discussed in the preceding sections that a local distortion is associated with each electronic state in the insulating phase even above T_{JT}. The local distortion will reduce the interionic electron transfer to stabilize the insulator phase. This idea seems to be essentially the same as the polaron effect proposed by Millis et al [14]. Conversely the electron transfer will suppress not only the cooperative effect but also the local distortions in the itinerant phase. However, it will be too haste to conclude a complete quenching of the Jahn-Teller coupling in the metallic phase; a dynamical process may contribute to its stabilization. Several people including Goodenough [15] will discuss this subject at the present symposium. I would like to conclude my discussion by emphasizing that one can discuss the crystal distortions in the insulating phase within a general framework without introducing ad hoc models.

References

[1] J.Kanamori, J.Appl.Phys.Suppl.**31,** 14S (1960).
[2] J.Kanamori, M.Kataoka and Y.Itoh, J.Appl.Phys. **39**, 688 (1969);M.Kataoka and J.Kanamori, J.Phys.Soc.Jpn. **32**, 113 (1972).
[3] M.Kataoka, J.Phys.Soc.Jpn. **36**, 456 (1974).
[4] M.Kataoka and Y.Endoh, J.Phys.Soc.Jpn. **48**, 912 (1980).
[5] P.J.Wojtowicz, Phys.Rev.**116**, 32(1959).
[6] Y.Tokura, in this Proceedings.
[7] J.H.Van Vleck, J.Chem.Phys. **7**, 472 (1939).
[8] U.Öpik and M.H.L.Pryce, Proc. Roy. Soc.(London) **A238**, 425 (1957).
[9] A.D.Lieh and C.J.Ballhausen, Ann.Phys.(N.Y.) **3**, 304 (1958).
[10] G.A.Gehring and K.A.Gehring, Rep.Prog.Phys.**38**, 1 (1975).
[11] J.Rodrígues-Carvajal, M.Hennion, F.Moussa. A.H.Moudden, L.Pinsard and A.Revcolevschi, Phys. Rev. **B 57**, R3189 (1998).
[12] J.Kanamori, Prog. Theoret. Phys. **17,** 177(1957); ibid. 197 (1957).
[13] J.B.Goodenough, "Magnetism and the Chemical Bond" John Wiley & Sons , New York(1963).
[14] A.J.Millis, Boris I.Shraiman, and R.Mueller, Phys.Rev.Letters 77, 175 (1996).
[15] J.B.Goodenough, in this Proceedings.

Present Status of the First-Principles Electronic Structure Calculations for the Strongly Correlated Transition-Metal Oxides

Kiyoyuki Terakura[1], Zhong Fang[2] and Igor V. Solovyev[2]

[1] JRCAT, National Institute for Advanced Interdisciplinary Research, 1-1-4 Higashi, Tsukuba, Ibaraki 305-8562, Japan

[2] JRCAT, Angstrom Technology Partnership, 1-1-4 Higashi, Tsukuba, Ibaraki 305-0046, Japan

Abstract: Intensive efforts have been made to improve the description of the strongly correlated transition-metal oxides by the first-principles electronic structure calculations. Two subjects in this context will be discussed in this article taking the classical and prototypical materials, MnO and FeO, as examples. One is the analysis of pressure induced phase transition and the other is the comparison among some different levels of approximate methods in the band-structure calculations.

1. Introduction

The first-principles electronic structure calculations (FPESC) are recently playing important roles in a wide range of fields, physics, chemistry, materials science and even biology. Such a situation was realized by significant progress in two fundamental aspects in FPESC. One is the simplification in treating the electron-electron interaction as formulated in the density functional theory (DFT) [1, 2, 3, 4] and another is the first-principles molecular dynamics (FPMD) method introduced by Car and Parrinello [5, 6]. Quantitative analyses of experimental observations and highly reliable predictions of structures and electronic properties are commonly made in various situations. Nevertheless, the particular field of transition-metal oxides (TMO) seems to be the field where FPESC has not been playing important roles.

Several attempts have been made in order to improve such a situation in the study of TMO. They are, for example, generalized gradient approximation (GGA) in DFT [2, 3, 4], LDA+U method [7, 8, 9], GW approximation [10, 11], and the optimized effective potential (OEP) method [12, 13]. In the following, we perform an empirical calculation of the electronic structure of MnO [14], a simple and prototypical example of Mott insulators, with the experimental spin wave dispersion as a guide and compare this result with the calculations based on some different approximations just mentioned. Though we admit that the present state of FPESC for TMO is not yet satisfactory, we also would like to demonstrate that FPESC can still make significant contributions to some aspects of the physics of TMO. A good example is the analysis of the high-pressure experiment on TMO. A favorable feature in this case is that under high pressure the electron correlation effects become less important because of the increased band width. We present in the following our recent analyses of the high-pressure phases of MnO and FeO [15, 16].

Before presenting the results of our calculations, we mention the essence of the methodology. The study of the pressure effects on MnO and FeO was performed by the FPMD method with the plane wave basis pseudopotential formalism. This approach is useful for the structural optimization under pressure because the Hellmann-Feynman forces acting on atoms and the stress can be calculated efficiently. However, the LDA+U calculation as well as the empirical calculation of the spin wave was performed with the LMTO method.

2. Pressure-induced structural phase transitions of MnO and FeO

MnO and FeO have been regarded as typical examples of Mott insulators and the study of their pressure effects has also a long history. However, remarkable progress has been made only very recently. At normal pressure, both MnO and FeO take the rock-salt type (B1) crystal structure with the type II antiferromagnetic (AF) ordering. Below Néel temperature, the lattice distorts rhombohedrally in both systems: compressed and elongated along the $<111>$ direction for MnO and FeO, respectively. We call this rhombohedrally distorted rock-salt structure simply rB1.

Figure 1 shows the phase diagram of FeO in the $p - T$ plane [17]. We note the following two important features. First, the rB1 phase persists up to higher temperature with increasing pressure in the low pressure regime. Second, on further compression, the system undergoes a phase transition from rB1 to the NiAs (B8) structure. We would like to make a few comments on the first feature.

1. At room temperature, FeO is paramagnetic at normal pressure and takes the B1 structure. By increasing pressure, the crystal distorts to the rB1 structure at 16 GPa [18] and the Mössbauer measurement indicates that the rB1 phase is antiferromagnetic. Although there is no magnetic measurements at higher pressure and higher temperature regions, the boundary between the rB1 and B1 phases may probably correspond to Néel temperature T_N. The boundary is approximately expressed as $T(p) = 14p + 76$ with the pressure p given in units of GPa. The data of T_N and its pressure derivative dT_N/dp as functions of the Fe deficiency x ($Fe_{1-x}O$) [19] shown in Fig.2 are at least not inconsistent with this conjecture: dT_N/dp sharply increases as x decreases and can be as large as 14 (K/GPa) as x approaches 0.04.

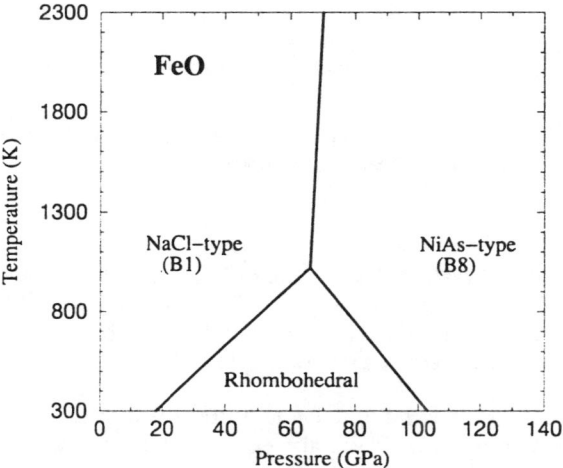

Figure 1: A schematic phase diagram of FeO. Figure 1 of ref.17 was slightly simplified.

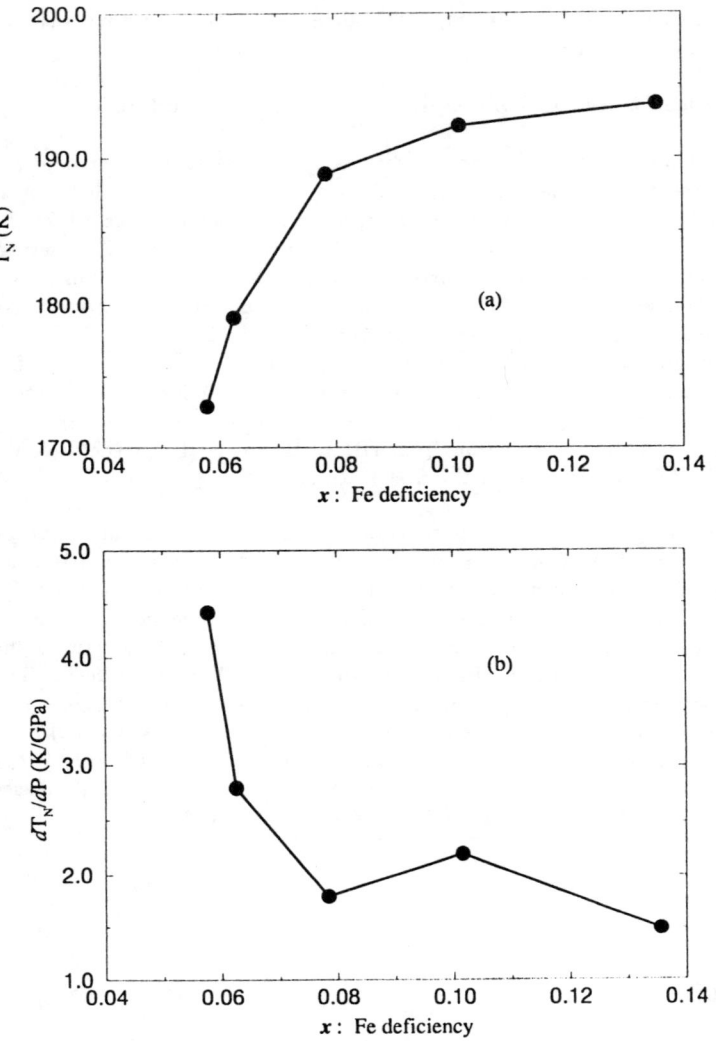

Figure 2: (a): Néel temperature T_N and (b): its pressure dependence dT_N/dp of $Fe_{1-x}O$ as functions of Fe deficiency x. The original data come from ref.19.

2. There is an experimental observation that the elongation of the lattice along the $<111>$ axis is enhanced upon compression at room temperature [18] and this trend was semiquantitatively reproduced by the GGA based band calculation which treats the system at zero temperature [16, 20].

Fei and Mao [17] concluded from their x-ray diffraction measurement that the high-pressure phase of FeO is of the NiAs type (B8). It was also experimentally known that this phase at high presssure and high temperature is metallic. As a simple extrapolation of most of the TM compounds with the B8 structure, a natural idea of the B8 FeO

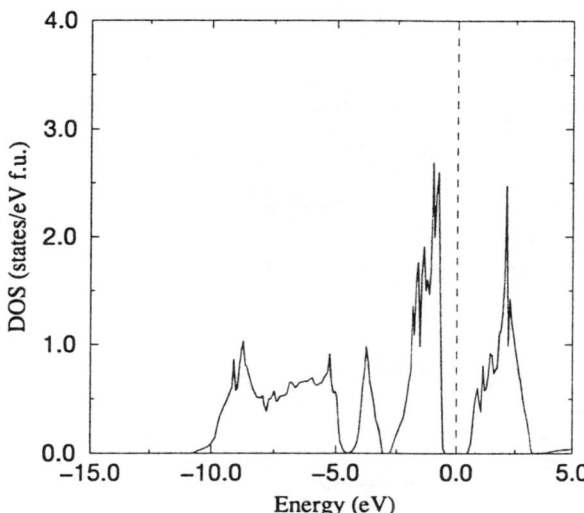

Figure 3: Density of states for the antiferromagnetic FeO with the inverse NiAs stucture. The experimental lattice constant at 96 GPa and 800 K [17] was used in the calculation.

is that Fe occupies the Ni site and O the As site. We call this structure the normal NiAs (nB8). Although nB8 FeO gives correct diffraction peak positions, it has been a puzzling problem that this structure does not reproduce the observed intensity profile. An interesting feature of the B8 structure is that the exchange of the Fe and O positions leads to a crystal structure inequivalent to the original one. The new structure is called the inverse NiAs (iB8). Our FPESC shows that iB8 FeO has a well defined band gap of about 1 eV even at 100 GPa as shown in Fig.3 and that this band gap produces special stability of this structure among others (see Fig.4a). Furthermore, the intensity profile of the x-ray diffraction for this structure is consistent with experiment [15, 16].

The above results were obtained with GGA and they may be valid in the high pressure range where the band width becomes large enough to break the Mott insulator condition. However, there are two problems in the calculated results. First, the insulating nature of the iB8 FeO is not consistent with the metallic behavior observed experimentally. As for this problem, we pointed out two possible origins [15]. One is that FeO has generally a large (several %) Fe deficiency which will cause carrier doping in the band insulating state of iB8 FeO. Another is that metallic nB8 FeO may coexist with the iB8 phase at the high temperature range where the metallic behavior was observed. The second problem seems to be more serious. The GGA level calculation makes iB8 FeO more stable than rB1 FeO even at normal pressure. We pointed out that the problem is due to the poor description of the rB1 phase of FeO at normal pressure by GGA, which makes FeO metallic incorrectly. The failure of a simple one-electron picture in predicting the insulating property of FeO can be easily understood in the following way. In the undistorted B1 structure of FeO, one third of the minority-spin t_{2g} band is filled. Elongation of the lattice in the $<111>$ direction splits the triply degenerate t_{2g} states into an A_1 state and an E_g state with the latter being half filled. Therefore, in the one-electron approximation, FeO is metallic even in the

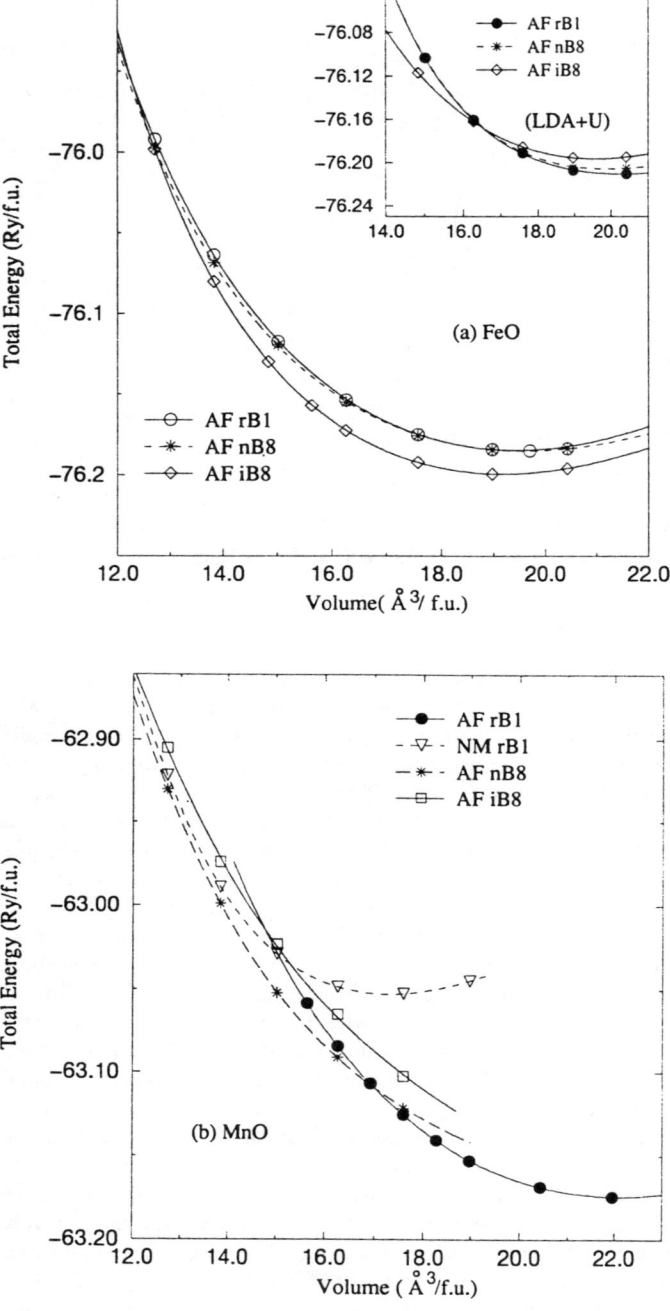

Figure 4: Total energies per formula unit of some different crystal structures as functions of the volume. (a): FeO and (b): MnO.

AF state with rhombohedral lattice distortion. Insulating FeO is realized by orbital polarization induced by the electron-electron interaction. The spin-orbit interaction fixes the particular orbital to be occupied. We demonstrated that the inclusion of the spin-orbit interaction and the local Coulomb interaction U can make FeO insulating and reproduce the orbital angular momentum properly [21]. It is interesting to note that the inconvenient aspect of the structural stability at normal pressure range can also be removed at the same time. (See the inset of Fig.4a.)

As for MnO [22, 23], the most stable high-pressure structure is nB8 and the AF and F orderings are nearly degenerate as shown in Fig.3b. However, the static compression measurement suggests multiple-phase coexistence even at 130 GPa [23]. We concluded that the metastable low-spin rB1 phase is the one which coexists with the nB8 phase. An interesting feature is that the lattice is elongated along the < 111 > direction in the low-spin rB1 phase while it is compressed in the high-spin rB1 phase [15, 16]. Anyway, our calculation suggests that the rB1 phase at high pressure is only metastable. This explains the experimental observation that the diffraction peaks corresponding to the nB8 phase grow while those of the rB1 phase shrink after laser annealing at 137 GPa. Our band calculation showes that both low-spin rB1 and nB8 phases are metallic. Experimentally, metallic reflection was observed at high pressure.

3. Empirical band structure of MnO giving correct spin-wave dispersion

As the ground state of MnO at normal pressure has an electronic configuration of high-spin d^5, the electron distribution is basically spherical and there is no complication coming from the orbital polarization as mentioned above for FeO. Because of this simplicity, MnO has been used as a good test case for various theoretical treatments. If we pay attention only to the enegy gap, different methods give a wide range of values: 1.3 eV by the LSDA calculation [24], 13 eV by the Hartree-Fock calculation [25], 4.2 eV by a model GW calculation [10], and $5 \sim 6$ eV by the recent optimized effective potential (OEP) method [13]. By the photoemission and BIS measurements, the band gap of MnO may be around 4 to 5 eV [26]. Although the experimental electronic excitation spectrum provides us with useful information to compare with the theoretical energy bands, orbital energies obtained by the DFT method and also by the Hartree-Fock method do not necessarily correspond to excitation spectrum. This is the reason why we decided to use different experimental information, here the spin wave disperion, as a guide for judging the validity of the theoretical energy band because the exchange coupling depends on the on-site exchange splitting Δ_{ex} and the charge-transfer energy Δ_{ct} through virtual electronic excitations.

The magnetic coupling between the second neighbor Mn atoms is governed by the superexchange interaction mediated by the oxygen atoms which, in the second order perturbation theory, is given by the following expression [27].

$$J_2 = -\frac{t_{2,pd}^4}{\Delta_{ct}^2}\left(\frac{1}{\Delta_{ex}} + \frac{1}{\Delta_{ct}}\right), \quad (1)$$

where the effective pd hopping integral for the octahedral oxygen coordination around Mn is defined as

$$t_{2,pd}^4 = (pd\sigma)^4 + (pd\pi)^4. \quad (2)$$

On the other hand, the first neighbor exchange coupling is governed not only by the superexchange but also by the direct exchange. As for the superexchange part, the above $t_{2,pd}^4$ has to be replaced with

Table I. Magnetic moments and exchange parameters in LSDA, LDA+U and OEP as well as those obtained by fitting the experimental spin-wave dispersion data (fit).

	μ (μ_B)	J_1 (meV)	J_2 (meV)
LSDA	4.50	-13.2	-23.5
LDA+U	4.68	-5.0	-13.2
OEP[a]	4.80	-8.9	-12.0
fit	4.84	-4.8	-5.6
Expt.	4.79[b], 4.58[c]	-4.8[d]	-5.6[d]

[a] Ref. 13.

[b] B. E. F. Fender, A. J. Jacobson and F. A. Wegwood, J. Chem. Phys. **48**, 990 (1968).

[c] A. K. Cheetham and D. A. O. Hope, Phys. Rev. B **27**, 6964 (1984).

[d] Ref. 28.

The experimental parameters J_1 and J_2 are multiplied by $S^2 = (5/2)^2$.

$$t_{1,pd}^4 = 2(pd\sigma)^2(pd\pi)^2 + (pd\pi)^4. \qquad (3)$$

On the other hand, the direct exchange part is given by

$$-\frac{2t_{dd}^2}{\Delta_{ex}}, \qquad (4)$$

where t_{dd} is the direct hopping integral between the nearest neighbor Mn d orbitals. A nice aspect of the spin-wave dispersion of MnO for our purpose is that only J_1 and J_2 are required to describe the spin-wave dispersion [28, 29]. Therefore, the parameters Δ_{ct} and Δ_{ex} can be uniquely fixed by the information of the spin-wave dispersion.

Like frozen phonon calculations, we adopted the frozen spiral spin idea for the spin-wave calculation. For matallic magnets, this prescription gives very good results [30]. For convenience of the practical calculations, we introduced, instead of Δ_{ct} and Δ_{ex}, a different set of parametes, $\Delta_\uparrow$ and $\Delta_\downarrow$, as the rigid shifts of the spin-up and spin-down Mn-d potentials [14]. With the proper choice of these parameters, the calculated spin-wave disperion fits perfectly the observed one. Table I lists the theoretical J_1 and J_2 obtained by using some different approximations in the band structure calculation and also experimental data.

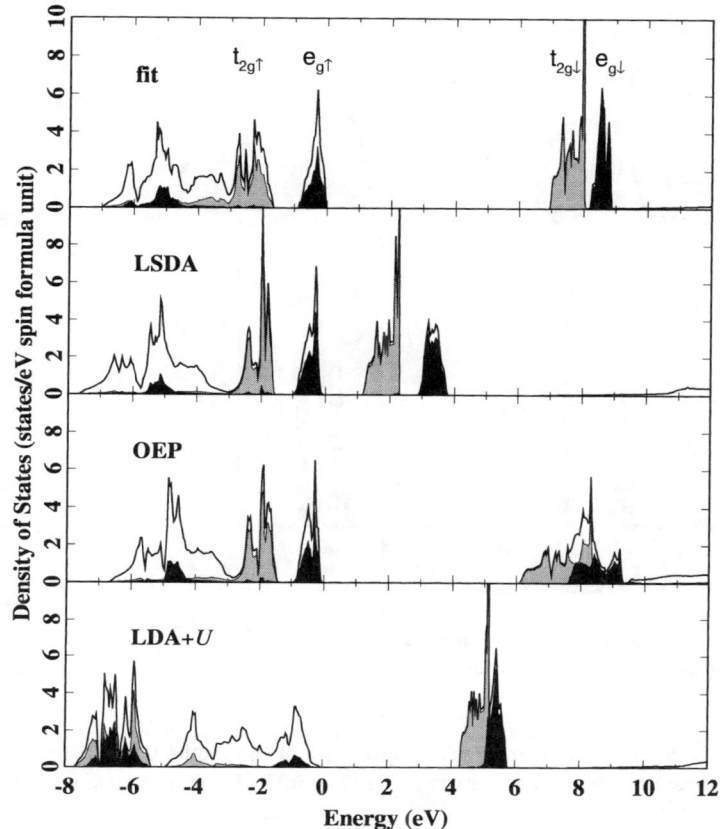

Figure 5: The densities of states of MnO obtained by some different approaches. The topmost one designated as "fit" is the present result. The black erea denotes the component of the e_g orbitals and the gray area that of the t_{2g} orbitals.

The most important outcome of this calculation is the electronic structure obtained by using $\Delta_\uparrow$ and $\Delta_\downarrow$ fixed in the above procedure. The result is shown in Fig.5 together with those based on some different approximations. As shown in Fig.6, the present result agrees very well with the photoemission and the inverse-photoemission measurements [26]. If we derive Δ_{ex} and Δ_{ct} from the calculated results, they are 10.6 eV for Δ_{ex} and 10.7 eV for Δ_{ct}. Furthermore, from Δ_{ex}, we estimate the Coulomb interaction parameter U to be about 8 eV assuming that J is about 0.8 eV. The usual LDA+U method with $U_{\text{eff}}(= U - J) \sim 7$ eV, brings the occupied d bands below the oxygen p band [7]. Therefore, the present analysis suggests that we need not only U but also an additional energy shift of the d states in order to reproduce the properties of MnO properly.

Despite the fact that the band structure obtained by OEP gives a reasonable values for Δ_{ct} and Δ_{ex}, the exchange couplings are appreciably overestimated. This may be caused by the large width of the unoccupied d band. We speculate that the use of local effective one-electron potential in the OEP method may have some limitation in

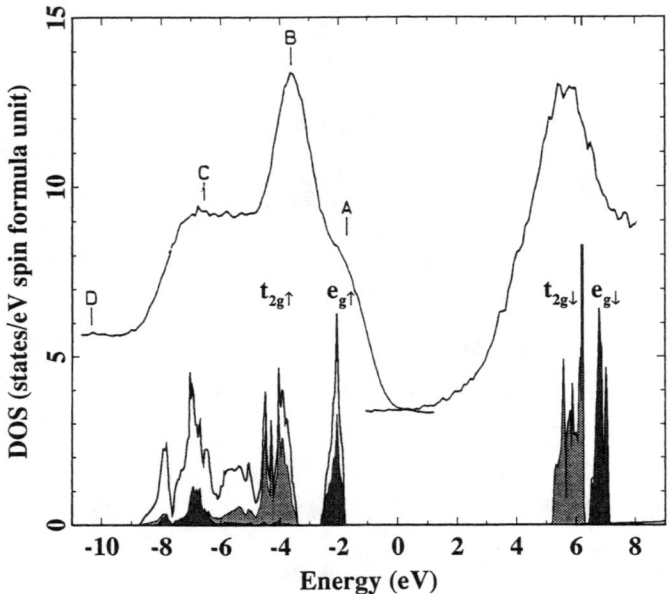

Figure 6: The present result of the band structure of MnO is compared with the photoemission and BIS measurements [26].

its applicability. Both our present empirical approach and the model GW calculation which adopt non-local one-electron potential make the width of the unoccupied d band nearly the same as that for the occupied d band. A natural extension of the present analysis is the theoretical determination of the parameters $\Delta_\uparrow$ and $\Delta_\downarrow$ without using experimental information. One possible way for this purpose is to borrow the basic idea of OEP [12].

In connection with this analysis, we would like to make a brief comment on the problem of LaFeO$_3$. We found that our calculations always underestimate the magnetic moment of Fe seriously: the experimental value is $4.6 \pm 0.2\,\mu_B$ while theoretical one is only 3.6 μ_B with LDA and 4.04 μ_B even with the LDA+U with U_{eff} of 7 eV [9]. In most cases, the LDA+U method overestimates the magnetic moment. Therefore this disagreement has been puzzling for us. However, we also noted that the peak position in the optical conductivity of LaFeO$_3$ is higher in experiment than in the calculated result [9]. Again the LDA+U method cannot make any significant improvement. As an attempt, we introduced $\Delta_\uparrow$ and $\Delta_\downarrow$ in the calculation of LaFeO$_3$ and adjusted these parameters to reproduce the peak position in the optical conductivity. Although we cannot determine two parametes uniquely by using only one experimental datum, we found that the magnetic moment of Fe is also significanlty enhanced to have a good agreement with the experimental value within a reasonable range of these parameters.

4. Concluding Remarks

We have first demonstrated that the FPESC can make important contributions to the analysis of the high-pressure phases of transition metal oxides, which are regarded as strongly correlated systems at normal pressure. Our study has revealed that the high-pressure phase of FeO (MnO) takes the inverse (normal) NiAs structure. To the best of our knowledge, there is no other transition-metal compounds which take the inverse NiAs structure. An attempt was also made to analyze the spin-wave dispersion of MnO with two adjustable parametes in our band structure calculation. We demonstrated that this procedure gives us a very reasonable electronic structure which agrees well with experiments. This analysis has clarified some problems in the attempts of first-principles methods for the electronic structure calculation of the transition-metal oxides.

Acknowledgement

The present work was partially supported by the New Energy and Industrial Technology Development Organization (NEDO).

References

[1] As a review of the density functional theory: R. O. Jones and O. Gunnarsson, Rev. Mod. Phys. **61**, 689 (1989).

[2] A. D. Becke, Phys. Rev. A **38**, 3098 (1988).

[3] J. P. Perdew, in *Electronic Structure of Solids'91*, ed. by P. Ziesche and H. Eschrig (Akademie Verlag, Berlin, 1991).

[4] M. Levy and J. P. Perdew, Phys. Rev. B **48**, 11638 (1993); Int. J. Quantum Chem. **49**, 539 (1994).

[5] R. Car and M. Parrinello, Phys. Rev. Lett. **55**, 2471 (1985).

[6] K. Terakura, in *Computational Physics as a New Frontier in Condensed Matter Research*, (Phys. Soc. Jpn., 1995), p.1.

[7] V. I. Anisimov, J. Zaanen and O. K. Andersen, Phys. Rev. B **44**, 943 (1991).

[8] I. V. Solovyev, P. H. Dederichs and V. I. Anisimov, Phys. Rev. B **50**, 16861 (1994).

[9] I. V. Solovyev, N. Hamada and K. Terakura, Phys. Rev. B **53**, 7158 (1996).

[10] S. Massidda, A. Continenza, M. Posternak and A. Baldereschi, Phys. Rev. Lett. **74**, 2323 (1995).

[11] F. Aryasetiawan and O. Gunnarsson, Rep. Prog. Phys. **61**, 237 (1998).

[12] J. D. Talman and W. F. Shadwick, Phys. Rev. A **14**, 36 (1976).

[13] T. Kotani, to be published in J. Phys.: Condens. Matter.

[14] I. V. Solovyev and K. Terakura: submitted to Phys. Rev. B.

[15] Z. Fang, K. Terakura, H. Sawada, T. Miyazaki and I. V. Solovyev, Phys. Rev. Lett. **81**, 1027 (1998).

[16] Z. Fang, I. V. Solovyev, H. Sawada and K. Terakura, submitted to Phys. Rev. B.

[17] Y. W. Fei and H. K. Mao, Science **266**, 1678 (1994).

[18] T. Yagi, T. Suzuki and S. Akimoto, J. Geophys. Res. **90**, 8784 (1985).

[19] S. Tamura, High Temperatures−High Pressures **22**, 399 (1990).

[20] D. G. Isaak, R. E. Cohen, M. J. Mehl and D. J. Singh, Phys. Rev. B **47**, 7720 (1993).

[21] J. Kanamori, Progr. Theor. Phys. **17**, 177 (1957); Progr. Theor. Phys. **17**, 197 (1957).

[22] Y. Noguchi, K, Kusaba, K. Fukuoka and Y. Syono, Geophys. Res. Lett. **23**, 1469 (1996).

[23] T. Kondo, T. Yagi and Y. Syono, Rev. High Pressure Sci. Technol. **7**, 148 (1998).

[24] K. Terakura, T. Oguchi, A. R. Williams and J. Kübler, Phys. Rev. B **30**, 4734 (1984).

[25] M. D. Towler, N. L. Allan, N. M. Harrison, V. R. Saunders, W. C. Mackrodt and E. Apra, Phys. Rev. B **50**, 5041 (1994).

[26] J. van Elp et al., Phys. Rev. B **44**, 1530 (1991).

[27] J. Zaanen and G. A. Sawatzky, Can. J. Phys. **65**, 1262 (1987).

[28] M. Kohgi, Y. Ishikawa and Y. Endoh, Solid State Commun. **11**, 391 (1972).

[29] G. Pepy, J. Phys. Chem. Solids **35**, 433 (1974).

[30] O. N. Mryasov, A. I. Liechtenstein, L. M. Sandratskii and V. A. Gubanov, J. Phys.: Condens. Matter **3**, 7683 (1991).

Aspects of Coupled Spin-Orbital Degrees of Freedom in d- and f-Electron Systems

Hiroyuki Shiba and Ryousuke Shiina

Department of Physics, Tokyo Institute of Technology, Oh-okayama, Tokyo 152-8551

Abstract. The long-range order related to orbital degrees of freedom is discussed by taking two examples: one from a $4f$-electron system CeB_6 and the other from $LaMnO_3$. It is pointed out for CeB_6, where an antiferro-quadrupole order occurs, that the magnetic field induces a large octupole moment and it plays an important role. For the latter case an effective Hamiltonian is carefully derived to discuss the electronic mechanism of coupling between spin and orbital degrees of freedom by taking $LaMnO_3$ as an example.

1 Introduction

The orbital order in $3d$ compounds is an old subject[1], which has revived recently in connection with $3d$ transition-metal oxides. A similar problem is also being studied in $4f$ or $5f$ compounds under the name of quadrupole order[2]. The purpose of this paper is to put these two streams close to each other in order to emphasize common physics.

In $3d$ ions having orbital degeneracy (like Mn^{3+} in a cubic crystal-field) the orbital degrees of freedom are important. Although the spin-orbit coupling is weak for quenched orbital moment, the Coulomb interaction among $3d$ electrons together with the orbital dependence of transfer integrals leads to a coupling between the spin and orbital orders[1]. The orbital degrees of freedom are coupled with atomic displacements (via Jahn-Teller mechanism), when they order, in particular[3]. Thus spin, orbital and lattice displacements are interrelated: When one of them orders or becomes disordered, it necessarily affects other degrees of freedom. Currently many transition-metal oxides are being examined in the light of orbital order or orbital fluctuations.

In $4f$ or $5f$ compounds, the spin-orbit coupling is strong; therefore, the total angular momentum J, instead of spin and orbital angular momenta, is a good quantum number. Local degrees of freedom of f electrons are most conveniently described by multipole moments. When the crystal field leaves a number of low-lying energy levels, long-range orders other than magnetic order are possible in general. The antiferro-quadrupole order is one of such possibilities that we encounter most frequently except for the dipole order. Many materials are being studied in connection with the antiferro-quadrupole order or multipole order.

Let us note at this point that the orbital order and the quadrupole order are conceptually the same. In this paper we wish to study typical cases chosen from f and d electron systems and compare them with each other.

2   Multipole Order in CeB$_6$

CeB$_6$ is one of dense Kondo materials and has a long history in research[4, 5, 6, 7, 8, 9, 10]. It has a simple cubic structure: Ce ions, each of which has one $4f$ electron, sit on the corner sites, while B$_6$ ions occupy the body-center sites. CeB$_6$ has plural ordered phases due to $4f$ electrons, in which our interest lies in the nature of "phase II" under the applied magnetic field, which is believed to be an antiferro-quadrupole (AFQ) phase. There are some mysteries in phase II.

(1) The transition temperature continuously increases; no sign of decrease is observed up to available largest field 15T. It strongly suggests that the main origin of phase II is electronic and the coupling with lattice displacements must be secondary.

(2) The magnetic field induces antiferromagnetic moments. However their nature is controversial. A triple-q structure was proposed from NMR[6], while the neutron diffraction suggested a field-induced antiferromagnetism with $(1/2, 1/2, 1/2)$[9].

The reason we take up this problem is that we believe there is simple and unique physics behind those mysteries. The crystal field ground state of $4f$ electron in CeB$_6$ is Γ_8 quartet, which is well separated from the excited state (Γ_7 doublet), giving an ideal Γ_8 system. Therefore we think that all these mysteries should be explained from the Γ_8 quartet with $J = 5/2$.

Let us note first that the local degrees of freedom of the Γ_8 quartet can be decribed by 4×4 hermitian matrices. Considering $J = 5/2$ in CeB$_6$, one obtains 15 nonvanishing moments (except for the trivial unit matrix) shown in Table 1. (The bars on the products in the Table means the symmetrized products: for example, $\overline{J_x J_y} \equiv J_x J_y + J_y J_x$.)

Notice that 3 types of octupole moments are present within $J = 5/2$ Γ_8-quartet in addition to two types of quadrupoles: Γ_4^+ (O_2^0, O_2^2) and Γ_5^+ (O_{xy}, O_{yz}, O_{zx}). Which multipole moments order depends very much on the nature of interaction between moments on different Ce ions.

2.1   Relation between AF Quadrupole Order and Field-Induced Moments

Before discussing quantitative problems, let us consider general properties based exclusively on symmetry[11]. Suppose that the long-range order of O_{zx} with wave vector $Q = (1/2, 1/2, 1/2)$ is realized. The uniform magnetic field applied along z induces a uniform dipole moment J_z. Then, it is clear from Landau theory of phase transition that a combination of J_z and AF O_{zx} leads to antiferro (AF) J_x and AF octupole moments T_x^α, T_x^β. This argument

Table 1. Multipole moments in $J = 5/2$ Γ_8-quartet

moment	representation	operators
dipoles	Γ_4^-	J_x, J_y, J_z
quadrupoles	Γ_3^+	$\frac{1}{2}(2J_z^2 - J_x^2 - J_y^2) \equiv O_2^0$
		$\frac{\sqrt{3}}{2}(J_x^2 - J_y^2) \equiv O_2^2$
	Γ_5^+	$\frac{\sqrt{3}}{2}\overline{J_y J_z} \equiv O_{yz}$
		$\frac{\sqrt{3}}{2}\overline{J_z J_x} \equiv O_{zx}$
		$\frac{\sqrt{3}}{2}\overline{J_x J_y} \equiv O_{xy}$
octupoles	Γ_2^-	$\frac{\sqrt{15}}{6}\overline{J_x J_y J_z} \equiv T_{xyz}$
	Γ_4^-	$\frac{1}{2}(2J_x^3 - \overline{J_x J_y^2} - \overline{J_z^2 J_x}) \equiv T_x^\alpha$
		$\frac{1}{2}(2J_y^3 - \overline{J_y J_z^2} - \overline{J_x^2 J_y}) \equiv T_y^\alpha$
		$\frac{1}{2}(2J_z^3 - \overline{J_z J_x^2} - \overline{J_y^2 J_z}) \equiv T_z^\alpha$
	Γ_5^-	$\frac{\sqrt{15}}{6}(\overline{J_x J_y^2} - \overline{J_z^2 J_x}) \equiv T_x^\beta$
		$\frac{\sqrt{15}}{6}(\overline{J_y J_z^2} - \overline{J_x^2 J_y}) \equiv T_y^\beta$
		$\frac{\sqrt{15}}{6}(\overline{J_z J_x^2} - \overline{J_y^2 J_z}) \equiv T_z^\beta$

Table 2. Relation between AF quadrupole order and induced moments

field	AF quadrupole moment	induced AF dipoles	induced AF octupoles
$(0,0,1)$	O_2^0	J_z	T_z^α
	O_2^2	–	T_z^β
	O_{xy}	–	T_{xyz}
	O_{yz}	J_y	T_y^α, T_y^β
	O_{zx}	J_x	T_x^α, T_x^β
$(1,1,1)$	$O_2^0, 2O_{xy} - O_{yz} - O_{zx}$	$2J_z - J_x - J_y$	$2T_z^\alpha - T_x^\alpha - T_y^\alpha, T_x^\beta - T_y^\beta$
	$O_2^2, O_{yz} - O_{zx}$	$J_x - J_y$	$2T_z^\beta - T_x^\beta - T_y^\beta, T_x^\alpha - T_y^\alpha$
	$O_{xy} + O_{yz} + O_{zx}$	$J_x + J_y + J_z$	$T_x^\alpha + T_y^\alpha + T_z^\alpha, T_{xyz}$
$(1,1,0)$	O_2^0, O_{xy}	$J_x + J_y$	$T_x^\alpha + T_y^\alpha, T_x^\beta - T_y^\beta$
	O_2^2	$J_x - J_y$	$T_x^\beta + T_y^\beta, T_x^\alpha - T_y^\alpha$
	$O_{yz} + O_{zx}$	J_z	T_{xyz}, T_z^α
	$O_{yz} - O_{zx}$	–	T_z^β

can be extended to any field direction for various AF quadrupole orders (see Table 2). It shows that the underlying AF quadrupole order can be inferred from induced AF dipole moments and octupole moments. Since induced AF dipole moments can be detected by neutron scattering, the relation in Table 2 can be used to identify AF quadrupole order, which is difficult to detect directly.

As a corollary of this result, one can similarly discuss what moments are induced additionally by AF magnetic order having the same AF wave vector, if it coexists with AFQ order. In that case it is generally expected that a uniform moment is necessarily induced, leading to a canted AF magnetic order; the direction of the uniform moment depends on the quadrupole order. We believe this phenomenon to occur in TmTe below T_N[12][13].

2.2 Interaction between Multipole Moments

Since the exact form of intersite interaction among multipole moments is difficult to determine microscopically, one has to guess which interaction is important. There are two limiting cases from which one can start, when an AF quadrupole moment orders first as the temperature decreases. One is that the interaction between quadrupole moments is large and other interactions can be ignored as a first approximation. Another case is that the interactions among all multipole moments are equally important. We believe the latter is more suitable in the case of CeB$_6$ for the following reasons. First, the AF quadrupole transition temperature increases almost linearly with the magnetic field, which is expected from the latter model. Second, CeB$_6$ is a metal, in which the interaction mediated by conduction electrons must be important and gives coupling among higher multipoles with nearly the same magnitude[14].

Assuming only the nearest neighbor interaction among multipoles and including the Zeeman term, we obtain the following Hamiltonian:

$$\mathcal{H} = \sum_{(ij)} \sum_{\alpha\beta} X_i^\alpha D^{\alpha\beta} X_j^\beta + g\mu_B \sum_j \boldsymbol{J}_j \cdot \boldsymbol{H}, \tag{1}$$

where $X_i^\alpha (\alpha = 1, \cdots, 15)$ represents 15 moments in Table 1 of Ce at the i-th site. The summation is taken over the nearest neighbor pairs. If we take an equal coupling for all the multipoles for simplicity (i.e., $D^{\alpha\beta} = D\delta_{\alpha\beta}$ for $\alpha, \beta = 1, \cdots, 15$; $D > 0$), we arrive at the SU(4) model, which was first proposed by Ohkawa[15]. The SU(4) model is too symmetric and leads to high degeneracy, which is unrealistic for CeB$_6$[11]. As a minimal modification, let us enhance the coupling for Γ_5-type quadrupoles (O_{xy}, O_{yz}, O_{zx}) slightly as $D^{\alpha\beta} = D(1+\lambda)\delta_{\alpha\beta}$ for O_{xy}, O_{yz}, O_{zx} and $D^{\alpha\beta} = D\delta_{\alpha\beta}$ otherwise. λ is an increment of the coupling; for CeB$_6$ $\lambda \sim 0.2$ is probably appropriate.

One obtains from the model (1) an AF order of O_{xy}, O_{yz}, or O_{zx} with $(1/2, 1/2, 1/2)$ at $H = 0$. At $H \neq 0$, depending on the direction of the magnetic field, either O_{xy} (for $H \parallel [001]$), $(O_{xy} + O_{yz} + O_{zx})/\sqrt{3}$ (for $H \parallel [111]$), or $(O_{yz} + O_{zx})/\sqrt{2}$ (for $H \parallel [110]$) is realized. Table 2 uniquely determines the direction of the AF dipole moment, which is consistent with available neutron experiments[9].

The most important consequence of (1) is that the magnetic field induces a fairly large AF octupole moment T_{xyz} for any direction of the field[16]. Actually this is the main stabilization mechanism of AFQ order in the magnetic field. The AF dipole moment is small and even vanishes for $H \parallel [001]$, when the AF order of O_{xy} is realized (see Table 2). If we take into account the AF octupole moment, the discrepancy between NMR and neutron scattering can be resolved as shown in [17].

The octupole moments appear to be important also in "phase III", which is a mixed phase of AF magnetic order and AF quadrupole order with different wave vectors[18][19]. There are other problems in CeB$_6$, which need to be studied further. One is the fluctuation spectrum of multipole moments in the ordered state[20] and the effect of fluctuations on the phase transition[21].

The present analysis on CeB$_6$ suggests that high multipole moments beyond quadrupoles, which are often overlooked, sometimes play an important role in f-electron systems. This is related to the problem of hidden order parameter.

3 Orbital Order in Manganites

Let us consider next the orbital order in $3d$ electrons, choosing cubic LaMnO$_3$ as a typical case[22]. In Mn^{3+}, one of doubly degenerate e_g orbitals is occupied by one electron, which is Hund-coupled with 3 t_{2g} electrons. There are $2\times 2 (=4)$ degrees of freedom within e_g. Similarly to Γ_8 quartet in CeB$_6$ there are 16 local operators to describe the degrees of freedom in e_g, which are summarized in Table 3 except for the trivial identity operator. ℓ is the orbital angular momentum operator whose magnitude is 2. Since the orbital dipole moment is quenched within e_g, magnetic moments are carried only by spins. τ_z and τ_x are Γ_3-type quadrupoles; τ_y is the octupole. σ is the Pauli matrix to describe spin degrees of freedom. The orbital order, if it exists, corresponds to an arrangement of Γ_3-type quadrupoles (τ_z, τ_x). $\sigma\tau_i$ ($i = z, x, y$) are coupled spin-orbital operators. For Mn^{3+} in a cubic field the Hund coupling between t_{2g} and e_g electrons has to be taken into account.

Table 3. Local operators in e_g orbital

operator	time reversal symmetry
$\tau_z \equiv -(1/6)(2\ell_z^2 - \ell_x^2 - \ell_y^2)$	+
$\tau_x \equiv (\sqrt{3}/6)(\ell_x^2 - \ell_y^2)$	
$\tau_y \equiv -(\sqrt{3}/18)\ell_x \ell_y \ell_z$	−
σ	−
$\tau_z \sigma$	−
$\tau_x \sigma$	
$\tau_y \sigma$	+

In contrast to f-electrons, where it is difficult to determine the interaction microscopically, the interaction between moments in LaMnO$_3$ can be described well by an extended Hubbard model $\mathcal{H} = \mathcal{H}_0 + \mathcal{H}_t$, where

$$\mathcal{H}_0 = U \sum_{i\tau} n_{i\tau\uparrow} n_{i\tau\downarrow} + U' \sum_{i\sigma\sigma'} n_{ia\sigma} n_{ib\sigma'} - J \sum_i \mathbf{s}_{ia} \cdot \mathbf{s}_{ib}$$
$$- K \sum_i (\mathbf{s}_{ia} + \mathbf{s}_{ib}) \cdot \mathbf{S}_i, \qquad (2)$$

$$\mathcal{H}_t = \sum_{\langle ij \rangle} (t_{ij}^{\tau\tau'} c_{i\tau\sigma}^\dagger c_{j\tau'\sigma} + \text{H.c.}). \qquad (3)$$

Here $\mathcal{H}_0$ is an atomic Hamiltonian for $3d$ electrons in Mn: U (U') and J (K) are the Coulomb interaction on the same (different) orbital and the Hund coupling on e_g orbitals (between e_g and t_{2g}), respectively[23]. The electrons on t_{2g} are described by S ($S = 3/2$). $t_{ij}^{\tau\tau'}$ is the orbital-dependent hopping integral between nearest neighbor Mn^{3+} ions via oxygen[1]. Although approximate, $\mathcal{H}_0$ describes fairly well the Coulomb interaction among $3d$ electrons, if the parameters are chosen adequately.

The Mott-insulating state can be most conveniently studied by using an effective Hamiltonian derived from (2) and (3) by assuming the Coulomb interaction is much larger than the hopping. Since $(3d)^4$ in Mn^{3+} form the total spin $I = 2$ due to the Hund coupling K, the effective Hamiltonian should be expressed in terms of I. The result is as follows[22]:

$$\mathcal{H} = \sum_\alpha \sum_{(ij)_\alpha} \left[-J_1(\mathbf{I}_i \cdot \mathbf{I}_j + 6)\frac{1}{2}(1 - \tau_i^\alpha \tau_j^\alpha) + J_2(\mathbf{I}_i \cdot \mathbf{I}_j - 4)\frac{1}{2}(1 - \tau_i^\alpha \tau_j^\alpha) \right.$$
$$\left. + J_3(\mathbf{I}_i \cdot \mathbf{I}_j - 4)\frac{1}{4}(1 + \tau_i^\alpha)(1 + \tau_j^\alpha) \right] + \sum_{(ij)} J_t \mathbf{I}_i \cdot \mathbf{I}_j, \quad (4)$$

where α denotes z, x and y directions of nearest neighbor pairs: $\tau^\alpha = \tau_z$ (for $\alpha = z$), $-\frac{1}{2}\tau_z - \frac{\sqrt{3}}{2}\tau_x$ (for $\alpha = x$) and $-\frac{1}{2}\tau_z + \frac{\sqrt{3}}{2}\tau_x$ (for $\alpha = y$). J_1, J_2 and J_3 are given by

$$J_1 = \frac{1}{10} \frac{t^2}{U' - \frac{1}{4}J}, \quad J_2 = \frac{3}{80} \frac{t^2}{U' - \frac{1}{4}J + \frac{5}{2}K} + \frac{1}{16} \frac{t^2}{U' + \frac{3}{4}J + \frac{3}{2}K},$$
$$J_3 = \frac{1}{4} \frac{t^2}{U + \frac{3}{2}K}. \quad (5)$$

J_t is the superexchange interaction due to t_{2g} electrons. The coupling between orbital and magnetic orders is evident in (4).

Quite recently Feiner and Oleś[24] have derived the same effective Hamiltonian as (4) starting from more complete Coulomb integrals instead of (2). The difference appears only in the expression for $J_1 \sim J_3$. According to Feiner and Oleś

$$J_1 = \frac{1}{10} \frac{t^2}{\varepsilon(^6A_1)}, \quad J_2 = \frac{3}{80} \frac{t^2}{\varepsilon(^4A_1)} + \frac{1}{16} \frac{t^2}{\varepsilon(^4E)}, \quad J_3 = \frac{1}{8}\left(\frac{t^2}{\varepsilon(^4E)} + \frac{t^2}{\varepsilon(^4A_2)} \right)$$
(6)

are obtained. Here ε is the excitation energy for each multiplet. Incidentally Ishihara et al.[23] derived a different effective Hamiltonian from (2); their result is not correct simply because the energy spectrum of (2) is not properly evaluated.

Clearly the energy denominator for J_1 is the smallest among (5) as

$$U' - J < U' - \frac{1}{4}J + \frac{5}{2}K, \; U' + \frac{3}{4}J + \frac{3}{2}K, \; U + \frac{3}{2}K. \quad (7)$$

The possible phases of (4) have been determined with the mean-field theory

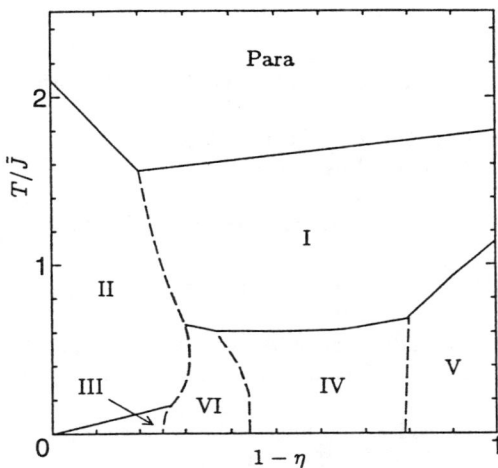

Fig. 1. Phase diagram for the model (4). The solid and broken lines represent 2nd and 1st order transitions, respectively.

by assuming for simplicity that the terms on the right-hand side of (7) are all equal and $(U' - J)/\eta$ $(0 < \eta < 1)$; we believe that $\eta \sim 0.5$ is appropriate for LaMnO$_3$. The effect of J_t has been checked by changing $\delta \equiv 8J_t/5J_1$; it has turned out that the results are rather insensitive to δ so that we have fixed to $\delta = 0.4$. These values are consistent with more reliable estimation by Feiner and Oleś[24]. The result is shown in Fig.1, where the nature of the phases I~VI is as follows: I (pure orbital ordering), II (G-type AF magnetic order without orbital order), III (G-type AF magnetic order accompanied by orbital order), IV (A-type AF magnetic order with orbital order), V (ferromagnetic order with orbital order) and VI (C-type AF magnetic order with orbital order). These phases can be described by a set of order parameters, which are mutually coupled due to the symmetry. The value of $\eta \sim 0.5$, which is presumably appropriate for LaMnO$_3$ naturally leads to the A-type antiferromagnetic state as observed experimentally. The anisotropy in the spin-wave stiffness constant is also reasonable.

So far we have discussed the insulating phase. What is left is the nature of metallic ferromagnetic phase, which shows up under sufficient doping on LaMnO$_3$, i.e. La$_{1-x}$Sr$_x$MnO$_3$ for instance. Experimentally there is no evidence of orbital order in the metallic phase. The nature of orbital degrees of freedom in the ferromagnetic metallic state is one of theoretical problems currently studied[25][26][27].

The authors wish to thank O. Sakai, P. Thalmeier and T. Nishitani for their contribution on the work reported here.

References

1. K. I. Kugel and D. I. Khomskii: Sov. Phys. JETP **37**, 725 (1973); Sov. Phys. Uspekhi **25**, 231 (1982)
2. See for instance P. Morin and D. Schmitt: *Ferromagnetic Materials* Vol.5, ed. by K. H. J. Buschow and E. P. Wohlfarth (Elsevier, 1990), p.1.
3. J. Kanamori: this Proceedings.
4. T. Fujita, M. Suzuki, T. Komatsubara, S. Kunii: T. Kasuya and T. Ohtsuka: J. Phys. Soc. Jpn. **35**, 569 (1980).
5. M. Kawakami, S. Kunii, K. Mizuno, M. Sugita: T. Kasuya and K. Kume: J. Phys. Soc. Jpn. **50**, 432 (1981).
6. M. Takigawa, H. Yasuoka, T. Tanaka and Y. Ishizawa: J. Phys. Soc. Jpn. **52** (1983) 728; M. Takigawa, Ph. D. Thesis (University of Tokyo, 1982, unpublished)
7. E. Zirngiebl, B. Hillebrands, S. Blumenröder, G. Güntherodt, M. Loewenhaupt, J. M. Carpenter, K. Winzer and Z. Fisk: Phys. Rev. B**30**, 4052 (1984).
8. N. Sato, S. Kunii, I. Oguro, T. Komatsubara and T. Kasuya: J. Phys. Soc. Jpn. **53**, 3967 (1984).
9. J. M. Effantin, J. Rossat-Mignod, P. Burlet, H. Bartholin, S. Kunii and T. Kasuya: J. Mag. Mag. Mater. **47&48** (1985) 145; W. A. C. Erkelens, L. P. Regnault, P. Burlet and J. Rossat-Mignod: J. Mag. Mag. Mater. **63&64**, 61 (1987).
10. B. Lüthi, S. Blumenröder, B. Hillebrands, E. Zirngiebl, G. Güntherodt and K. Winzer: Z. Phys. B**58**, 31 (1984).
11. R. Shiina, H. Shiba and P. Thalmeier: J. Phys. Soc. Jpn. **66**, 1741 (1997).
12. R. Shiina and H. Shiba: to be published in Physica B.
13. T. Sakakibara: private communication.
14. H. Teitelbaum and P. Levy: Phys. Rev. B**14**, 3058 (1976).
15. F. J. Ohkawa: J. Phys. Soc. Jpn. **52** (1983) 3897; ibid. **54** (1985) 3909.
16. R. Shiina, O. Sakai, H. Shiba and P. Thalmeier: J. Phys. Soc. Jpn. **67**, 941 (1998).
17. O. Sakai, R. Shiina, H. Shiba and P. Thalmeier: J. Phys. Soc. Jpn. **66**, 3005 (1997).
18. O. Sakai, R. Shiina, H. Shiba and P. Thalmeier: preprint.
19. M. Sera and S. Kobayashi: preprint.
20. P. Thalmeier, R. Shiina, H. Shiba and O. Sakai: J. Phys. Soc. Jpn. **67**, 2363 (1998).
21. G. Uimin, Y. Kuramoto and N. Fukushima: Solid State Commun. **97**, 595 (1996).
22. R. Shiina, T. Nishitani and H. Shiba: J. Phys. Soc. Jpn. **66**, 3159 (1997).
23. S. Ishihara, J. Inoue and S. Maekawa: Phys. Rev. B**55**, 8280 (1997).
24. L. F. Feiner and A. M. Oleś: preprint.
25. S. Ishihara, M. Yamanaka and N. Nagaosa: Phys. Rev. B**56**, 686 (1997).
26. A. Takahashi and H. Shiba: Euro. Phys. J. (in press).
27. P. Horsch, J. Jaklic and F. Mack: cond-mat/9807255.

Part III

Manganites

Anisotropic Charge Dynamics in Layered Manganite Crystals

Y. Tokura[1,2], T. Kimura[2], and T. Ishikawa[1]

[1] *Department of Applied Physics, University of Tokyo, Tokyo 113, Japan*

[2] *Joint Research Center for Atom Technology (JRCAT), Tsukuba 305, Japan*

Abstract

Anisotropic charge dynamics has been investigated for single crystals of layered manganites, $La_{2-2x}Sr_{1+2x}Mn_2O_7$ ($0.3 \leq x \leq 0.5$). Remarkable variation in the magnetic structure as well as in the charge-transport properties is observed with changing doping-level x. A crystal with $x=0.3$ behaves like a 2-dimensional ferromagnetic metal at the temperature region between $\sim$90 K and $\sim$270 K, and shows the interplane tunneling magnetoresistance at lower temperatures which is sensitive to the interplane magnetic coupling between the adjacent MnO_2 bilayers. Optical probe for these layered manganites has also clarified highly anisotropic and incoherent charge dynamics.

I. INTRODUCTION

The recent observations of the large negative magnetoresistance (MR) effect have shed renewed light on the study of perovskite manganites, producing a great deal of interest in underlying physics. Recently, extensive studies [1–6] have also been performed in the so-called Ruddlesden-Popper (RP) structure series for manganese oxides which are formulated as $(RE, AE)_{n+1}Mn_nO_{3n+1}$ (RE and AE being trivalent rare earth or divalent alkaline earth ions, respectively). The basic structure in this homologous series is based on alternate stacking of rock-salt-type block layers $(RE, AE)_2O_2$ and n MnO_2-sheets along the c-axis, as shown in the insets to Fig. 1. Among them, the $n=2$ member of the RP series, $RE_{2-2x}AE_{1+2x}Mn_2O_7$, shows a wide variety of physical properties like the $n=\infty$ (perovskite) analogs, including large MR [3] and magnetostriction [4,7] effects related to paramagnetic insulator$-$to$-$ferromagnetic metal transition. The charge-ordering transition has also been observed in the bilayered manganite with the carrier concentration $x=0.5$ [8]. One of the

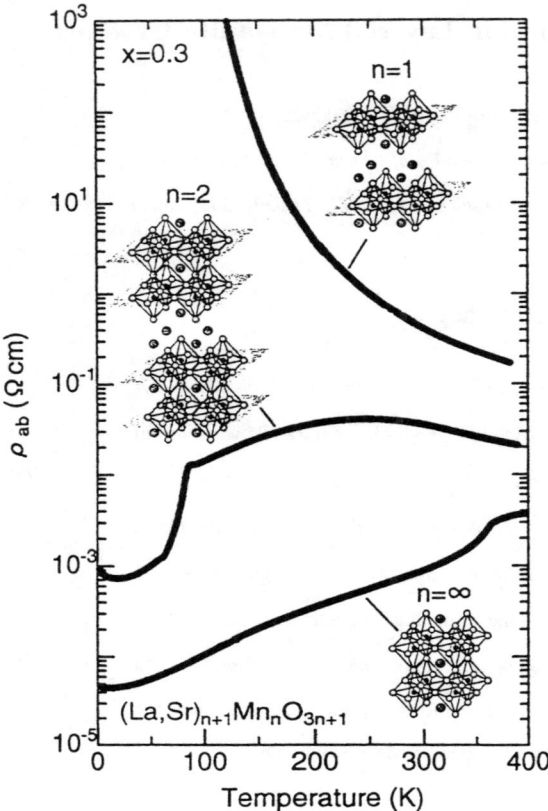

FIG. 1. Temperature dependence of inplane resistivity (ρ_{ab}) for $n=1$, 2, and ∞ members of the series of Ruddlesden-Popper phases $(La,Sr)_{n+1}Mn_nO_{3n+1}$. The nominal hole concentration is fixed at $x=0.3$. Inset: The crystal structure of $(La,Sr)_{n+1}Mn_nO_{3n+1}$ ($n=1$, 2, and ∞). Shaded planes represent MnO_2 layers.

most distinctive features for the bilayered manganite is its anisotropic characters (the resistivity ratio $\rho_c/\rho_{ab} > 10^2$) due to the layered structure.

The interplane tunneling magnetoresistance (TMR) effect arising from such a large anisotropy in the layered manganites may provide a novel approach to the large MR attainable at low magnetic fields [9]. The bilayered compound can be viewed as composed of ferromagnetic-metallic (FM) MnO_2 bilayers with intervening insulating (I) $(La,Sr)_2O_2$ blocks. In other words, the bilayered manganite intrinsically contains the infinite arrays of FM/I/FM tunneling junctions in its crystal structure. In such a quasi-two-dimensional (quasi-2D) FM, the interplane as well as inplane charge dynamics (and hence the MR characteristics) is expected to critically depend on the interlayer magnetic coupling between the FM MnO_2 bilayers. Here we show the results of a systematic study on the magneto-transport and optical properties of single crystals of the bilayered manganite, $La_{2-2x}Sr_{1+2x}Mn_2O_7$, with various carrier concentrations.

II. RUDDLESDEN-POPPER SERIES OF HOLE-DOPED MANGANITES

A series of $La_{2-2x}Sr_{1+2x}Mn_2O_7$ ($n=2$) single crystals as well as $La_{1-x}Sr_{1+x}MnO_4$ ($n=1$) and $La_{1-x}Sr_xMnO_3$ ($n=\infty$) with various doping levels ($0.3 \leq x \leq 0.5$) was grown by the floating zone method as reported previously [3]. Figure 1 shows the temperature dependence of the ρ_{ab} in the series of RP phases ($n=1,2,$ and ∞) with the nominal hole concentration fixed at $x=0.3$. The ρ_{ab} as well as its temperature dependence changes from metallic to semiconducting, with decreasing the number (n) of MnO_2 sheets per unit cell.

In the $n=\infty$ compound, the steep drop of the resistivity at $\sim$360 K corresponds to the onset of ferromagnetic ordering which is driven by the double exchange interaction. The $n=1$ phase with an isolated MnO_2 sheet, on the other hand, does not undergo the ferromagnetic-metallic transition and remains insulating down to the lowest temperature. In the $n=2$ compound, ρ_{ab} shows a semiconducting temperature dependence above the room temperature. With decreasing temperature, ρ_{ab} shows a broad maximum around $\sim$270 K, and then shows a metallic temperature dependence in contrast to a semiconducting behavior of ρ_c. This result implies that the short-range ferromagnetic correlation which extends only within MnO_2 bilayers evolves with decrease of temperature below $\sim$270 K [9]. With further decreasing temperature the long-range spin-ordering takes place around $T_c \approx 90$ K where both ρ_{ab} and ρ_c steeply decrease into a low-temperature region. Below T_c the spin correlation extends over the adjacent MnO_2 bilayers, and reduces the spin scattering of the conduction electron in the transport process along the c-axis as well as the ab-plane.

III. PRESSURE-ENHANCED INTERPLANE TUNNELING MAGNETORESISTANCE

Let us focus on the pressure and magnetic field effects on the resistivity in $La_{2-2x}Sr_{1+2x}Mn_2O_7$ ($x=0.3$). Figures 2(a) and (b) show the temperature profiles of ρ_{ab} and ρ_c at several pressures under magnetic fields of 0 and 3 T, respectively. All these data

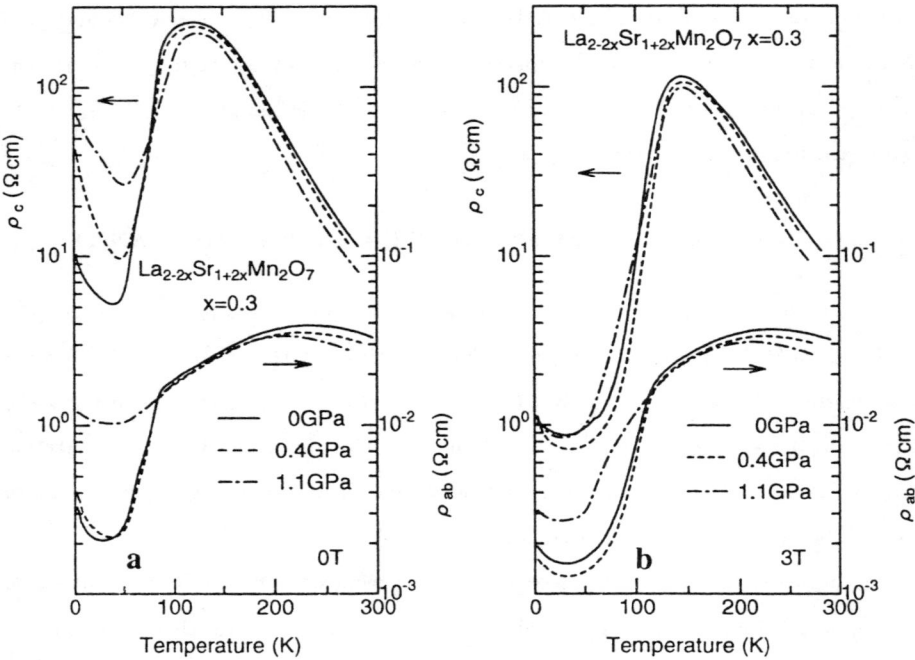

FIG. 2. Temperature dependence of inplane (ρ_{ab}) and interplane (ρ_c) resistivity under magnetic fields of 0 T (a) and 3 T (b) with $H \| c$ at pressures of 0, 0.4, and 1.1 GPa in the La$_{2-2x}$Sr$_{1+2x}$Mn$_2$O$_7$ ($x=0.3$) crystal.

were taken in the warming run. In the temperature range above T_c, the both ρ_{ab} and ρ_c monotonically decrease with increasing pressure under the magnetic fields of 0 T and 3 T. A similar reduction of resistivity has also been observed in $n=\infty$ phase, La$_{1-x}$Sr$_x$MnO$_3$ [10]. The results indicate that the application of pressure enhances the transfer interaction for e_g electrons and causes the increase in conductivity as far as the temperature-region above T_c is concerned.

The most striking feature is found in the temperature region below T_c. Both the ρ_{ab} and ρ_c are remarkably increased by applying pressure at zero magnetic field, as displayed in Fig. 2(a). In other words, the steep drops of ρ_{ab} and ρ_c are suppressed by applying pressure, which signals the pressure-suppression of the long-range spin-ordering. Under the pressure of 1.1 GPa the ρ_{ab} shows no drop due to the spin-ordering but dull temperature dependence characteristic of the 2D metallic state down to the lowest temperature. A similar feature is observed in the pressure effect in ρ_c. Compared with the magnetic susceptibility data suggestive of the evolution of the interplane antiferromagnetic (AF) coupling below $T_N \approx 60$ K as well as of its critical pressure-surpression [11], the pressure-induced change in the resistivity below T_c may have an intimate connection to the change of the interplane magnetic coupling. In other words, the short-range 2D spin ordering appears to extend down to the lowest temperature under high pressure, which sensitively affects the charge-transport. By contrast, under the magnetic field of 3 T the pressure effects on ρ_{ab} and ρ_c become considerably small below T_c, as seen in Fig. 2(b). By applying high magnetic fields, the 3D ferromagnetic spin-arrangement should be realized even at high pressures. This may explain the rather weak pressure-dependence of resistivity at high magnetic fields. For an origin of the pressure effect on the interlayer coupling, we may need to consider the orbital degrees of freedom of the e_g-like conduction electrons.

The field dependence of ρ_c has been measured with $H \| c$ at 4.2 K under the pressures of 0 and 0.7 GPa, as displayed in Fig. 3. For these measurements, the crystals were slowly cooled from room temperature to 4.2 K at 7 T, and then magnetic fields were swept cyclically. At ambient pressure the ρ_c rapidly decreases with increasing magnetic fields, and becomes nearly constant above the saturation field $H_{sat} \approx 0.2$ T. The interplane MR,$[\rho_c(0$ T$)-\rho_c(0.8$ T$)]/\rho_c(0.8$ T$) \sim 490$ %, is much larger than the inplane MR, $[\rho_{ab}(0$ T$)-\rho_{ab}(0.8$ T$)]/\rho_{ab}(0.8$ T$) \sim 25$ % [11]. This has been ascribed to the presence of the interplane tunneling MR process for the current-perpendicular-to-plane (CPP) configuration. Namely, the c-axis transport of the spin-polarized electron is blocked at the insulating block due to the AF-type coupling between the adjacent MnO_2 bilayers, but the magnetization process removes such AF-coupling boundaries and allows the interplane tunneling of spin-polarized electrons. In addition, we have observed the non-linear $I-V$ characteristics in ρ_c, suggestive of a FM/I/FM tunneling process [13,14].

As noted in Fig. 3, the magnitude of the interplane MR is drastically enhanced up to $\sim 4,000$ % by applying pressure. This is because the ρ_c value at zero field remarkably increases by high pressure while remaining nearly pressure-independent at magnetic fields more than H_{sat} [see and compare Fig. 2(a) with (b)]. H_{sat} is increased slightly with the increase of pressure. Such an enhanced MR effect can be closely related to the reduction of the interplane magnetic coupling. The weak pressure-dependent ρ_c at $H \geq H_{sat}$ suggests that the pressure effect is small in the absence of the magnetic domain boundaries. By contrast, at low temperatures ($\leq T_c$) ρ_c shows strong pressure-enhancement at zero field.

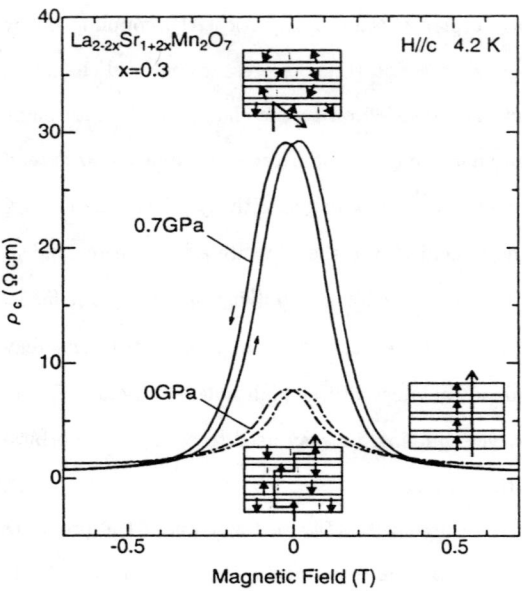

FIG. 3. Normalized interplane resistivity as a function of a magnetic field parallel to the c axis at 4.2 K in a $La_{2-x}Sr_{1+2x}Mn_2O_7$ ($x=0.3$) crystal. Inset shows schematic spin arrangement; (a) At ambient pressure, the MnO_2 bilayers exhibit the interlayer antiferromagnetic ordering, but include the ferromagnetic one as defects. The spin-polarized electrons may move along the c-axis in an inhomogeneous path, through the ferromagnetically-ordered region, although the antiferromagnetically-ordered region serves as a barrier. (b) The application of pressure suppresses the interlayer magnetic coupling, and make the conducting pathes in ferromagnetically-ordered region extinguish. (c) Applications of a magnetic field aligns all the magnetic moments irrespective of pressure. Spin-polarized electrons can move along the c-axis over the whole region of the specimen.

This indicates that conduction electrons moving along the c-axis in the pressure-induced 2D FM state suffer from even stronger scattering than those in the weakly AF-coupled state at ambient pressure.

In the real material, magnetic domains must be present within each ferromagnetic MnO_2 bilayer below T_N at ambient pressure. The spin arrangement for this concept are illustrated in the insets of Fig. 3. The regions of the interplane AF alignment are therefore incorporated with local regions of the interplane F alignment [inset (a) to Fig. 3]. The regions of the interplane F alignment which remain as defects in the globally AF state, are hence likely to reduce ρ_c and lead to the relative decrease in the interplane MR. (If this is the case, the microscopic current path for the measurement of ρ_c must be highly inhomogeneous.) The suppression of the interplane coupling by pressure may decouple these defects of the F alignment, as well as the global AF-coupled state, and cause the increase of the resistivity at zero field [inset (b) to Fig. 3]. At $H \geq H_{sat}$ the decoupled spin domain of each MnO_2 bilayer is aligned along the field direction, which can be viewed as the transition from the 2D ferromagnetic (paramagnetic along the c-axis) to the 3D ferromagnetic state [inset (c) to Fig. 3]. This arises from the fact that the 2D FM state is a highly diffuse metal as seen in $\rho_{ab}-T$ curves at 1.1 GPa of Fig. 2(a). The pressure-enhanced inplane MR thus reflects the deconfinement transition of the spin-polarized carriers, being reminiscent of the spin-valve effect [15].

IV. POSSIBLE ENHANCEMENT OF MR DUE TO AF FLUCTUATIONS

The metallic regime in ρ_{ab} above T_c disappears by degrees with increasing the nominal hole concentration. We display in Fig. 4(a) and (b) the temperature profiles of the ρ_{ab} in $x=0.3$ and 0.4 crystals under several magnetic fields. In the $x=0.4$ crystal, the both ρ_{ab} and ρ_c show a steep increase toward T_c with a large activation energy($\approx$30meV), which contrasts with the metallic ρ_{ab} in the $x=0.3$ crystal. The MR ratio in the $x=0.4$ crystal immediately above T_c [$\rho_{ab}(0\ T)$-$\rho_{ab}(7\ T)$]/$\rho_{ab}(7\ T)\approx 2\times 10^4$ %] is much larger than that in the $x=0.3$ crystal [$\rho_{ab}(0\ T)$-$\rho_{ab}(7\ T)$]/$\rho_{ab}(7\ T)\approx 3\times 10^2$ %]. One of the possible origins for this anomalous behavior of the $x=0.4$ crystal is related to the charge-/orbital-ordering which takes place in specimens with the hole concentration near $x=0.5$. The semiconducting resistivity above T_c has been observed also in pseudo-cubic perovskite manganites at a hole concentration near $x=0.5$ with controlled one-electron band width [16]. In these systems, the AF charge-ordering instability competes with the ferromagnetic double-exchange interaction near above T_c, and may be relevant to the semiconducting behavior. The competition gives rise to the phase instability and the extremely large MR at temperatures immediately above T_c due to the field suppression of AF fluctuation. For bilayered compounds, recent neutron scattering measurements revealed that the long-lived AF clusters within bilayers coexist with ferromagnetic critical fluctuations above T_c for the $x=0.4$ crystal [17]. The evolution of the semiconducting behavior and the enhanced MR effect with increasing the hole concentration

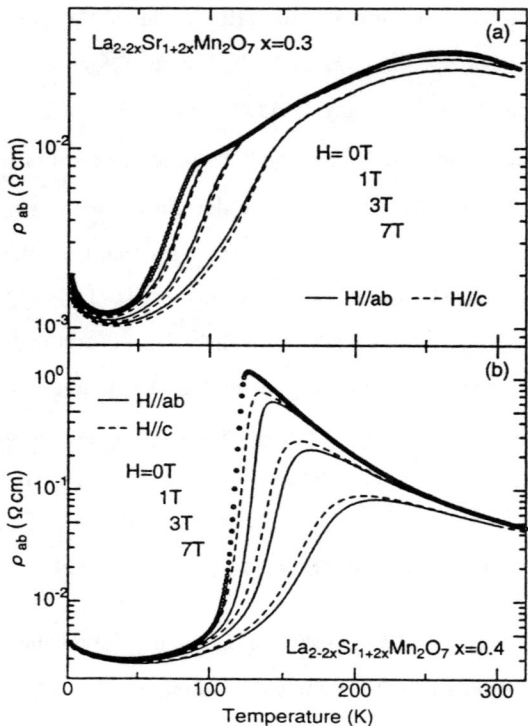

FIG. 4. Temperature dependence of ρ_{ab} under various magnetic fields with two field orientations ($H\|c$ and $H\perp c$), in $La_{2-2x}Sr_{1+2x}Mn_2O_7$ [x=0.3 (a) and 0.4 (b)] crystals. The appreciable differences near T_c between the field directions parallel to the c-axis and the ab-plane are related to the easy axis of magnetization ($H\|c$ for x=0.3 crystal, $H\perp ab$ for x=0.4 crystal).

arises from the localization effect due to the inplane AF correlation, that is perhaps related to the aforementioned charge-/orbital-ordering.

V. OPTICAL PROBE OF CHARGE DYNAMICS

To investigate the anisotropic and temperature(T)-dependent charge dynamics in the layered manganite crystal with the ferromagnetic ground state, we have measured optical conductivity spectra for a single crystal of $La_{2-2x}Sr_{1+2x}Mn_2O_7$ (x=0.4).

We show in the lower panel of Fig. 5 the T dependence of the optical conductivity spectra for $E \parallel ab$ and $E \parallel c$ for a $La_{2-2x}Sr_{1+2x}Mn_2O_7$ (x=0.4) which were deduced by

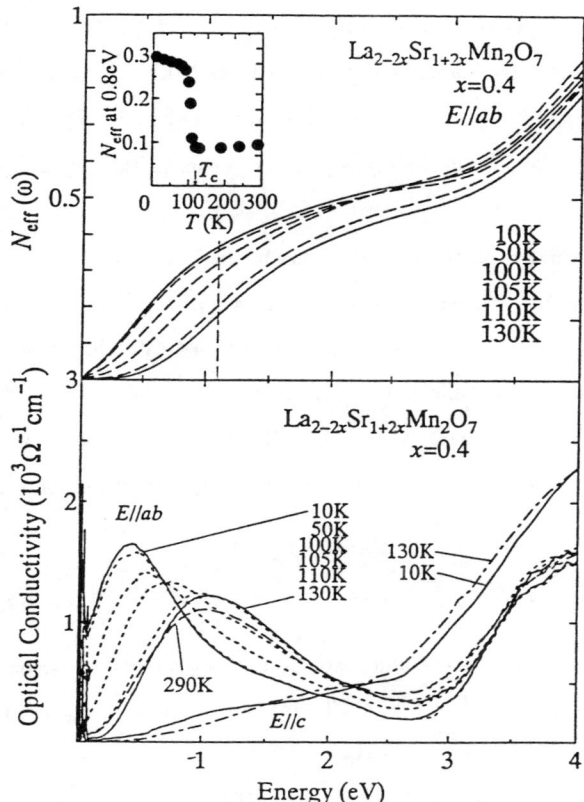

FIG. 5. Upper panel: Temperature-variation of spectra of the effective number of electrons $N_{eff}(\omega)$, which were deduced by integrating the observed $E \parallel ab$ $\sigma(\omega)$. The inset shows the temperature dependence of N_{eff} at 0.8eV as a measure of the kinetic energy of the conduction electrons(see the text). Lower panel: Temperature dependence of $E \parallel ab$ and $E \parallel c$ optical conductivity spectra for a $La_{2-2x}Sr_{1+2x}Mn_2O_7$ ($x=0.4$) crystal with the 3D spin-ordering temperature $T_c=121K$. Solid lines represent the spectra at 10K and 130K.

Kramers-Kronig analysis of the reflectivity data. At a glance, one may notice a conspicuous T-dependent change for the in-plane ($E \parallel ab$) spectra up to 3 eV. By contrast, the $E \parallel c$ spectra shows a minimal T-dependent change, except for slight accumulation of the spectral weight below 1.5 eV with decreasing T below T_c. Both the $E \parallel ab$ and $E \parallel c$ spectra show an onset around 3 eV, peaking at 4 eV. This peak can be assigned to the charge-transfer type transition between the O$2p$ and Mn t_{2g}-like (down-spin) states [18,19]. The spectra below 3 eV are dominated by the intra- and interband transitions relevant to O$2p$ and Mn

63

e_g-like states. The in-plane conductivity spectrum shows a broad peak around 1eV above T_c, forming the pseudogap ($\approx$ 0.2eV) structure. As T decreases from 290 K to just above T_c, the 1eV-peak height increases and the gap feature becomes clearer, although the overall feature is rather unchanged. Thus the optical-conductivity spectra confirms the insulating behavior above T_c.

With decreasing T below T_c a large spectral change occurs in the $E \parallel ab$ spectra over an energy range of 0–3 eV. Importantly, the transferred spectral weight does not form the Drude-like coherent peak centered at $\omega = 0$, while peaking around 0.4 eV. This signals unconventional charge dynamics in the metallic state of this layered manganite. Here, let us first argue the overall spectral-weight transfer as observed. To estimate the transferred spectral weight, we have calculated the effective number of electrons (N_{eff}), which is defined as

$$N_{\text{eff}}(\omega) = \frac{2m}{\pi e^2 N} \int_0^\omega \sigma(\omega') d\omega'. \tag{1}$$

Here, N represents the number of Mn atoms per unit volume. The N_{eff} spectra at various temperatures are shown in the upper panel of Fig. 5. Taking account of the increasing experimental error in $\sigma(\omega)$ with ω, we may view that the $N_{\text{eff}}(\omega)$ curves tend to merge into a nearly single line above 3 eV. In other words, the T-dependent distribution of the spectral weight ($N_{\text{eff}} \approx 0.5$) occurs nearly below 3 eV, which is relevant to the doped holes or the band-filling e_g-like electrons. The inset to Fig. 5 (upper panel) shows the T dependence of N_{eff} at $\omega_c = 0.8$ eV, an appropriate cut-off energy for the estimate of the transferred low-energy spectral weight. Between 290 K and 120 K $N_{\text{eff}}(\omega_c)$ is nearly constant. Once the compound undergoes the 3D spin ordering transition, $N_{\text{eff}}(\omega_c)$ show a conspicuous increase in accord with the occurrence of the metallic conduction.

Similar spectral change with temperature or with evolution of the ferromagnetic magnetization was also observed for the underdoped pseudocubic perovskite manganite crystals such as barely metallic $La_{1-x}Sr_xMnO_3$ around x=0.2 [20,21] and insulating (semiconducting) $Nd_{1-x}Sr_xMnO_3$ (x =0.3) [22]. Such a large spectral change as observed in the perovskite manganites has been argued in various contexts: (1) the T-dependent change of the density of state for the spin-polarized e_g-state conduction bands with a large exchange splitting exceeding the band width [20,21,23–25], (2) the Jahn-Teller polaron band with T-dependent *effective* coupling strength [26,22], and (3) the T-dependent change of the interband transitions between the different-orbital branches at non-specific k points [27,28]. All the respective scenarios may be relevant, more or less, to the observed features, perhaps depending on the energy region.

The overall feature of the T-dependent change of the in-plane low-energy conductivity spectrum is similar to the case of the 3D analog, $La_{1-x}Sr_xMnO_3$, yet some distinct features can be noticed for this quasi-2D ferromagnet in addition to the inherent anisotropic behavior. The spectral weight is rapidly accumulated with decrease of T, but forms the broad peak around 0.4 eV. The formation of such a midinfrared peak implies the small-polaronic conduction in this system. According to a conventional small-polaron model [29], the peak energy of the $\sigma(\omega)$ corresponds approximately to twice the polaron binding energy. In fact, the 290 K (or higher-temperature) in-plane spectrum can be crudely fitted with a set of polaron parameters (binding energy E_b= 0.535 eV, hopping energy J= 0.59 eV). However, such a calculation completely fails to reproduce the shape of the lower-temperature $\sigma(\omega)$ spectrum that shows a subsisting low-energy (down to ω=0 eV) tail as well as a pronounced peak structure around 0.4eV. As a more suitable model, we may consider the dynamical mean-field calculation (infinite-dimensional approach) on the dynamic Jahn-Teller effect [the above scenario (2)] by Millis et al. [26]. In fact, the calculated Jahn-Teller polaron spectra with the electron-phonon coupling strength $\lambda \approx 1.08$ [26] can qualitatively reproduce the observed temperature-dependent feature that the 1 eV-peak around and above T_c is gradually shifted to a lower energy (down to 0.4 eV) with increasing spectral weight. (In this calculation, J_H is assumed infinite, so that the higher-energy scale structure than the Jahn-Teller coupling energy ($\approx$ 1 eV) cannot be discussed.) To be more quantitative about the singular shape of the low-temperature $\sigma(\omega)$ for the *metallic* ground state, however, the consideration of the collective nature of the dynamic Jahn-Teller distortion or the short-range orbital ordering would be necessary for this quasi-2D system beyond the infinite-dimensional approach [26].

In the present quasimetallic layered manganites (x=0.4) the carrier motion is diffuse down to near zero frequency even in the fully spin-polarized ground state. Namely, the strong scattering of the spin-polarized carriers by bosonic excitations such as phonons and orbital excitations gives rise to a broad midinfrared peak around 0.4 eV in the optical-conductivity spectra, while the lower-energy spectral weight (Drude weight) is extremely small. The orbital fluctuations originating from the pseudo degeneracy of the e_g-like state or the associated dynamical Jahn-Teller distortion forming the polaronic carriers is likely responsible for the diffuse charge transport and incoherent low-energy spectral feature as revealed in this study.

VI. ACKNOWLEDGMENT

We would like to thank Y. Tomioka, T. Okuda, A. Asamitsu, H. Kuwahara, T. Katsufuji, R. Kumai, T. G. Perring, G. Aeppli, and M. Tamura for their collabolation and enlightening discussions. This work was supported in part by NEDO and the Ministry of Education, Japan.

REFERENCES

[1] R. A. M. Ram, P. Ganguly, and C. N. R. Rao, J. Solid State Chem. **70**, 82 (1987).

[2] Y. Moritomo, Y. Tomioka, A. Asamitsu, Y. Tokura, and Y. Matsui, Phys. Rev. B **51**, 3297 (1995).

[3] Y. Moritomo, A. Asamitsu, H. Kuwahara, and Y. Tokura, Nature **380**, 141 (1996).

[4] J. F. Mitchell, D. N. Argyriou, J. D. Jorgensen, D. G. Hinks, C. D. Potter, and S. D. Bader, Phys. Rev. B **55**, 63 (1997).

[5] P. D. Battle, M. A. Green, N. S. Lasky, J. E. Millburn, M. J. Rosseinsky, S. P. Sullivan, and J. F. Vente, Chem. Commun., 767 (1996); P. D. Battle, M. A. Green, N. S. Lasky, J. E. Millburn, P. G. Radaelli, M. J. Rosseinsky, S. P. Sullivan, and J. F. Vente, Phys. Rev. B **54**, 15967 (1996).

[6] H. Asano, J. Hayakawa, and M. Matsui, Appl. Phys. Lett. **71**, 844 (1997).

[7] D. N. Argyriou, J. F. Mitchell, C. D. Potter, S. D. Bader, R. Kleb, and J. D. Jorgensen, Phys. Rev. B **55**, R11965 (1997).

[8] J. Q. Li, Y. Matsui, T. Kimura, and Y. Tokura, Phys. Rev. B **57**, R3205 (1998).

[9] T. Kimura, Y. Tomioka, H. Kuwahara, A. Asamitsu, M. Tamura, and Y. Tokura, Science **274**, 1698 (1996).

[10] Y. Moritomo, A. Asamitsu, and Y. Tokura, Phys. Rev. B **51**, 16491 (1995).

[11] T. Kimura, A. Asamitsu, Y. Tomioka, and Y. Tokura, Phys. Rev. Lett. **79**, 3720 (1997).

[12] S. Ishihara, S. Okamoto, and S. Maekawa, J. Phys. Soc. Jpn. **66**, 2965 (1997).

[13] M. Julliere, Phys. Lett. A **54**, 225 (1975).

[14] S. Maekawa and U. Gäfvert, IEEE Trans. Magn. **MAG-18**, 707 (1982).

[15] B. Dieny, V. S. Speriosu, S. S. P. Parkin, B. A. Gurney, D. R. Wilhoit, and D. Mauri, Phys. Rev. B **43**, 1297 (1991).

[16] H. Kuwahara, Y. Tomioka, Y. Moritomo, A. Asamitsu, M. Kasai, R. Kumai, and Y. Tokura, Science **272**, 80 (1996); S. Tokura, H. Kuwahara, Y. Moritomo, Y. Tomioka, and A. Asamitsu, Phys. Rev. Lett. **76**, 3184 (1996); H. Kuwahara, Y. Moritomo, Y. Tomioka, A. Asamitsu, M. Kasai, R. Kumai, Y. Tokura, Phys. Rev. B **56**, 9386 (1997).

[17] T. G. Perring, G. Aeppli, Y. Moritomo, and Y. Tokura, Phys. Rev. Lett. **78**, 3197 (1997).

[18] T. Arima, Y. Tokura, and J. B. Torrance, Phys. Rev. B **48**, 17006 (1993); T. Arima and Y. Tokura, J. Phys. Soc. Jpn. **64**, 2488 (1995).

[19] Y. Moritomo, T. Arima, and Y. Tokura, J. Phys. Soc. Jpn. **64**, 4117 (1995).

[20] Y. Okimoto, T. Katsufuji, T. Ishikawa, A. Urushibara, T. Arima, and Y. Tokura, Phys. Rev. Lett. **75**, 109 (1995).

[21] Y. Okimoto, T. Katsufuji, T. Ishikawa, T. Arima, and Y. Tokura, Phys. Rev. B **55**, 4206 (1997).

[22] S. G. Kaplan, M. Quijada, H. D. Drew, D. B. Tanner, G. C. Xiong, R. Ramesh, C. Kwon, and T. Venkatesan, Phys. Rev. Lett. **77**, 2081 (1996).

[23] N. Furukawa, J. Phys. Soc. Jpn. **64**, 2734 (1995).

[24] N. Furukawa, J. Phys. Soc. Jpn. **64**, 3164 (1995).

[25] D.S. Dessau, C.-H. Park, T. Saitoh, P. Villella, Z.-X. Shen, Y. Moritomo, Y. Tokura, and N. Hamada, preprint.

[26] A. J. Millis, R. Mueller, and B. I. Shraiman, Phys. Rev. B **54,** 5405 (1996).

[27] H. Shiba, R. Shiina, and A. Takahashi, J. Phys. Soc. Jpn. **66,** 941 (1997).

[28] P.E. de Brito and H. Shiba, Phys. Rev. B **57,** 1539 (1998).

[29] M. I. Klinger, Phys. Status Solidi **11,** 499 (1965); **12,** 765 (1965); H. G. Reik and D. Hesse, J. Phys. Chem. Solids **28,**581 (1967).

Roles of Orbitals in Transition Metal Oxides

Y. Endoh[1], K. Hirota[1], Y. Murakami[2], H. Nojiri[3], S. Ishihara[3],
S. Maekawa[3], H. Kimura[1], T. Fukuda[4], N. Okamoto[3]

[1] Department of Physics, Graduate School of Science, Tohoku University,
Aramaki Aoba, Aoba-ku, Sendai, 980-8578, Japan, CREST
[2] Photon Factory, Institute of Materials Structure Science, KEK
1-1, Ohomachi, Tsukuba, Ibaraki, 305-0801, Japan, CREST
[3] Institute of Material Research, Tohoku University,
Katahira, Aoba-ku, Sendai, 980-8577, Japan, CREST
[4] SPring8, Japan Atomic Energy Research Institute,
Mihara, Mikazuki-cho, Sayo-gun, Hyogo, 679-2740, Japan

Abstract Neutron and synchrotron x-ray scattering experiments with additional bulk measurements such as the magnetization, and resitivity under high magnetic field have been performed from single crystals of the colossal magneto resistance (CMR) manganese oxides of $La_{1-x}Sr_xMnO_3$. Systematic experimental investigations show decisively that the complicated phase diagram around $x \approx 0.12$ has been comprehended as the first order phase transition from metal to insulator applying magnetic field driven by the orbital order. The ferromagnetic insulating phase is not originated by the Jahn Teller interaction but is by the electron correlation, since the crystal symmetry is pseudo cubic. The novel CMR mechanism is proposed based on the present experimental investigations.

1 Introduction

Discovery of the high temperature superconductivity (HTS) in copper oxides by Bednortz and Muller [1] in 1986 stimulated the elucidation on inherent relation between magnetism and conductivity in the strongly correlated electron systems. Doping of holes in quantum antiferromagnetic insulating copper oxides, such as La_2CuO_4, gives rise to the Metal-Insulator (MI) transition and at the same time, it induces the novel HTS in a certain doping concentration range. However, it is still a big issue whether quantum spin fluctuations persisting in such a novel superconducting Cu oxides is a necessary ingredient for the realization of the novel HTS.

Meanwhile, the MI transition accompanied with the colossal magnetoresistance (CMR) effect in the manganese oxides has been elucidated along this line established in the studies for the strongly correlated electron systems [2]. Namely, $LaMnO_3$ of the parent material of CMR is recognized as the Mott Insulator (Charge Transfer type like La_2CuO_4). The double exchange interaction in these doped manganese oxides was first proposed in 1950s [3], but the presently discussed CMR effect cannot be comprehensive within this simple context of the double exchange mechanism. Furthermore unusual properties of the CMR effect are regarded as a cooperative phenomena associated with a structural change due to a tiny atomic displacement, competing magnetic interactions and charge fluctuations between different valencies of manganese cations. In this respect, Mn^{3+} orbitals are Jahn-Teller (JT) active [4], hence the degeneracy can easily lift by lowering the crystal symmetry costing the lattice distortion energies. Therefore it is believed that the CMR should result of the effect of the local distortion which is often defined as "polarons". The itinerancy of such polarons give rise to the conduction. However, there has been no definitive experimental result showing the existence of the polarons except speculative interpretation of the data of electrical conductivity. Nonetheless, it is well recognized that the JT interaction plays a crucial role to determine the superexchange interactions in these transition metal oxides.

$LaMnO_3$ undergoes the cooperative JT transition blow 780 K of the orthorhombic crystal structure (O*) and lifts the degenerated e_g orbital, which orders such that nearest neighbor z axis of $d_{3z^2-r^2}$ is alternately aligned with $90°$ or of antiferro-type order in the C plane. The type 1 antiferromagnetic structure in $LaMnO_3$ has been interpreted by the strong ferromagnetic superexchange interaction acting in the C plane with weak antiferromagnetic interaction perpendicular to the plane [5]. In this respect, if $d_{x^2-y^2}$ orbital of the e_g band is stabilized, oxygen ions at apex perpendicular to the basal plane approach closer to the central cation, hence the crystal undergoes to the tetragonal symmetry. In fact, the tetragonal structure in La_2CuO_4 of the parent compound of the HTS is well recognized as the JT distortion.

In this presentation, we focus on dynamical properties of magnetism or spin dynamics as well as the 'orbital' order (OO) in the CMR materials and discuss that the CMR effect can be comprehended by the role of the "orbital" degree of the freedom. In order to investigate this subject, we have conducted a decisive observation of the OO in the typical CMR manganese oxides. we therefore briefly describe both the experimental method and results [6]. Then we will argue that the electron correlations also acts a rather significant key role for the OO and eventually the CMR effect in the manganese oxides.

The format of the paper is as follows. Magnetic properties, in particular the dynamical features are summarized, taking the role of the OO into discussions of the infinite layered CMR manganese oxides. Then the experimental results will be followed by the description of the principle of detection of OO. The recent experiments from the $La_{0.88}Sr_{0.12}MnO_3$ single crystal will be presented, which exhibits the ferromagnetic insulator to metal transition at 145 K driven by the OO. Then the final section is devoted to discussions of our recent experimental results.

2 Magnetic properties in the CMR manganese oxides

The mother compound of CMR manganese oxides, $LaMnO_3$ consists of infinitely stacked layers of MnO_2 intervened by LaO layers, which was determined to be the type I antiferromagnetic insulator (CT type Mott Insulator) in 1955 [7]. A strongly anisotropic spin-wave dispersion curve at low temperatures represented as the quasi 2 dimensional (2D) ferromagnetic spin-wave reflects the ferromagnetic super-exchange interaction in the C plane assisted by the JT interaction as stated in the preceding section [5]. The spin wave dispersion relation of $LaMnO_3$ was then well interpreted by the spin Hamiltonian of the sum of the nearest neighbor ferromagnetic in the C plane and the weak antiferromagnetic interaction along the direction perpendicular to the plane with the single ion anisotropy. It should be noted that the fitted parameters are also compared with the recent theoretical calculation based on the Local Spin Density Approximation (LSDA) [8].

In this way, inelastic neutron scattering measurement proves the ferromagnetic superexchange interaction due to the OO of $d_{3x^2(3y^2)-r^2}$, but recently, as will be described in the following section, the occurrence of the OO has been directly detected by the synchrotron radiation experiment [9]. It should be noted here that appearance of the ferromagnetic superexchange interaction realized in $LaMnO_3$ has been a subtle issue, because the bond angle is far from $180°$ due to the crystal distortion of $LaMnO_3$, and the ground state of d orbital sensitively depends also on the local crystalline field around Mn sites.

The small doping of holes by substitution of Sr to La (x=0.05 in $La_{1-x}Sr_xMnO_3$) enhances the anisotropy in the spin-wave dispersion relation, namely the dispersion curve in the C plane becomes steeper, but on the other hand, it is nearly flat along the C axis [10]. The anisotropy energy or a spin-wave gap decreases, which is not compatible with the fact of spin canting towards the perpendicular direction of the C plane, together with the smaller JT distortion.

Further increase of holes, ($x \geq 0.1$), makes the drastic change in magnetism; the spin-wave dispersion relation becomes rather isotropic with no appreciable energy gap in spin-wave excitations, which implies that the dimensional cross-over occurs in the spin dynamics around $x \approx 0.1$ [10]. It should be noted that the electric conduction in $La_{0.9}Sr_{0.1}MnO_3$ is not completely transformed to be metallic but rather complicated showing characteristic temperature dependence as shown in the last section in this paper. The magnetic transition temperature monotonically increases upon doping beyond $x \approx 0.3$ suggesting the exchange interaction stronger. In fact, the spin-wave stiffness constants, $\mathscr{D}$ derived from the quadratic dispersion relation in a small q region: $w_q = \mathscr{D}q^2$ coincide with T_C in this doping range x ($x \leq 0.3$) [11].

The CMR effect in $La_{0.7}Ca_{0.3}MnO_3$ is enhanced near the the Curie temperature (T_C), where thermal evolution of spin dynamics has been reported to show an anomalous feature [15]. Spin excitations remain to be well defined but the ferromagnetic order disappears rather discontinuously. It is unusual and characteristic that a so called elastic diffusive component is strongly enhanced near the Curie temperature. These characteristic features in the $La_{0.7}Ca_{0.3}MnO_3$ should be compared with those of $La_{1-x}Sr_xMnO_3$ ($x \approx 0.2 - 0.3$) near T_C, which behaves rather normal like those in conventional metallic ferromagnets [12]. Spin excitations there usually change the feature in a quite narrow temperature range close to T_C: they are all inelastic in the collective mode below T_C, and then at and above T_C, they condensate to the diffusive mode, where the center of the excitation spectrum always corresponds to zero energy, $\omega=0$. Scattering intensities or $S(q=0,\omega=0)$ of such a quasielastic mode critically diverges at T_C, and therefore this type of the quasielastic scattering is defined as the critical scattering.

The energy spectrum of magnetic critical scattering or the dynamical structure factor can be derived by using dynamical scaling hypothesis applicable to the localized spin system [13]. The $S(q,\omega)$ near T_C ($T>T_C$) can be treated as the double Lorenzian form as shown in the following equation for such isotropic Heisenberg ferromagnets.

$$S(q,\omega) = 2k_B T \chi(0) \frac{1}{\pi} \frac{\kappa_1^2}{\kappa_1^2 + q^2} \cdot \frac{\Gamma}{\Gamma^2 + \omega^2} \tag{1}$$

Here, $\chi(0)$, κ_1 and Γ are defined, respectively, the static magnetic susceptibility, the inverse correlation length and the energy width determined by the diffusion constant. Γ and κ_1 in equation (1) is further rewritten by using the heuristic relations of the dynamical scaling law, just shown below,

$$\Gamma = A q^{2.5} f\left(\frac{\kappa_1}{q}\right)$$

$$\kappa_1 = \kappa_0 \left(\frac{1 - \frac{T}{T_C}}{T_C}\right)^{\nu}$$

here, $f(x) = 1$ $x = 0$
   $f(x) \approx x^{0.5}$ $x = \infty$

Then, equation (1) is further converted to the following equation.

$$S(q,\omega) \propto \frac{T_C}{A} \kappa_0 \frac{q^{\frac{5}{2}}}{q^5 + \left(\frac{\omega}{A}\right)^2} \tag{2}$$

In our experience, the dynamical scaling function $f(q,\kappa_1)$ is valid for the most cases of ferromagnetic insulator and even for the ferromagnetic transition metals [14].

As stated in this section, $LaMnO_3$ is in fact understood to be the 2D ferromagnet, rather than the 3 dimensional (3D) antiferromagnet. The $La_{1-x}Sr_xMnO_3$ with $x \geq 0.1$, however, can be regarded as rather 3D Heisenberg (isotropic) ferromagnets. The dimensional cross-over from 2D to 3D in the spin dynamics therefore, appears in $La_{1-x}Sr_xMnO_3$ at $x \approx 0.1$, which we will focus on in the later section. In this respect, other CMR compounds consisting of the elements of smaller tolerance factors show unusual spin dynamics near T_C though the hole concentration x is the same.

The double exchange interaction has been believed to be responsible for the transition to ferromagnetic metal from the antiferromagnetic insulator. We emphasize, however, that the fact of dimensional cross-over in spin-wave excitations near at $x \approx 0.1$ in the non metallic state of $La_{1-x}Sr_xMnO_3$ is incompatible to the simple concept of the double exchange mechanism, which should be coincident with the metallic transition in the manganese oxides. Naturally, we must look for a novel concept, which should also interpret the CMR mechanism in manganese oxides.

3 Detection of orbital ordering by synchrotron x-ray scattering

As described in preceding sections, we speculate that the drastic change in both magnetic properties and conductivity accompanied by the crystal distortion might be controlled by a "hidden" parameter of the "orbital" degree of freedom. Then we might directly investigate whether the OO occurs accompanied with these phase transitions applying the recently established experimental method by using the synchrotron x-ray scattering technique.

By tuning the synchrotron radiation energies at the resonant transition energy between 1s core and 4p unoccupied electronic level corresponding to the energy of K edge, we could observe the orbital ordering structure. Murakami et al. first demonstrated both the charge (CO) and orbital order (OO) state in the single layered $La_{0.5}Sr_{1.5}MnO_4$ where Mn^{3+} and Mn^{4+} are mixed equally (x=0.5) [15]. Prior to this demonstration, the detailed magnetic structure below $T_N=110K$ and the lattice modulation due to the charge ordering was determined by neutron diffraction [16]. We briefly describe here the principle of this decisive experiment determining the OO followed by the results.

The most important character of the synchrotron radiation source is not only the brightness of the coherent x-ray beam but the energy tunability by continuous scans of the monochromator. Detection of the CO can be made by using a small (4eV) energy difference between the Mn^{3+} and Mn^{4+} K absorption edge at around 6.55 KeV, which also shows a high performance of the energy resolution of the synchrotron radiation.

How the scattering is enhanced near the K-absorption edge can be expressed in the following. The X-ray scattering factor is generally given in equation (3).

$$f(E) = f_o + f'(E) + if''(E) \qquad (3)$$

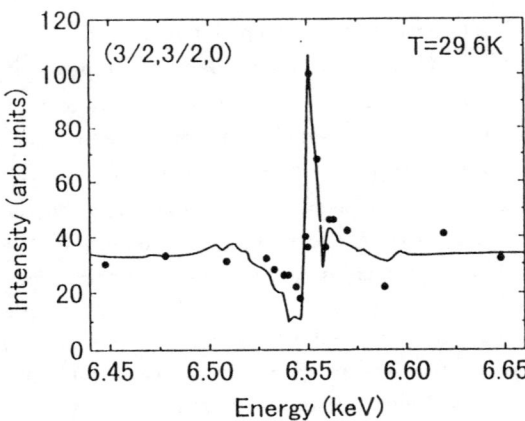

Fig 1 Energy dependence of charge ordering superlattice reflection (3/2, 3/2, 0) of $La_{0.5}Sr_{1.5}MnO_4$ near the manganese K edge. This scan was performed at T=29.6 K. Line drawn in the figure was calculated of the result of eq. (4) using determined f' and f" from the absorption experiments [15].

Here, f_o, $f'(E)$ and $f''(E)$ are respectively, the Thomson scattering factor, real and imaginary part of anomalous scattering factor. The latter two factors are energy dependent, which is largely enhanced near the K absorption edge. For instance, the $f''(E)$ term for Mn^{3+} and Mn^{4+} can be directly deduced from the absorption spectra of $LaSrMnO_4$ (Mn^{3+}) and $La_{0.5}Sr_{1.5}MnO_4$ (Mn^{3+}/Mn^{4+}=1). The energy dependence of real part, $f'(E)$, is deduced from the Kramers-Kronig transformation of the imaginary part. The evidence of the charge ordered state is then easily examined by applying this anomalous scattering.

The structure factor of the superlattice reflections caused by the CO, $(h,k,0)$ with h,k= half integers, is given by the following formula.

$$f_{(hk0)} \propto (f'_{3+}(E) - f'_{4+}(E)) + i(f''_{3+}(E) - f''_{4+}(E)) + C \qquad (4)$$

C is the energy independent scattering factor from the difference of the Thomson scattering related to the different valence contribution, which is readily separated from the energy scans of radiation photons. The result is shown in Fig.1, where the solid line is calculated by using the result of f"(E). The observed data are perfectly explained by the calculation, which gives rise to the direct evidence of the charge ordering of Mn^{3+} and Mn^{4+}. This result accords to the previously determined structure by neutron diffraction measurements.

The polarization dependence of the anomalous scattering factor arises in the anisotropy of the charge, which may be determined either as the bond or orbital. This fact was first pointed out by Dmitrienko, who defined the scattering of anomalous

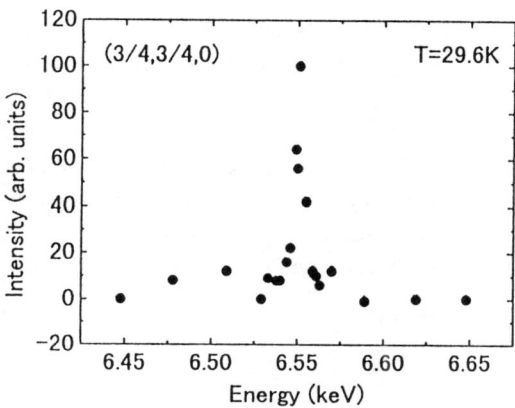

Fig.2 Energy dependence of orbital ordering superlattice reflection (3/4, 3/4, 0) of $La_{0.5}Sr_{1.5}MnO_4$ near the manganese K edge measured at T = 29.6 K. Temperature dependence shown in the insert indicates the simultaneous occurrence of both CO and OO below T_{CO} = 217 K [15].

tensor of charge susceptibility (ATS scattering), since the anomalous scattering amplitudes of f'(E) and f"(E) are written as the tensor [17]. Such a polarization dependent photo-absorption is readily known as the dichroism or the birefringence, in electromagnetic forces which rotates the polarization of the light. The essence of the polarization dependent phenomenon is the same as that of the anisotropy of the charge susceptibility caused through the resonant scattering, which is enhanced near the absorption edge. The ATS scattering means that the scattering appears at the reflection forbidden by the atomic configuration of the crystal. Taking $La_{0.5}Sr_{1.5}MnO_4$, in addition to a class of the forbidden (h,k,0) reflections with h,k of the half integer corresponding to the CO structure of Mn^{3+} and Mn^{4+}, the (3/4,3/4,0) is also the forbidden reflection which is only allowed by the orbital order of $(3z^2- r^2)$ type in e_g band of Mn^{3+}. The ATS scattering due to the orbital order was clearly shown by the careful experiments; various search of energy scans, polarization dependence (σ to π), (σ to σ) or (π to π) and the special scan with rotating the crystal around the scattering vector (azimuthal scan), which should be constant for the charge scattering at least. In other words, the amplitude of ATS scattering becomes a tensor rather than a scalar reflecting the anisotropy of the orbitals, which is explained below in more detail. Here, the ATS scattering results from $La_{0.5}Sr_{1.5}MnO_4$ are shown in Fig. 2, where the anomalous scattering of the ATS appears at the resonant energy of 6.55 KeV near the K edge and at the forbidden reflection of either (3/4,3/4,0) or (5/4.5/4,0). The thermal evolution also shows that the CO and orbital order occur simultaneously at 217 K.

Then theoretical investigations of the ATS scattering clarified the detailed mechanism of this demonstration just described above [18], which guides the principle of the ATS scattering for the orbital order of 3d electrons or/and the quadrupolar order in some rare-earth and actinide compounds. The resonant elastic scattering near the K absorption edge is due to the elastic dipole transition. In other words, the synchrotron light excites electrons from 1s to 4p and the resonant light is scattered resonating with the loss process of excited electrons. The resonant X-ray scattering process is already formulated [19]. The anomalous scattering factor of $\Delta f'(E)$ and $\Delta f''(E)$ described above is written in the following formula, where the electronic system is excited from the initial state, $|0\rangle$ with energy ε_0 to the intermediate state, $|1\rangle$ with ε_1, and is finally relaxed to the final state, $|f\rangle$ with ε_f,

$$\Delta f_{\alpha\beta} \propto \frac{e^2}{mc^2} \sum_1 \left\{ \frac{\langle f|j_{i\alpha}(-k')|1\rangle\langle 1|j_{i\beta}(k'')|0\rangle}{\varepsilon_0 - \varepsilon_1 - \omega_{k'} - i\delta} \right.$$

$$\left. + \frac{\langle f|j_{i\alpha}(k'')|1\rangle\langle 1|j_{i\beta}(-k')|0\rangle}{\varepsilon_0 - \varepsilon_1 + \omega_{k'} - i\delta} \right\} \quad (5)$$

$$j_{i\alpha} = \frac{eA_\sigma(\vec{k})}{m} \sum_\sigma P^+_{i\alpha\sigma} s_{i\sigma}$$

α,β : polarization of photons
$\omega_{k'(k'')}$: incident (scattered) photon energy with momentum $k'(k'')$
δ : constant
$A_\sigma(\vec{k})$: coupling constant   (4p (operator $P^+_{i\alpha\sigma}$) → 1s($s_{i\sigma}$))

Here the real and imaginary parts of $\Delta f_{\alpha\beta}$ are respectively $\Delta f'(E)$ and $\Delta f''(E)$. Regarding to this formula, the dichroism or birefringence of the light occurs by the same type of microscopic elastic dipole transition. Then a question arises why the ATS scattering occurs due to the 3d orbital order, or how the 3d orbital ordering reflects the anomalous scattering through 1s - 4p transition process. According to the detailed electronic structure of the MnO_6 cluster, it was found that the 4p levels above the Fermi level lift the degeneracy mainly by the strong Coulomb interaction between 3d and 4p electrons in the same manganese cation, which should be defined the intra-atomic Coulomb interaction. The p_z orbital lifts from both p_x and p_y, when e_g orbitals are polarized in the basal plane. In fact, the calculation showed that the energy difference of the split 4p levels due to the intra-atomic Coulomb interaction is 1.2 ± 0.6 eV, which is enough to induce the anomalous scattering in the resonant process. Then the scattering tensor can be calculated by assuming the 3d orbital ordering. The intensity of the forbidden interaction is also given in the following formula.

$$\overset{t}{f} = \begin{vmatrix} \Delta f_{xx} & 0 & 0 \\ 0 & \Delta f_{yy} & 0 \\ 0 & 0 & \Delta f_{zz} \end{vmatrix} \quad (6)$$

$$I \propto |\Delta f_{xx} - \Delta f_{zz}|^2 \quad (7)$$

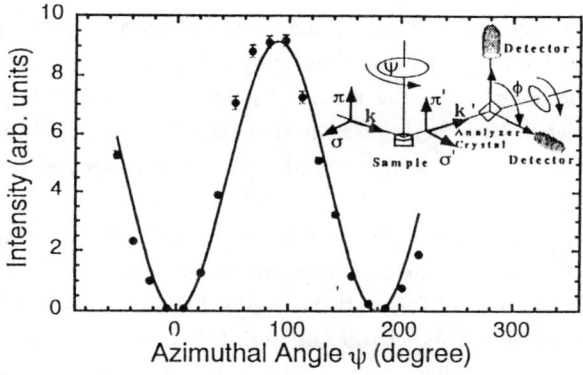

Fig.3 Experimental configuration for the detection of orbital ordering is shown in the insert, in which φ and ψ are defined as polarization analyzing angle and azimuthal angle, respectively. Azimuthal angle (ψ) dependence of scattering intensities of the orbital ordering reflection (3,0,0) of LaMnO$_3$ is proportional to sin2ψ, which was given by theory [9,18]

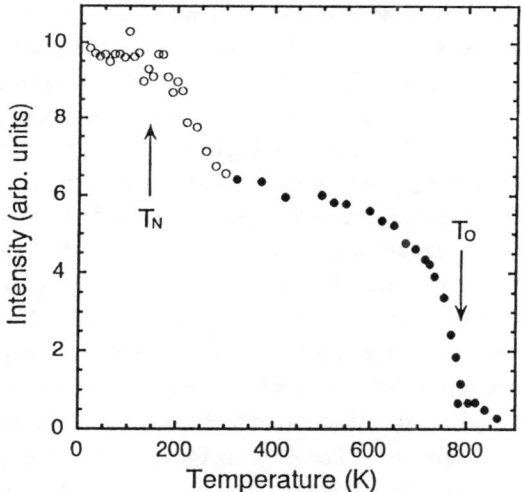

Fig.4 Temperature dependence of scattered intensities of the orbital ordering reflection (3,0,0) of LaMnO$_3$ at the x-ray energy of 6.555 KeV. T_O = 80 K and T_N = 140 K are respectively the structural change from cubic to orthorhombic induced by the Jahn Teller coupling and Neel temperature [9].

The $\Delta f''_{xx(zz)}$ for either occupied $d_{3z^2-r^2}$ or $d_{x^2-y^2}$ is the basis of the experimental observation of ATS scattering. Hence, the anisotropy of the d orbital reflects the tensor form in eq.(6) as well as the azimuthal angle dependence of the ATS as shown in Fig. 3.

Along with this theoretical study, the ATS scattering measurements have been carried out to observe the OO in LaMnO$_3$ [9]. All the experimental result revealed the antiferro-type of d$_{3z^2-r^2}$ orbital order where the z axis of polarization alternately directs in the C plane as predicted. New feature here is that the OO is nearly of the second ordered transition associated with the orthorhombic distortion at around 780 K as shown in Fig.4.

Now, we understand completely the microscopic mechanism of the ATS scattering and then we recognize that this method gives a decisive experimental probe to investigate the OO, hopefully the orbital fluctuations by the inelastic scattering. Since the microscopic coupling among spins, charges and orbitals acts an essential role to drive the macroscopic CMR characters in the manganese oxides, the combination of both neutron scattering and synchrotron x-ray scattering is highly desirable for thorough explorations for the microscopic mechanism of the CMR effect. In particular, the orbital component should be investigated experimentally in detail.

4 Ferromagnetic metal-insulator transition in La$_{0.88}$Sr$_{0.12}$MnO$_3$

La$_{1-x}$Sr$_x$MnO$_3$ is recognized to be simplest among many CMR manganese oxides and also it is the most extensively investigated material so far. Even so, the phase diagram of this system is already complicated, in particular near x ≈ 0.1, where several phase boundaries of insulator-metal, lattice symmetry as well as magnetic structure are entangled on the temperature (T) - doping concentration (x) diagram [20]. In other words, if we understand correctly this complexity in the phase diagram, we might find a possible clue to the CMR mechanism in the manganese oxides.

The CO, lattice distortion and magnetic properties are known to couple with each other around x ≈ 0.125 (1/8). We could grow a high quality single crystal by the conventional floating zone method using a lump image focusing furnace. After characterizing this single crystal, which is determined x=0.12, we systematically investigated both bulk and microscopic characters . Here, we summarize the essence of our experiments showing an unusual phase transition from ferromagnetic metal to insulator transition by either applying magnetic field or lowering temperature [6,21].

First, we confirmed that sequential phase transitions occur varying temperature [6,20], which is shown in Fig.5. Upon cooling from high temperatures, the crystal symmetry undergoes from pseudo-cubic to orthorhombic at T$_H$=291 K. Then ferromagnetic long range order (lro) occurs below T$_C$=172 K. The crystal symmetry changes again from orthorhombic to another pseudo cubic at T$_L$=145 K accompanied with the jump in magnetic structure. Above T$_L$=145 K, only (2,0,0) ferromagnetic component appears, which indicates simple 3D isotropic ferromagnetic order below T$_C$=172 K. Below T$_L$=145 K, very small antiferromagnetically ordered component appears, which implies that the magnetic moment simply incline from the principal axis by few degrees, probably 3 - 4 degrees. The electric resistivity jumps by as much as 10 times at T$_L$=145 K seen in Fig.6. Note that the conductivity below T$_C$ is rather

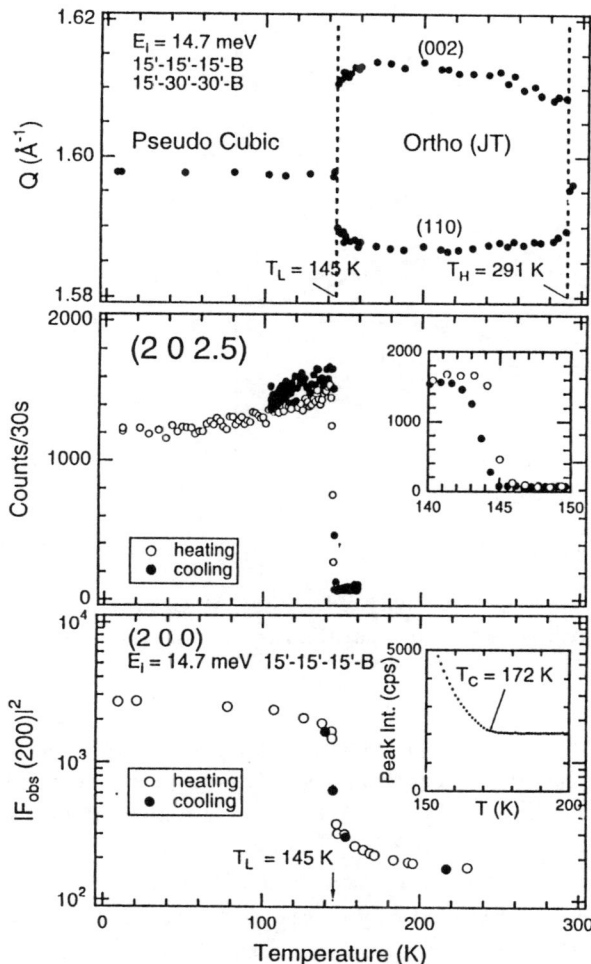

Fig.5 Temperature dependence of neutron scattering intensities from various reflections in $La_{0.88}Sr_{0.12}MnO_3$ showing sequential phase transitions at T_H = 290 K, T_C = 172 K and T_L = 145 K. Upper panel shows the structural transitions, middle is the CO transition and the lowest shows the first order magnetization jump at TL and ferromagnetic order at T_C.

metallic (dR/dT>0) though the conductivity itself is not as good as is in the normal metals. The conductivity is apparently isotropic which reflects the isotropic ferromagnetic nature determined both bulk magnetic properties and spin dynamics [10].

The most fascinating feature is that the resitivity jumps more in enhanced manner under the applied field near above T_L as shown in Fig.6. It means that at certain tmeperatures in this temperature range of $T_L<T<T_C$, there appears a striking phenomenon of the large positive magnetoresistance effect in the CMR manganese

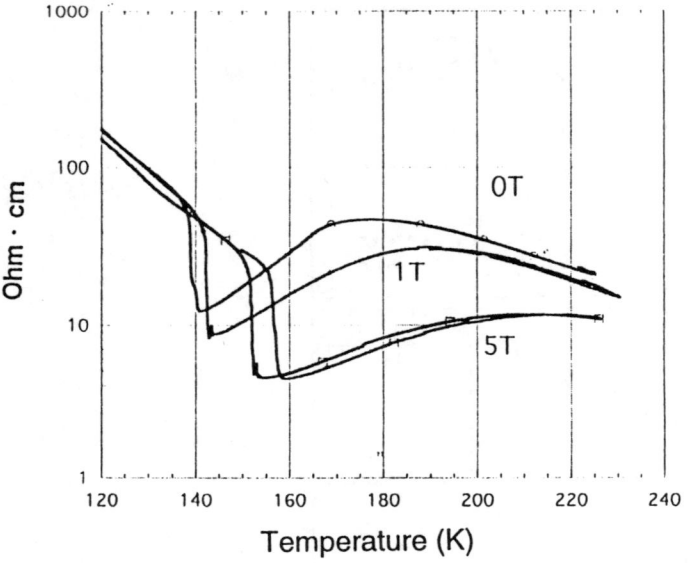

Fig.6 Temperature dependence of the resistivity of $La_{0.88}Sr_{0.12}MnO_3$ in H=0, 1T and 5T. Note that the the resistivity is plotted in logarithmic scale and the resistivity has no apparent anisotropy.

oxides. Furthermore, the crystal undergoes the transition to be less distorted below T_L seen in Fig.5. Therefore we speculated that this phase transition in the magnetic field should be continued from the zero field. In fact, we could confirm it by the extensive experiments at various magnetic field showing the metal-insulator transition temperature, T_L goes up with applied magnetic field ($dT_L/dH > 0$).

The synchrotron x-ray scattering as well as neutron scattering experiments show that the CO, OO and detailed structure of ferromagnetic lro simultaneously changes at T_L. Namely both CO and OO appears below T_L observing the distinct Bragg peaks at both forbidden reflections of $(h,k,l+1/2)$ with integer h,k,l and (h,k,l) with $h+k+l$=odd integer, respectively. The transition is definitely of the first order with distinct thermal hysteresis as well as the abrupt jump in intensity at T_L. This CO reflection can be expected to be observed by the model proposed by Y. Yamada [22], which indicates that Mn^{4+} cations localize at the center surrounded 8 Mn^{3+} units in every other C layers. In order to confirm that the Bragg peak at (0,3,0) is originated by OO, we have continued scans searching energy, azimuthal angle as well as polarization dependence. All of the results show consistently that $3D$ OO appears below T_L as depicted in Fig.7. We emphasize here that the OO of the Mn^{3+} cation sites presented here occurs in the undistorted crystal of the cubic or pseudo cubic symmetry, whereas the previously observed or predicted OO arises in the distorted lattice driven by the JT interaction. We already confirmed that the orthorhombic distortion disappears accompanied with the appearance of the charge ordering

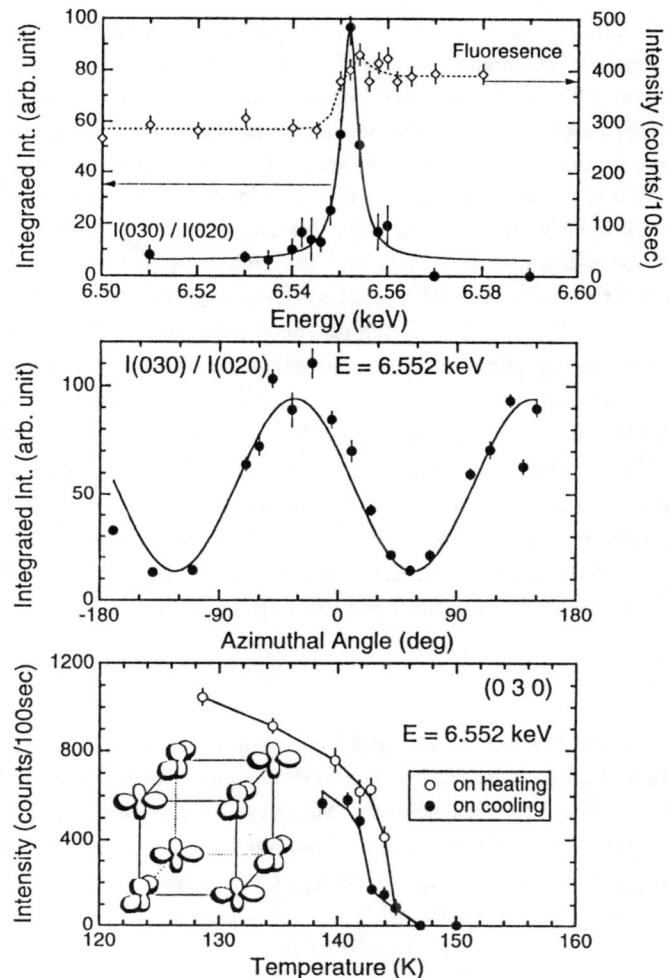

Fig.7 The experimental results of orbital ordering of $La_{0.88}Sr_{0.12}MnO_3$ below T_L = 145 K. Upper panel shows the energy dependence of the orbital ordering reflection at (0,3,0) indicating the resonance peak at 6.552 KeV, middle shows the azimuthal angle dependence of peak intensities and the lowest is the temperature dependence, indicating the transition is of the first order. The insert is theoretically proposed orbital order pattern.

reflection at $(h,k,l+1/2)$ under the magnetic field in the temperature range of $T_L<T<T_C$. Judging from an important fact of the isotropic $3D$ ferromagnetic nature, the antiferro-type OO is speculated not to be the same pattern in $LaMnO_3$, but rather unique orbital state of hybridization of $d_{z2-x2\ (y2-x2)}$ and $d_{3x2-r2\ (3y2-r2)}$.

Now, we interpret this experimental fact in terms of the recent theoretical result that two kinds of ferromagnetic phase can exist in different carrier concentration regions in this vicinity [6,23]. The result is based on the model Hamiltonian where both spin and orbital degrees of freedom are treated on an equal footing. The phase diagram is derived at T=0 with varying the exchange parameter of superexchange interaction of t_{2g} band. This mean field calculation also suggests the coexistence of two ferromagnetic phases associated with two different orbital structure in a certain x range between 0.08 and 0.42. This model calculation contains the common aspects with experimental evidence showing that the stabilization of the ferromagnetic insulator phase by both applying magnetic field and lowering temperature for the x=0.12 crystal : (1) both ferromagnetic ordering and antiferro-type orbital ordering are cooperatively stabilized in 3D manners, (2) the magnetic moment is enlarged by changing dominant magnetic coupling from double exchange to superexchange interaction. Then more importantly, this concept of the microscopic phase separation or the phase mixture for the intermediate range, $T_L<T<T_C$, can easily be extended to the CMR effects due to the fact that it occurs near the transition of the paramagnetic insulator to the ferromagnetic metal where the orbitals remain to be disordered. We argue therefore the present systematic investigations definitely show an important role of the orbital degree of the freedom on the metal-insulator transition in the CMR manganese oxides, which should also act a key role to the CMR effect.

Acknowledgement

Authors acknowledge dedicating assistance of H.Fujioka and M.Onodera in the single crystal preparation of manganese oxides. The work has partly been supported by a Grant in Aid for Scientific Research from the Ministry of Education, Science, Sports and Culture of Japan in addition to the Core Research for Evolutional Science and Technology (CREST) by the Japan Science Technology Corporation.

References

[1] J.G.Bednorz and K.A.Müller, Z. Phys., **B 64**, 189 (1986)
[2] Y.Tokura, A.Urushibara, Y.Moritomo, T.Arima, A.Asamitsu, A.Kido and
    H.Furukawa, J. Phys. Soc. Jpn., **63**, 3931 (1994)
[3] C.Zener, Phys. Rev., **82**, 403 (1951);
    P.W.Anderson and H.Hasegawa, Phys. Rev., **100**, 671 (1955);
    P.G.de Gennes, Phys. Rev., **118**, 141 (1960)
[4] J.B.Goodenough, Phys. Rev., **100**, 564 (1955);
    J.Kanamori, J. appl. Phys. **31**, 14S (1960)
[5] K.Hirota, H.Kaneko, A.Nishizawa and Y.Endoh, J. Phys. Soc. Jpn., **65**, 3736 (1996)
[6] Y.Endoh, K.Hirota, Y.Murakami, T.Fukuda, H.Kimura, H.Nojiri, K.Kaneko,
    S.Ishihara, S.Okamoto and S.Maekawa; unpublished
[7] E.O Wallan and W.C.Kohler, Phys. Rev., **100**, 545 (1955)

[8] I.Solvyev, N.Hamada and K.Terakura, Phys. Rev. Lett., **76**, 4825 (1996)

[9] Y.Murakami, I.Koyama, M.Tanaka, H.Kawata, J.P.Hill, D.Gibbs, M.Blume, T.Arima, Y.Tokura, K.Hirota and Y.Endoh, Phys. Rev, Lett., **81**, 582 (1998)

[10] K.Hirota, H.Kaneko, A.Nishizawa, Y.Endoh, M.C.Martin and G.Shirane, Physica B **237 - 238**, 36 (1997)

[11] Y.Endoh and K.Hirota, J.Phys. Soc. Jpn., **66**, 2264 (1997)

[12] M.C.Martin, G.Shirane, Y.Endoh, K.Hirota, Y.Moritomo and Y.Tokura, Phys. Rev. B **53**, R14285 (1996)

[13] P.Resibois and C.Piette, Phys. Rev. Lett., **24**, 514 (1970)

[14] Y.Endoh, Magnetism In Metals, Matematisk-fysiske Meddeleleser, **45**, 149 (1997)

[15] Y.Murakami, H.Kawada, H.Kawata, M.Tanaka, T.Arima, Y.Morotomo and Y.Tokura, Phys. Rev, Lett., **80**, 1932 (1998)

[16] B.Steinlieb, J.P.Hill, U.C.Wildgruber, G.M.Luke, B.Nachumi, Y.Moritomo and Y.Tokura, Phys. Rev. Lett., **76**, 2169 (1996)

[17] V.E.Dmitrienko, Acta Cryst., A **39**, 29 (1983); D.H.Templeton and L.K.Templeton, Acta Cryst., A **41**, 133 (1985)

[18] S.Ishihara and S.Maekawa, Phys. Rev. Lett., **80**, 3799 (1998)

[19] M.Blume, *Resonant Anomalous x-ray Scattering -Theory and Applications-* edited by G.Materik, C.J.Sparks and K.Fisher; Elsevier Science B.V. (1994)

[20] H.Kawano, R.Kajimoto, M.Kubota and H.Yoshizawa, Phys. Rev. B **53**, R14712 (1996)

[21] H.Nojiri, unpublished

[22] Y.Yamada, O.Hino, S.Nohdo, R.Kanao, T.Inami and S.Katano, Phys. Rev. Lett., **77**, 904 (1996)

[23] S.Okamoto, unpublished

Theory of Anomalous X-ray Scattering in a Variety of Orbital Ordered Manganites

Sumio Ishihara and Sadamichi Maekawa

Institute for Materials Research, Tohoku University, Sendai, 980-8577 Japan

Abstract. Colossal magnetoresistive manganites are widely recognized as a typical example of spin, charge and orbital coupled systems. Recently, the anomalous X-ray scattering by utilizing the synchrotron radiation has been applied to manganites in order to directly observe the orbital order. We theoretically study the anomalous X-rat scattering as a probe to detect the orbital state in manganites. We identify an origin of the anisotropy of the atomic scattering factor in the orbital ordered state which causes the scattering in the orbital superlattice reflection point. By utilizing the derived general formula of the polarization dependence of the scattering intensity, we propose a method to determine several types of the orbital order. The recent development of the anomalous X-ray scattering experiments in orbital ordered manganites is briefly reviewed.

1 Introduction

Since the discovery of High-Tc superconducting cuprates, physical and chemical properties in several transition metal oxides have been reinvestigated in modern view points. A typical example is perovskite manganites $R_{1-x}A_xMnO_3$ (R = La, Pr, Sm, A = Sr, Ca) and their related compounds where the colossal magnetoresistance was recently discovered [1, 2, 3, 4]. It is widely known that the conventional double exchange scenario can not explain the unique magneto transport phenomena observed in the metallic phase near the Mott insulator. One of the additional ingredient required is the orbital degree of freedom in a Mn^{3+} ion. The electron configuration in the ion is represented as $(t_{2g})^3(e_g)^1$ where one electron occupies the doubly degenerate e_g orbital. Therefore, the ion has the orbital degree of freedom, as well as spin and charge ones. Roles of the orbital in the magnetic properties in undoped manganites, where the long range orbital order is realized, were pointed out by Goodenough and Kanamori in 1950s [5, 6]. Furthermore, in doped case, it was proposed that the optical and transport properties are affected by the orbital states [7, 8, 9]. However, the experimental technique which directly detects the orbital order was limited to the case where the polarized neutron scattering is utilized [10]. Recently, Murakami *et al.* are successful in observation of the orbital order by applying the anomalous X-ray scattering in single layered manganites [11]. This unique experimental method was recognized immediately as a powerful technique to study the orbital state and was applied to several types of manganites thereafter [12, 13]. In this paper, we identify an origin of the anomalous X-ray scattering in orbital ordered state [14] and propose a method to determine several types of orbital order [15]. We also briefly review the experimental results of the anomalous X-ray scattering in orbital ordered manganites.

2 Overview of the Experimental Results

The polarized anomalous X-ray scattering was observed for the first time in sodium uranyl acetate in 1982 by Templeton and Templeton [16]. They measured the anomalous part of the scattering factor near the U L-edge. The anomalous part depends on the polarization of the incident and scattered X-ray, although the crystal structure is cubic, where the microscopic dichroism is not expected. It was concluded that this dichroism in the X-ray region is originated from the anisotropic configuration of the chemical bond in the crystal. Furthermore, through a prediction and an observation of the new extinction rule for the forbidden reflection [17], it was recognized that the anomalous X-ray scattering reflects the anisotropy in the microscopic electronic structure.

Recently, this experimental technique has been applied to observe orbital structure, which is flexibly changed by changing temperature, carrier concentration, an applied pressure, an external magnetic field etc. in comparison with chemical bond. Murakami et al. carried out the X-ray diffraction measurement in single layered manganites $La_{0.5}Sr_{1.5}MnO_4$ [11] where the charge and spin orderings have already been confirmed by the electron and neutron diffraction experiments, respectively. The orbital order is realized in the MnO_2 plane and supports the CE-type spin ordering. They observed the scattering intensity at (3/4 3/4 0) reflection point where it should disappear when the electronic structure in all Mn^{3+} ions is equivalent. The result implies the alternate alignment of the different types of the orbital in the [110] direction. The scattering intensity shows the resonant behavior near the K-edge of a Mn^{3+} ion with changing the incident photon energy. It suggests that the electric dipole transition between Mn $1s$ and Mn $4p$ orbitals causes the scattering and the conventional origins of the forbidden reflection originated from the normal part of the scattering factor, such as the asphericity of atomic electron density, the anharmonic thermal motion of atoms, the simultaneous (Renninger) reflection and so on, are excluded. The polarization analyses are carried out by rotating the sample crystal with respect to the linear polarized incident light. The scattering intensity shows a squared sinisoidal curve in contrast to that at the fundamental Bragg reflection which is independent of the rotation angle. The result implies that the scattering factor has a tensor character with respect to the polarization. It is also revealed that the orbital and charge orderings occur concomitantly at $210K$ through the measurement of the temperature dependence. They also detected the orbital ordering in the undoped manganites $LaMnO_3$ [12].

Furthermore, this unique method was applied to doped perovskite manganites $La_{0.88}Sr_{0.12}MnO_3$ [13]. In this compound, an existence of the two kinds of the ferromagnetic phases, which are termed F_1- ($T < 145$) and F_2- ($145 < T < 172$) phase, hereafter, was confirmed by the high magnetic field measurement and the neutron scattering experiments. F_1- and F_2-phases are identified as the insulating and metallic ferromagnetic phases, respectively. The structural phase transition from O' to O^* phases occurs at $145K$ with decreasing temperature. In the anomalous X-ray scattering experiment, it is surprising that the scattering intensity at the (030) reflection point, which corresponds to the orbital superlattice reflection point, is observed in the F_1-phase, where the static Jahn-Teller (JT) distortion is released. On the other hand, in the high temperature phase (F_2-phase), where the JT distortion remains, the scattering intensity disappears. It is concluded that the orbital order observed in the F_1-phase is not one supported by the JT distortion but induced by the superexchange interaction between orbital degrees in the nearest neighboring sites.

3 Anisotropy of the Atomic Scattering Factor in the Orbital Ordered State

The X-ray scattering factor is represented by the normal and anomalous parts of the atomic scattering factor,

$$F_{\alpha\beta}(\mathbf{k}',\mathbf{k}'') = N \sum_{i \in cell} e^{i(\mathbf{k}'-\mathbf{k}'')\cdot \mathbf{r}_i} \left(f_{0i}\delta_{\alpha\beta}(\mathbf{k}',\mathbf{k}'') + \Delta f_{i\alpha\beta}(\mathbf{k}',\mathbf{k}'') \right), \quad (1)$$

where $\mathbf{k}'$ and $\mathbf{k}''$ are the incident and scattered photon momenta, respectively, and N is the number of unit cell in the crystal. The normal term $f_{0i}(\mathbf{k}',\mathbf{k}'')$ describes the scattering from the electronic charge density. On the other hand, the anomalous term $\Delta f_{i\alpha\beta}(\mathbf{k}',\mathbf{k}'')$ is given by the second order perturbation with respect to the interaction between electronic current and photon as follows,

$$\Delta f_{i\alpha\beta}(\mathbf{k}',\mathbf{k}'') = \frac{m}{e^2} \sum_l \left\{ \frac{\langle f|j_{i\alpha}(-\mathbf{k}')|l\rangle\langle l|j_{i\beta}(\mathbf{k}'')|0\rangle}{\varepsilon_0 - \varepsilon_l - \omega_{k''} - i\delta} + \frac{\langle f|j_{i\beta}(\mathbf{k}'')|l\rangle\langle l|j_{i\alpha}(-\mathbf{k}')|0\rangle}{\varepsilon_0 - \varepsilon_l + \omega_{k'} - i\delta} \right\}, \quad (2)$$

where $|0\rangle$, $|f\rangle$ and $|l\rangle$ are the initial, final and intermediate electronic states in the perturbational process, respectively, and ε_0 and ε_l are the energies in the corresponding states. δ is a damping constant. In the present case where the incident photon is tuned near the K-edge, the current operator describes the dipole transition between Mn $1s$ and $4p$ orbitals given by $j_{i\alpha}(\mathbf{k}) = \frac{e}{m}\sum_\sigma (A_\alpha(\mathbf{k})P^\dagger_{i\alpha\sigma}s_{i\sigma} + h.c.)$. Here, $P_{i\alpha\sigma}$ and $s_{i\sigma}$ are the annihilation operators of electron in Mn $4p$ and Mn $1s$ orbitals, respectively, with spin σ and Cartesian coordinate α. It is worth noting the following points: (1) In the condition where the incident photon energy is far from the absorption edge, the anomalous term is neglected and the normal one dominates the scattering factor. On the other hand, near the edge, the anomalous term becomes comparable to the normal one. (2) The anomalous term has a tensor character with respect to the incident and scattered photon polarization, in contrast to the normal part which is a scalar. The anisotropy of the atomic scattering factor is the origin of the forbidden reflection at the orbital superlattice reflection point. (3) Eq.(2) is the same as the scattering factor in the conventional Raman scattering. However, the microscopic anisotropy due to the chemical bond, the orbital structure etc. are detectable in the present case, since the wave length is much smaller than that in the conventional Raman scattering. On the other hand, the microscopic dicroism and birefringence observed by the conventional Raman scattering experiment are reflected from the global crystal symmetry and the macroscopic anisotropy. (4) The anomalous term is almost independent of $\mathbf{k}'$ and $\mathbf{k}''$ because the wave length of photon is much larger than the average radius of the Mn $1s$ orbital. It is in contrast with the normal part which gradually decreases with increasing the momentum transfer $|\mathbf{k}' - \mathbf{k}''|$.

Next, we focus on the anistropy of the anomalous part of the atomic scattering factor and examine how the ordering of Mn 3d orbitals reflects on the transition between Mn 1s and Mn 4p orbitals. In order to study the problem, we consider the electronic structure in a MnO_6 small cluster. In a Mn ion, the following orbitals are introduced: $\{1s, 3d_\gamma$ ($\gamma = \gamma_{\theta+}, \gamma_{\theta-}$), $4p_\gamma$ ($\gamma = x, y, z$) $\}$, where $|3d_{\gamma_{\theta+}}\rangle = \cos(\theta/2)|3d_{3z^2-r^2}\rangle + \sin(\theta/2)|3d_{x^2-y^2}\rangle$ and $|3d_{\gamma_{\theta-}}\rangle$ is its counterpart. Six 2p orbitals in the O sites, which contribute to the σ-bond with the Mn 3d orbitals, are also considered. These O 2p orbitals are recombined as $\{2p_{\gamma_{\theta+}}, 2p_{\gamma_{\theta-}}, 2p_x, 2p_y, 2p_z, 2p_{r^2}\}$, where $x^2 - y^2$ etc represent the irreducible representation in O_h group. When we consider the electron transfer between Mn 3d and O 2p orbitals and that between Mn 4p and O 2p orbitals, it is easily shown that the Mn 3d and 4p orbitals are decoupled. It is concluded that the electron transfer does not cause the anisotropy of the atomic scattering factor.

One of the promising candidate which causes the anisotropy is the orbital dependence of the Coulomb interactions. The Coulomb interaction between Mn 3d and Mn 4p electrons transfers an information of the orbital order to the Mn 4p electronic state. It is given by

$$V(3d_{\gamma_{\theta\pm}}, 4p_\gamma) = F_0(3d, 4p) \pm 4F_2(3d, 4p) \cos\left(\theta + m_\gamma \frac{2\pi}{3}\right), \qquad (3)$$

where $m_x = +1$, $m_y = -1$, and $m_z = 0$. $F_n(3d, 4p)$ is the Slater-integral between 3d and 4p electrons defined by $F_0(3d, 4p) = F^{(0)}(3d, 4p)$ and $F_2(3d, 4p) = \frac{1}{35}F^{(2)}(3d, 4p)$. In the case where $3d_{3z^2-r^2}$ orbital is occupied, energy of $4p_z$ orbital is higher than that of $4p_{x(y)}$ by $6F_2(3d, 49)$. As a result, the scattering near the K-edge is dominated by the transition between Mn 1s and Mn $4p_{x(y)}$

orbitals. An additional origin of the anisotropy is the inter-site Coulomb interaction between Mn 3d and O 2p electrons. Since the electron hybridization between Mn 3d and O 2p is significant, $|3d^1_{3z^2-r^2}\rangle$ state is strongly mixed with $|3d^1_{3z^2-r^2}3d^1_{x^2-y^2}2p_{x^2-y^2}\rangle$ state. $|2p_{x^2-y^2}\rangle$ describes the state where a hole occupies the O $2p_{x^2-y^2}$ orbital. The inter-site Coulomb interaction between Mn 4p and O 2p electrons splits 4p orbital and induces the anisotropy of the scattering factor. The explicit formula of the interaction is given by

$$V(2p_{\gamma\theta\pm}, 4p_\gamma) = -\varepsilon \mp \frac{\varepsilon\rho^2}{5}\cos\left(\theta + m_\gamma\frac{2\pi}{3}\right). \qquad (4)$$

Here, $\varepsilon = Ze^2/a$ and $\rho = \langle r_{4p}\rangle/a$ with $Z = 2$ and $\langle r_{4p}\rangle$ being the average radius of Mn 4p orbital. The energy of $4p_z$ orbital is higher than that of $4p_{x(y)}$ by $3/10\varepsilon\rho$. Therefore, the above two mechanisms induces the anisotropy cooperatively. The JT type lattice distortion characterized by the difference of the lattice distortion in a MnO$_6$ octahedron $\delta a = a_z - a_{x(y)}$ also brings about the anisotropy in the 4p orbital. Although this mechanism competes with that originated from the Coulomb interaction, it is supposed to be neglected in La$_{0.5}$Sr$_{1.5}$MnO$_4$ and La$_{0.88}$Sr$_{0.12}$MnO$_3$ where the remarkable JT distortion is not observed as mentioned previously.

With taking into account the orbital dependence of the Coulomb interactions discussed above, we calculate the atomic scattering factor as a function of the occupied orbital and the polarization. In the numerical calculation, the following Hamiltonian is taken,

$$H = H_0 + H_t + H_{3d-4p} + H_{4p-2p} + H_{core} + H_{3d-3d}, \qquad (5)$$

where H_0 describes the energy level in the each orbital. The second and third terms are the Coulomb interaction terms between Mn 3d and Mn 4p orbitals and between Mn 3d and O 2p orbitals, respectively. H_{core} is the core hole potential between Mn 3d (4p) electron and Mn 1s hole created in the intermediate state.

H_{3d-3d} describes the Coulomb interaction between Mn 3d e_g electrons and the Hund coupling between e_g and t_{2g} spins. The configuration interaction method is adopted in the above Hamiltonian to calculate the atomic scattering factor. In Fig. 1, the real and imaginary parts of the atomic scattering factors are shown as a function of the incident photon energy in the case where $d_{3z^2-r^2}$ orbital is occupied. The edge of the main peak in $\Delta f''$ corresponds to the K-edge in the Mn^{3+} ion. The main and satellite peaks correspond to the transition from $|3d_{3z^2-r^2}\rangle$ to $|\underline{1s}\ 3d_{3z^2-r^2}3d_{x^2-y^2}4p_{x(y,z)}2p_{x^2-y^2}\rangle$ and $|\underline{1s}\ 3d_{3z^2-r^2}4p_{x(y,z)}\rangle$, respectively, although the both states are strongly mixed each other. As expected, the clear anisotropy is observed near the edge and $\Delta f''_{xx(yy)}$ component dominates the scattering factor near the edge. On the other hand, in the case where $d_{x^2-y^2}$ orbital is occupied, the anisotropy is entirely opposite to that in Fig. 1, that is, $\Delta f''_{zz}$ component is dominant. In Fig. 2, the amplitudes of the main peak in the imaginary part are presented

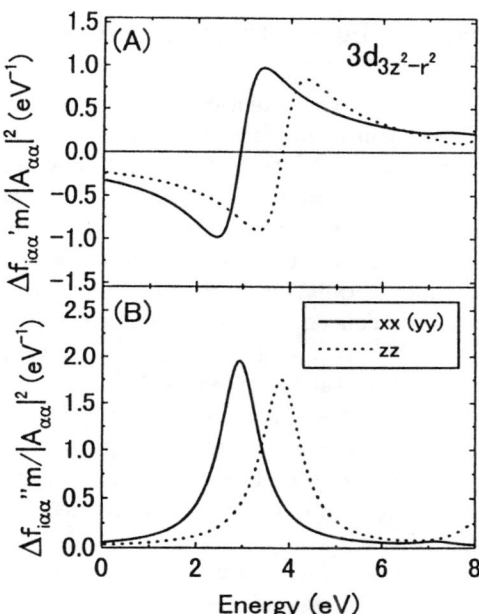

Fig. 1. The real and imaginary parts of the anomalous part of the atomic scattering factor in the case where $d_{3z^2-r^2}$ orbital is occupied. The origin of the energy is taken to be arbitrary.

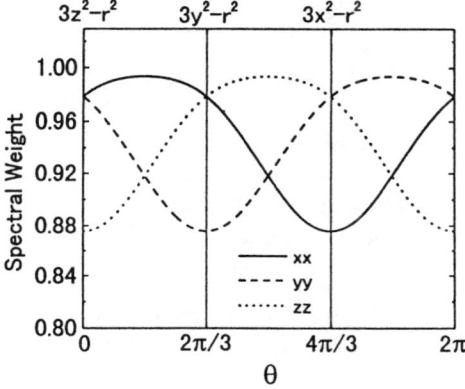

Fig. 2. The amplitude of the main peak in the atomic scattering factor as a function of the occupied orbital.

as a function of types of the occupied orbital. It is founded out that the amplitude strongly depends on the occupied orbital state. It is concluded that the scattering intensity at the orbital superlattice reflection point is attributed to the anisotropy of the scattering factor presented in these figures.

4 Polarization Dependence of the Scattering Intensity

In this section, we theoretically predict an experimental method which is utilized to identify several types of the orbital order in the anomalous X-ray scattering. In $La_{0.5}Sr_{1.5}MnO_4$, Murakami *et al.* tried to determine which type of the orbital order, $(3d_{3x^2-r^2}/3d_{3y^2-r^2})$ or $(3d_{z^2-x^2}/3d_{y^2-z^2})$, is realized [11]. Although they carried out the polarization dependence of the scattering intensity at the orbital superlattice point, they did not succeed in the identification. In the theoretical side, several types of the orbital ordering are predicted as a function of the carrier concentration, the external field, etc. [8, 9] Since the orbital order is strongly correlated to the spin structure, it is essential to identify the types of the orbital to understand the systematic change of the spin structures in manganites. Once a method to identify types of the orbital order is developed, this experimental technique will be applied to the wide range of compounds. We propose how to determine types of the orbital order by utilizing the polarization dependence in an adequate reflection point.

At first, we derive the general formulae of the scattering intensity as a function of the photon polarization. In the optical system where the X-ray scattering is carried out, the X-ray from the synchrotron source has an almost perfectly σ-polarization. The polarization analyses are performed by a rotation of the sample crystal around the scattering vector, which is termed the azimuthal scan and is characterized by the azimuthal angle ϕ, and a rotation of the analyzer crystal and the detector characterized by the analyzer angle ϕ_A. In this system, the scattering intensity is given by [18, 19]

$$I(\mathbf{k'}, \mathbf{k''}, \varphi, \varphi_A) = |F_A|^2 \left| \cos\varphi_A A_{\sigma\sigma}(\mathbf{k'}, \mathbf{k''}, \varphi) - \sin\varphi_A A_{\pi\sigma}(\mathbf{k'}, \mathbf{k''}, \varphi) \right|^2, \quad (6)$$

where F_A is the scattering factor in the analyzer crystal. The σ- and π-polarized components in the scattered photon are detected by the photon detector with $\varphi_A = 0$ and $\pi/2$, respectively. The scattering amplitude $A_{\lambda''\lambda'}(\mathbf{k''}, \mathbf{k'}, \varphi)$ is represented by the structure factor $F(\mathbf{k'}, \mathbf{k''})$ as follows,

$$A_{\lambda''\lambda'}(\mathbf{k''}, \mathbf{k'}, \varphi) = \frac{e^2}{mc^2} \sum_{\alpha''\alpha'} \epsilon^{(s)}_{\lambda''\alpha''} \left[U(\varphi) V F(\mathbf{k'}, \mathbf{k''}) V^\dagger U(\varphi)^\dagger \right]_{\alpha''\alpha'} \epsilon^{(i)t}_{\lambda'\alpha'}. \quad (7)$$

$\epsilon^{(i)}_{\lambda\alpha}$ and $\epsilon^{(s)}_{\lambda\alpha}$ are the polarization vectors of the incident and scattered photons, respectively, and their explicit forms are given by

$$\epsilon^{(l)}_{\lambda\alpha} = \begin{pmatrix} 1 & 0 & 0 \\ 0 & \pm\sin\theta & \cos\theta \end{pmatrix}, \quad (8)$$

where $+$ and $-$ are for $l = i$ and $l = s$, respectively. The unitary matrix $U(\varphi)$ and V describe the azimuthal rotation and the transformation from the coordinate in the crystallographic axis to that in the laboratory system, respectively.

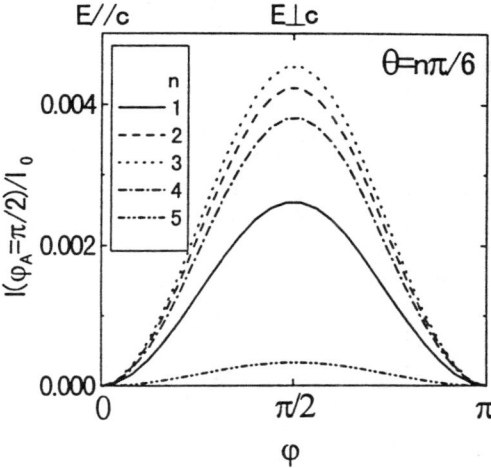

Fig. 3. The polarization dependence of the scattering intensity at the orbital superlattice reflection point in several orbital ordered case.

By using the formulae introduced above, we show the polarization dependence of the scattering intensity in the following three cases: (1) at the orbital superlattice reflection point, (2) at the fundamental reflection point, and (3) at the charge order reflection point in the charge ordered state. In Fig. 3, we show the scattering intensity at the orbital superlattice reflection point which corresponds to (300) and (3/4 3/4 0) points in $La_{0.5}Sr_{1.5}MnO_4$ and $LaMnO_3$, respectively. Two kinds of the orbital are considered and the orbital states are assumed as $\theta_A = -\theta_B$, where $\theta_{A(B)}$ describes the orbital in the $A(B)$ sublattice. The squared sinisoidal curve corresponds to that experimentally observed. At the reflection point, the scattering factor is given by $F_{\alpha\beta}/N = \Delta f_{\alpha\beta}(A) - \Delta f_{\alpha\beta}(B)$ where $\Delta f_{\alpha\beta}(A(B))$ is the atomic scattering factor in the $A(B)$-sublattice. Therefore, the scattering intensity reflects a difference of the scattering factor, termed the antiferro-component of the scattering factor. The scattering intensity becomes its maximum(minimum) at $\phi = \pi/2(\pi)$, where the electric field is parallel to the ab-plane (c-axis). It is interpreted that the a difference of the orbital state between two sublattices is remarkable in the ab-plane and it disappears along the c-axis. A remarkable discrepancy between the $(3d_{3x^2-r^2}/3d_{3y^2-r^2})$ and $(3d_{z^2-x^2}/3d_{y^2-z^2})$ cases is not found out. Therefore, we conclude that it is difficult to determine which orbital order is realized through the polarization analyses at this reflection point.

Next, we present the polarization dependence of the scattering intensity at the fundamental reflection point (Fig. 4) where the two kinds of the orbital are also considered. At the reflection point, the scattering factor is given by $F_{\alpha\beta}/N = (f_0(A) + f_0(B))\delta_{\alpha\beta} + \Delta f_{\alpha\beta}(A) + \Delta f_{\alpha\beta}(B)$. A sum of the anomalous part termed the ferro-component of the scattering factor reflects the scatter-

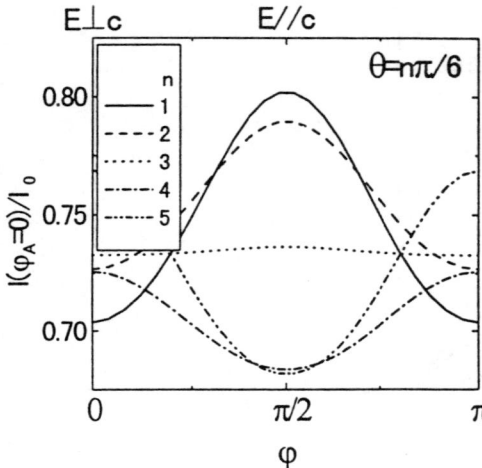

Fig. 4. The polarization dependence of the scattering intensity at the fundamental reflection point in several orbital ordered case.

ing intensity. A magnitude of the scattering intensity is almost dominated by the normal part of the scattering factor and the small azimuthal dependence is originated from the interference between the normal part and the real part of the anomalous one. It is found that in the $(3d_{3x^2-r^2}/3d_{3y^2-r^2})$ case, the phase of the modulation is opposite to that in the $(3d_{z^2-x^2}/3d_{y^2-z^2})$ case. It is attributed to the fact that the anisotropies of the ferro-component of the scattering factor are entirely different between the two orbital orders.

At last, in Fig. 5. we show the polarization dependence of the scattering intensity at the charge order reflection point in the charge ordered phase. We consider the state where Mn^{3+} and Mn^{4+} ions are alternately aligned in the MnO_2 plane. In the Mn^{3+} sublattice, the A- and B-orbitals are arranged along the [110] direction, as experimentally supposed in $La_{0.5}Sr_{1.5}MnO_4$. At this point, the scattering factor is given by $F_{\alpha\beta}/N = (f_0(A) + f_0(B) - 2f_0(4+))\delta_{\alpha\beta} + \Delta f_{\alpha\beta}(A) + \Delta f_{\alpha\beta}(B) - 2\Delta f(4+)\delta_{\alpha\beta}$, where $f_0(4+)$ and $\Delta f(4+)$ are the normal and anomalous parts in a Mn^{4+} ion. In this formula, it is worth to note that the normal part of the scattering factor is almost canceled out, $\Delta f(4+)$ is scalar and its magnitude is smaller than that in a Mn^{3+} ion near the Mn^{3+} K-edge. Therefore, the ferro-component of the anomalous scattering factor is obtained without disturbance of the normal part in the polarization analyses. As shown in Fig. 5, the polarization dependence is more remarkable in comparison with that in Fig. 3. We find the clear distinction between the $(3d_{3x^2-r^2}/3d_{3y^2-r^2})$ and $(3d_{z^2-x^2}/3d_{y^2-z^2})$ cases. The polarization dependence is opposite to that shown in Fig. 3, that is, the scattering intensity has its maximum when the electric vector is parallel to the c-axis in the $(3d_{3x^2-r^2}/3d_{3y^2-r^2})$ case. The discrepancy is attributed to the fact that the anomalous part dominates the scattering intensity in the present case.

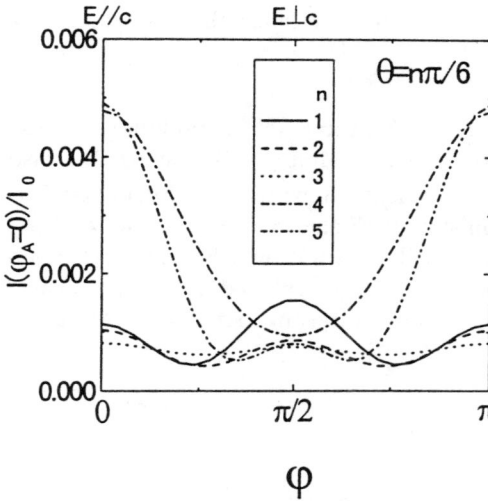

Fig. 5. The polarization dependence of the scattering intensity at the charge order superlattice reflection point in the charge ordered phase.

On the other hand, the interference between the normal and anomalous part becomes dominant in the case of Fig. 3. The polarization analyses in the charge ordered reflection point may be easier than that in the fundamental reflection to detect a type of the orbital order.

5 Summary and Discussion

Although it has been recognized that the orbital degree plays an important role in the electronic properties in transition metal compounds, it was considered as a hidden degree of freedom for a long time. The anomalous X-ray scattering has recently been developed as a new probe to directly detect the orbital order and shed light on the subject. In this paper, we briefly reviewed the recent experimental developments of the anomalous X-ray scattering which is applied to several orbital ordered manganites. In our theoretical studies, we revealed an origin of the anisotropy of the scattering factor and predicted a method to identify types of the orbital ordering. The anomalous X-ray scattering is promising as a probe to study the orbital ordering in not only manganites but other transition metal compounds. It is proposed that as one of the extension of observation of the orbital order, the method is applicable to detect the orbital excitation as a resonant inelastic X-ray scattering (RIXS). The propagation of the orbital excitation in the orbital degenerate system is termed the orbital wave which is theoretically predicted [20]. The mechanism to detect the orbital wave by RIXS is analogous to the Raman scattering to detect the two magnon excitation in the magnetic compound [14]. RIXS with the high energy resolution and the momentum analyses can detect the dispersion relation of the orbital wave since the wave length is comparable to the lattice constant.

ACKNOWLEDEGMENTS

We would like to thank Y. Endoh and Y.Murakami for providing the experimental data prior to publication and for valuable discussions. We also indebted to W.Koshibae for helpful comments. This work was supported by Priority Areas Grants from the Ministry of Education, Science and Culture of Japan, and CREST (Core Research for Evolutional Science and Technology Corporation) Japan. Part of the numerical calculation was performed in the HITACS-3800/380 supercomputing facilities in Institute for Materials Research, Tohoku University.

References

1. K. Chahara, T. Ohono, M. Kasai, Y. Kanke, and Y. Kozono, Appl. Phys. Lett. **62**, 780 (1993).
2. R. von Helmolt, J. Wecker, B. Holzapfel, L. Schultz, and K. Samwer, Phys. Rev. Lett. **71**, 2331 (1993).
3. Y. Tokura, A. Urushibara, Y. Moritomo, T. Arima, A. Asamitsu, G. Kido, and N. Furukawa, Jour. Phys. Soc. Jpn. **63**, 3931 (1994).
4. S. Jin, T. H. Tiefel, M. McCormack, R. A. Fastnacht, R. Ramesh, and L. H. Chen, Science, **264**, 413 (1994)
5. J. B. Goodenough, Phys. Rev. **100** 564 (1955).
6. J. Kanamori, J. Phys. Chem. Solids, **10** 87 (1959).
7. S. Ishihara, M. Yamanaka, and N. Nagaosa, Phys. Rev. B **56** 686 (1997).
8. S. Ishihara, S. Okamoto, and S. Maekawa, Jour. Phys. Soc. Jpn. **66** 2965 (1997).
9. R. Maezono, S. Ishihara, and N. Nagaosa, Phys. Rev. B **57** R13993 (1998).
10. Y. Ito, and J. Akimitsu, Jour. Phys. Soc. Jpn. **40** 1333 (1976).
11. Y. Murakami, H. Kawada, H. Kawata, M. Tanaka, T. Arima, H. Moritomo and Y. Tokura, Phys. Rev. Lett. **80** 1932 (1998).
12. Y. Murakami, J. P. Hill, D. Gibbs, M. Blume, I. Koyama, M. Tanaka, H. Kawata, T. Arima, T. Tokura, K. Hirota, and Y. Endoh, Phys. Rev. Lett. **81** 582 (1998).
13. Y. Endoh, K. Hirota, N. Nojiri, T. Fukuda, K. Kaneko, H. Kimura, Y. Murakami, S. Ishihara, S. Okamoto, and S. Maekawa , (unpublished).
14. S. Ishihara, and S. Maekawa, Phys. Rev. Lett. **80** 3799 (1998).
15. S. Ishihara, and S. Maekawa, Phys. Rev. B **58** Nov. 15 (1998).
16. D. H. Templeton, and L. K. Templeton, Acta Cryst A**38** 62 (1982), ibid. A**41** 133 (1985).
17. V. E. Dmitrienko, Acta Cryst A**39** 29 (1983).
18. T. Nagano, J. Kokubun, I. Yazawa, T. Kurasawa, M. Kuribayashi, E. Tsuji, K. Ishida, S. Sasaki, T. Mori, S. Kishimoto, and Y. Murakami, J. Phys. Soc. Jpn. **65**, 3060 (1996).
19. A. Kirfel, and A. Petcov, Acta. Cryst, A **47**, 180 (1991).
20. S. Ishihara, J. Inoue, and S. Maekawa, Physica C **263** 130 (1996), and Phys. Rev. B **55** 8280 (1997).

Nonlinear Optical Responses of Strongly Correlated Electronic Systems

E. Hanamura, Y. Tanabe* and M. Fiebig*†

Optical Sciences Center, The University of Arizona
1630 E University Blvd. Tucson, AZ 85721
*Department of Applied Physics, The University of Tokyo
7-3-1 Hongo, Bunkyo-ku, Tokyo 113, Japan
†Japan Science and Technology Corporation (JST)
1-4-25 Mejiro, Toshima-ku, Tokyo 171-0031, Japan

Abstract Two kinds of nonlinear optical responses, i.e., second harmonic generation (SHG) and third-order susceptibility (two-photon absorption) of strongly correlated electronic systems are presented. We clarify the microscopic mechanism of SHG due to electric and magnetic dipole moments and their interference in antiferromagnets. The third-order optical signals including two-photon absorption are theoretically shown to be enhanced by strong correlation among elementary excitations.

1 Introduction

Nonlinear optical responses are shown in this paper to give detailed information on electronic correlation effects, especially in antiferromagnetic insulators. The wavelength of the radiation field is much longer than the size of the unit cell and the correlation length so that the optical signal consists of some superposition of atomic signals depending upon the correlation. For example, in the crystal Cr_2O_3 of corundum structure, three d-electrons $(t_{2g})^3$ of Cr^{3+} ion compose the $|^4A_{2g}\rangle$ ground state due to Hund coupling. Below the Néel temperature $T_N = 307.6$ K, four ions in the unit cell have alternating spin directions along the C_3 axis and the crystal loses both time-reversal and spatial inversion symmetry. Then, SHG(χ^e) due to the electric dipole moment becomes finite below T_N although it vanishes above T_N. On the other hand, SHG(χ^m) by magnetic dipole moment is finite both above and below T_N.

In Section 2, the microscopic mechanism of SHG χ^e and χ^m processes and their interference are presented by taking into account the proper magnetic and crystalline structure and the propagation effect of the induced magnetic dipole moment. The third-order optical susceptibility $\chi^{(3)}$ describes two-photon absorption, differential transmission and phase-conjugated wave generation. We will show in Section 3 that these signals will possibly be enhanced by the correlation effects of electrons both in the ground and excited electronic states. From these examples, we will be able to understand that the nonlinear optical responses will play an important role of in the detection of the correlated electronic system.

2 Nonreciprocal Nonlinear-Optical Responses

For antiferromagnetic Cr_2O_3, SHG due to magnetic and electric dipole moments were observed separately and their interference were clearly observed [1,2]. Both SHG spectra and their interference were theoretically analyzed by taking account of the proper magnetic and crystalline structures and the propagation of excitations by superexchange interaction in the electronic excited states[3,4]. The latter are introduced from a standpoint which is different different from the one taken previously [3,4].

2.1 Interference between Magnetic and Electric Dipoles

The crystal and magnetic structures in the ordered phase $(R\bar{3}'c')$ is drawn in Fig.1. The phase difference $\Delta\theta \equiv \theta^m - \theta^e$ between the SH magnetic- and electric-dipole susceptibilities

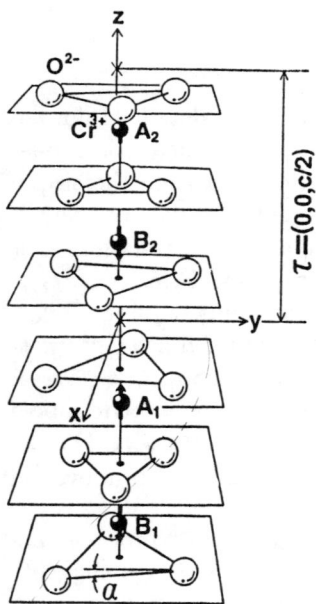

Fig. 1. The crystal and magnetic structure of an unit cell of Cr_2O_3 below Néel temperature

$$\chi^m = \chi'_m + i\chi''_m = |\chi^m|\exp(i\theta^m) \tag{1a}$$

and

$$\chi^e = \chi'_e + i\chi''_e = |\chi^e|\exp(i\theta^e) \tag{1b}$$

is defined as a function of $\omega_s = 2\omega$, with ω being the frequency of the incidential fundamental light wave. In order to obtain perfect interference, we must satisfy two conditions $|\chi^e| = |\chi^m|$ and $|\theta^m - \theta^e| = \pi/2$. These are derived as follows. When we take the z- and x- axes in the threefold (C_3) and the twofold (C_2) axes, the second-order susceptibilities χ^m and χ^e have the following symmetries:

$$\chi^m \equiv \chi^{mee}_{xxx} = -\chi^{mee}_{xyy} = -\chi^{mee}_{yxy} = -\chi^{mee}_{yyx}, \tag{2a}$$

$$\chi^e \equiv \chi^{eee}_{xxx} = -\chi^{eee}_{xyy} = -\chi^{eee}_{yxy} = -\chi^{eee}_{yyx}. \tag{2b}$$

Then, the source term for the second harmonic

$$\boldsymbol{S} = \mu_0\left(\nabla \times \frac{\partial \boldsymbol{M}}{\partial t} + \frac{\partial^2 \boldsymbol{P}}{\partial t^2}\right) \tag{3}$$

is expressed as follows:

$$\begin{pmatrix} S_x \\ S_y \\ S_z \end{pmatrix} = \frac{4\omega^2}{c^2}\begin{pmatrix} 2\chi^m E_x E_y - \chi^e(E_x^2 - E_y^2) \\ \chi^m(E_x^2 - E_y^2) + 2\chi^e E_x E_y \\ 0 \end{pmatrix}, \tag{3a}$$

or

$$\begin{pmatrix} S_+ \\ S_- \\ S_0 \end{pmatrix} = \frac{4\sqrt{2}\omega^2}{c^2}\begin{pmatrix} (-i\chi^m - \chi^e)E_-^2 \\ (i\chi^m - \chi^e)E_+^2 \\ 0 \end{pmatrix}, \tag{3b}$$

setting $S_\pm = (S_x \mp iS_y)/\sqrt{2}$ and $E_\pm = (E_x \mp iE_y)/\sqrt{2}$. From Eq.(3a), the spectra of χ^e and χ^m can be measured separately by observing the second harmonic linearly polarized along x and y axes, respectively, under the irradiation of fundamentals linearly polarized along x axis [2].

From Eq.(3b) the signal intensity $I \propto |S|^2$ for the circularly polarized light is expressed as

$$|S|^2 \propto (|\chi^m|^2 + |\chi^e|^2)(|E_+|^4 + |E_-|^4) - 2(\chi'_m\chi''_e - \chi''_m\chi'_e)(|E_+|^4 - |E_-|^4). \tag{4}$$

Since χ^e in Eq.(4) is linearly proportional to the magnetization of the sublattice, it is possible to observe the magnetic domains of the antiferromagnetic crystals by using circularly polarized fundamental waves. The second term $-2|\chi^m||\chi^e|\sin\Delta\theta$ cancels out the first term for E_- circularly polarized light but doubles that for E_+ one when $|\chi^m| = |\chi^e|$ and $\Delta\theta = \pi/2$. This is the case at $2\omega = 2.1\text{eV}$ of Cr_2O_3 [1] and this was used to observe the antiferromagnetic domains [2].

2.2 Is $|\chi^m| = |\chi^e|$?

Two photons with the fundamental frequency excite an electron from the $|^4A_{2g}\rangle$ ground state into the $|^4T_{2g}\rangle$ excited state which can be mediated by an electric-($\sim \chi^e$) or magnetic-($\sim \chi^m$) dipole interaction, respectively. The emission of the second harmonic $\omega_s = 2\omega$ can be induced by the magnetic dipole moment or by the electric dipole moment. In general, the expectation value of the electric dipolemoment operator is by a few orders of magnitude larger than that of the magnetic one if both are allowed, but the electric one is forbidden between the ground state $|^4A_{2g}\rangle$ and the excited state $|^4T_{2g}\rangle$ using as bases the eigenfunctions of cubic field. However, the odd-parity crystalline field and the spin-orbit interaction work as perturbation making the electric dipole transition allowed. This can be seen by an explicit investigation of the symmetry of the Cr_2O_3 crystal.

Four Cr^{3+} ions along the z axis constitute a unit cell and are called B_1, A_1, B_2, A_2 with down, up, down, up spin from bottom to top, in this order. Symmetry of the magnetic space group $R\bar{3}'c'$ requires that A_1 with its environment is carried into B_1 by $C_{2x}(\tau)$, into B_2 by ΘI and A_2 by $\Theta\sigma_d(\tau)$, where τ is the displacement vector $(0, 0, c/2)$, Θ the time-reversal operation, $C_{2x}(\tau)$ the rotation by π around the x axis followed by a displacement by τ, and $\sigma_d(\tau)$ the mirror reflection in the yz plane with a displacement by τ. The matrix elements of an operator $\hat{A}$ at different sites are correlated to each other by the equations:

$$\langle R\Psi|\hat{A}|R\Psi'\rangle = \langle\Psi|R^{-1}\hat{A}R|\Psi'\rangle, \tag{5a}$$

$$\langle\Theta R\Psi|\hat{A}|\Theta R\Psi'\rangle = \langle\Psi|\Theta^{-1}R^{-1}\hat{A}R\Theta|\Psi'\rangle^*, \tag{5b}$$

where R stands for the operations such as $C_{2x}(\tau)$ etc., acting upon both orbital and spin states. The matrices of M_x and P_x at the various sites are thus related to those at A_1 by

$$\begin{aligned} M_x[B_1] &= M_x[A_1], & P_x[B_1] &= P_x[A_1], \\ M_x[A_2] &= -M_x[A_1]^*, & P_x[A_2] &= -P_x[A_1]^*, \\ M_x[B_2] &= -M_x[A_1]^*, & P_x[B_2] &= -P_x[A_1]^*. \end{aligned} \tag{6}$$

We sum up the contributions from the four ions in the unit cell and keep the dominant term with the two-photon resonance. Here we introduce relaxation effects in order to satisfy the casuality and we obtain

$$\chi^m = \frac{4iNn}{\epsilon_0 c} \sum_i \rho_i \sum_{m,k} \frac{\text{Im}(MPP)_{imki}}{(\omega_{mi} - 2\omega - i\Gamma_{mi})(\omega_{ki} - \omega)}, \qquad (7)$$

and

$$\chi^e = \frac{4iN}{\epsilon_0} \sum_i \rho_i \sum_{m,k} \frac{\text{Im}(PPP)_{imki}}{(\omega_{mi} - 2\omega - i\Gamma_{mi})(\omega_{ki} - \omega)} \qquad (8)$$

per unit volume of the crystal, N being the number density of unit cells.

The basis functions to be employed in evaluating the matrix elements in Eqs. (7) and (8) should be eigenfunctions which diagonalize the spin-orbit interaction $\mathcal{H}_{so}$ and the lower-symmetry crystalline field:

$$V_{trig} = V(T_{2g}\,x_0) + V(T_{1u}\,a_0) + V(T_{1g}\,a_0) + V(T_{2u}\,x_0). \qquad (9)$$

However, using the eigenfunctions in the cubic crystalline field as bases, V_{trig} and $\mathcal{H}_{so}$ are taken into account as perturbations in Eqs. (7) and (8).

Muthukumar, Valenti, and Gros [6] proposed a microscopic theory to explain the observed non-reciprocal effect, assuming a $(CrO_6)_2$ cluster model with D_{3d} symmetry, which did not, however, reflect the actual symmetry of the environment of Cr^{3+} ion in the crystal. In their model the symmetry below T_N is D_{3d} which still contains the space-inversion operation, so that χ^e vanishes after the correct summation of all the contributions. We have shown [4,7] that if the correct symmetry is taken into account, both χ^m and χ^e vanish below T_N when contributions from the four Cr^{3+} ions in a unit cell are summed up, as long as conventional fields of even and odd parity are assumed as done in Ref. [6]. It is noted here that the symmetry determines not only the type and symmetry of the perturbations but also the complex properties of their matrix elements [7]. Only after having determined whether these matrix elements are real or imaginary, can we establish a correct microscopic model for χ^m and χ^e [3]. Then, we have found that the twist crystalline field ($T_{1g}a_0$ and $T_{2u}x_0$), together with $\mathcal{H}_{so}$ give finite contribution to χ^m and χ^e, and it turned out that $|\chi^m| \approx \|\chi^e|$, their absolute values being in good agreement with the observation [3].

2.3 Is the Phase Difference $\pi/2$?

In order to achive the complete interference between χ^m and χ^e, we must also solve the problem that their phase difference $\Delta\theta \equiv \theta^m - \theta^e$ should become approximately equal to $\pm\pi/2$. The expressions χ^m and χ^e of Eqs. (7) and (8) give almost the same phase as both relaxation rates are almost equal and the relative values of numerators at $^4T_{2g}x_\pm$ and $^4T_{2g}x_0$ are also similar for χ^m and χ^e.

Here we take into account the propagation effect of magnetic dipole moment through superexchange interaction in the electronic excited states. The interaction between the two Cr^{3+} ions i and j,

$$V = \sum_{<i,j>} K_{ij} L_i \cdot L_j \tag{10}$$

is taken into account by the mean field approximation, where L_i and L_j are orbital angular momentum vectors of the ions i and j, respectively, and K_{ij} will be at least of the same order of magnitude as the superexchange interaction between the ions i and j in their ground state. Then this magnetic-dipole–magnetic-dipole interaction is represented as local field corrections at $\omega_s = 2\omega$:

$$\frac{\bar{\chi}^m}{\bar{\chi}^e} = \frac{f_m(2\omega)}{f_e(2\omega)} \frac{\chi^m}{\chi^e}. \tag{11}$$

Here, $f_e(2\omega)$ is the Lorentz field correction, $\{\epsilon(2\omega)+2\}/3$ which is real. The local magnetic field correction $f_m(2\omega)$ consists of two contributions (1) due to the long-range magnetic dipolar interaction and (2) due to the neighboring ions expressed by Eq. (10). The former effect is negligible but the latter is evaluated in the mean field approximation as

$$f_m(2\omega) = \frac{1}{1 - \xi \, \chi_{\text{ion}}(2\omega)} \tag{12}$$

with

$$\xi = -\sum_j K_{ij}/(\mu_0 \mu_B^2) \tag{12a}$$

and

$$\chi_{\text{ion}}(2\omega) = \mu_0 \sum_i \rho_i \sum_m \frac{|\langle i|M_x|m\rangle|^2}{\omega_{mi} - 2\omega - i\Gamma}, \tag{12b}$$

where μ_0 is the magnetic permeability of the vacuum and μ_B the Bohr magneton. Then, we can evaluate the phase difference $\Delta\theta = \theta^m - \theta^e$ as

$$\tan \Delta\theta = \frac{\xi \, \text{Im}[\chi_{\text{ion}}(2\omega)]}{1 - \xi \, \text{Re}[\chi_{\text{ion}}(2\omega)]}. \tag{13}$$

Assuming 200 meV for the value of Γ, which is close to the linear absorption bandwidth, and $\sum_j K_{ij} \approx 65$meV, $|\Delta\theta|$ is estimated to be $\pi/3$ at the peak of $^4T_{2g}$ as shown in Fig. 2. The cross relaxation from $|^4T_{2g}x_\pm\rangle$ to $|^4T_{2g}x_0\rangle$ through phonon emission contributes to the phase difference spectrum as shown by the solid line in Fig. 2. The Figure is to be compared with the spectrum of the observed phase difference in Fig.3. By using the given numerical values, we can draw the theoretical (and experiental) interference spectrum as shown in Fig. 4.

The value $\sum_j K_{ij} \approx 65$meV is roughly three times the Zeeman energy in the molecular field estimated from the Néel temperature $T_N = 307.6$ K.

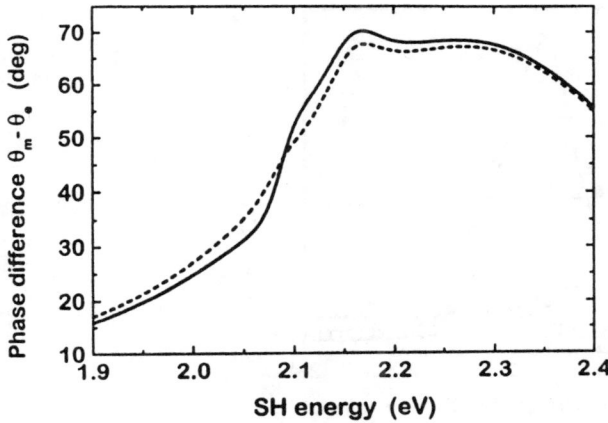

Fig. 2. Calculated phase difference $\theta^m - \theta^e$ around $^4T_{2g}$ with (solid line) and without (dashed line) the effect of cross relaxation

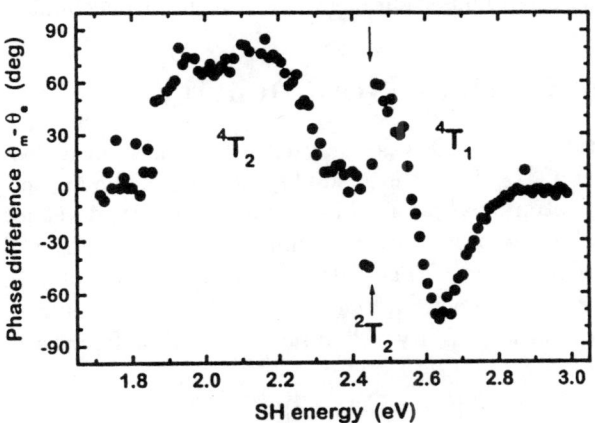

Fig. 3. phase difference $\theta^m - \theta^e$ derived from the experimental SH spectra Fig. 2 in Ref. [1]. The various $d - d$ transitions are labeled

This looks reasonable because the excitation propagation is brought about on the excited configuration $(t_{2g})^2 e_g$ involving the (one-electron) orbital e_g which extends more than the ground state $(t_{2g})^3$, while T_N is determined by the superexchange interaction among the configurations $(t_{2g})^3$.

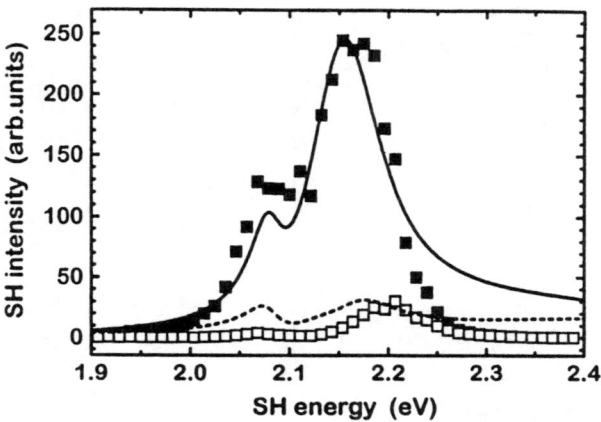

Fig. 4. Calculated SH spectra for I_+ (solid line) and I_- (dashed line) compared with the experimental data at 10K for I_+ (full squares) and I_- (open squares). The spectra for I_+ and I_- correspond to right and left circularly polarized incident light in the experiment, respectively, and describe the interference of χ^e and χ^m contributions. For the theoretical spectra, the effects of local-field correction and cross relaxation have been taken into account

3 Third-Order Optical Nonlinearity

As shown in the last section, the SHG was analyzed for the case of a two-photon resonant excitation of $^4T_{2g}$ from the ground state $^4A_{2g}$ of the antiferromagnetic Cr_2O_3. Two-photon absorption (TPA) was not observed yet for this transition. SHG is the second order optical nonlinearity described by $\chi^{(2)}$ while TPA is the third-order one $\chi^{(3)}$. The rate $w^{(2)}(\omega_t, \omega_p)$ of two-photon absorption with ω_p as pump frequency and ω_t as test photon frequency is expressed by the third-order susceptibility $\chi^{(3)}(\omega_t; \omega_t, -\omega_p, \omega_p)$ as follows:

$$\begin{aligned}
w^{(2)}(\omega_t, \omega_p) &= \frac{\partial}{\partial t}\mathrm{Tr}_m\left[\langle n_t-1, n_p-1|\rho(t)|n_t, n_p\rangle\right] \\
&= \frac{1}{\hbar}\mathrm{Im}\langle \vec{P}_t(t)\cdot \vec{E}_t\rangle_p \quad (14) \\
&= \frac{1}{\hbar}\mathrm{Im}\chi^{(3)}(\omega_t; \omega_t, -\omega_p, \omega_p)|E_p|^2|E_t|^2.
\end{aligned}$$

Here $\rho(t) \equiv \exp(-iH_t t/\hbar)\rho(0)\exp(iH_t t/\hbar)$ with H_t the total Hamiltonian including electronic interaction with external fields of pump $\boldsymbol{E}_p(\omega_p)$ and test $\boldsymbol{E}_t(\omega_t)$ beams, and trace in the first line is taken over the material system. The expectation value in the second line is taken under the presence of the pump field. The bottom line of Eq. (14) is in the lowest order in $\boldsymbol{E}_p$ and $\boldsymbol{E}_t$.

3.1 Conventional Two-Photon Absorption

Let us consider the conventional two-phonton absorption rate in which two photons with angular frequencies ω_p and ω_t are absorbed simultaneously and the crystal is excited into the elctronic excited states $|e\rangle$ from the ground state $|g\rangle$. This rate can be evaluated by perturbation theory [8] as

$$w^{(2)} = \frac{2\pi}{\hbar} \left(\frac{e}{m}\right)^4 \left(\frac{2\pi\hbar}{\kappa_t V \omega_t}\right) \left(\frac{2\pi\hbar}{\kappa_p V \omega_p}\right) n_t n_p$$

$$\times \left| \left\langle e \left| \epsilon_t \cdot \sum_j p_j \sum_i \frac{|i\rangle\langle i|}{E_{ig} - \hbar\omega_p} \epsilon_p \cdot \sum_{j'} p_{j'} \right| g \right\rangle \right. \quad (15)$$

$$\left. + \left\langle e \left| \epsilon_p \cdot \sum_j p_j \sum_i \frac{|i\rangle\langle i|}{E_{ig} - \hbar\omega_t} \epsilon_t \cdot \sum_{j'} p_{j'} \right| g \right\rangle \right|^2 \delta(E_{eg} - \hbar\omega_t - \hbar\omega_p).$$

For ordinary cases in which a single-electron picture is well justified, the first photon $\omega_p(\omega_t)$ excites an electron in the ground state $|g\rangle$ into the intermediate state $|i\rangle$ at the site j' and the second one $\omega_t(\omega_p)$ excites the electron in the intermediate state $|i\rangle$ into the final excited state $|e\rangle$ in the same $j = j'$. Therefore, $w^{(2)} \equiv N\sigma^{(2)}$, where N is the numbers of ions which contribute to TPA and $\sigma^{(2)}$ the single-ion cross section of TPA.

3.2 Giant TPA due to Bound States of Two Excitations

Once two excitations from the ground state form a bound state, a new channel of TPA, i.e., $\chi^{(3)}$- process is opened, and the TPA rate is extremely enhanced. This giant TPA was predicted in 1973 for an excitonic molecule (biexciton), i.e., a bound state of two Wannier excitons [9] and this TPA was experimentally confirmed for a biexciton in CuCl [10]. The origin of this enhancement is explained diagramatically in Fig. 5(b) in contrast to the conventional case of Fig. 5(a). Here, let us consider a system of Frenkel excitons. When two excitations form a bound state [11]:

$$|j, j'\rangle = F(j - j')|j\rangle|j'\rangle, \quad (16)$$

we can excite any ion j as the second excitation around the first excitation j' in Eq. (15) as long as $|j - j'|$ is within the extension of the bound state $F(j - j')$ in Eq. (16). These excited states can be described by the spin-1/2 operators as

$$|i\rangle \equiv S_i^+|g\rangle \quad \text{and} \quad |i, j\rangle \equiv S_i^+ S_j^+|g\rangle.$$

Such a bound state originates in dipolar interactions between Frenkel excitons described as [11]

$$H_j = \sum_{\langle i,j \rangle} J(i,j) S_{iz} S_{jz}. \quad (17)$$

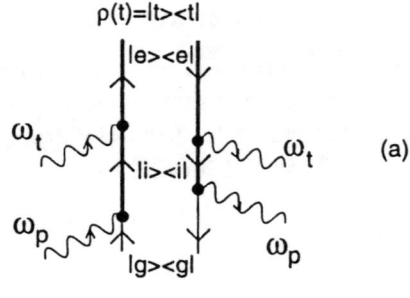
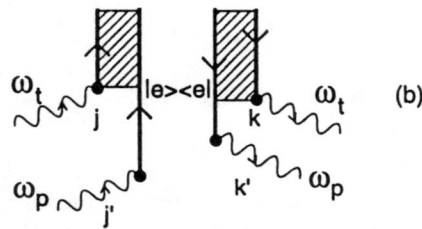
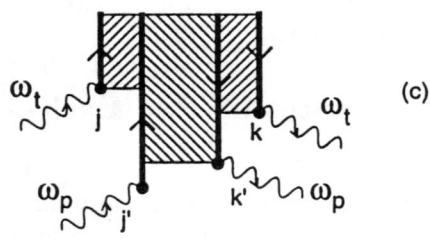

Fig. 5. Contributions to the third-order susceptibility describing two-photon absorption. (a) Diagram of conventional two-photon absorption. (b) Diagram of $\chi^{(3)}$ under two-photon resonant pumping of a bound state of two excitons, which results in giant two-photon absorption. (c) Diagram contributing to the colossal two-photon absorption in the strongly correlated electronic system with charge-transfer gap

The enhancement factor n_J is evaluated in terms of the envelope function $F(i-j)$ for the relative motion of two excitations as

$$n_J = \left[\sum_j F(i-j)\right]^2. \qquad (18)$$

The physical origin of this enhancement can be explained as follows: Both the pump and the test beams can excite potentially many ions as denoted by $\sum_j p_j$ and $\sum_{j'} p_{j'}$ in Eq. (15). When two excitations form the bound state described by Eq. (16), spatial freedom in choosing the second excitation j around the first one at j' gives the enhancement factor n_J in contrast to the case regarded in section 3.1.

3.3 Colossal TPA in Strongly Correlated Systems

In this subsection, we will present how the optical nonlinearity may be enhanced by the strong correlation of electrons in the ground state as well as in the electronic excited states. Let us consider the charge-transfer gap system in which an electron in the 2p orbital of oxygen is excited into the unoccupied 3d orbitial of a transition metal ion, say Cu^{2+}. Then, the annihilation operator $c_{i\sigma}$ of d-electron with spin σ at the i-th Cu^{2+} ion is presented by spinon annihilation (creation) $f_{i\sigma}(f_{i\sigma}^+)$, holon b_i, b_i^+ and doublon d_i, d_i^+ operators [12] as

$$c_{i\sigma} = f_{i\sigma}b_i^+ + \sum_{\sigma'} \xi_{\sigma\sigma'} f_{i\sigma'}^+ d_i, \qquad (19a)$$

with the constraint

$$\sum_\sigma f_{i\sigma}^+ f_{i\sigma} + b_i^+ b_i + d_i^+ d_i = 1, \qquad (19b)$$

and with $\xi_{\sigma-\sigma} = 1$ and $= 0$ for other cases. Then the transition dipole moment P can be described as

$$P = \mu \sum_{(i,j)} b_j^+ d_i^+ (f_{i\uparrow} f_{j\downarrow} - f_{i\downarrow} f_{j\uparrow}). \qquad (20)$$

The dipolar interaction with external field makes, e.g., the up-spin electron in the oxygen 2p orbital transfer to the d-orbital of the i-th metal ion as shown in Fig. 6. Then the doublon is formed and the spinon is annihilated at i-th ion, creating the singlet spin state. At the same time, the final state interaction between the up-spin d-electron at j-th ion and remaining down-spin 2p-electron at oxygen form the spin-singlet state [13], whose coupling force J_k is much larger than the transfer energy t. As a consequence, we have a bound state of a holon at j and a doublon at i, which can propagate over the crystal due to the translational symmetry.

For the third-order optical nonlinearity, dipolar interaction of Eq. (20) with the external field works four times as shown by Fig. 5(c). The holon at j and the doublon at $i = j \pm 1$ make up a bound state due to the strong attraction working between holon and doublon, which may be called charge-transfer exciton. First, the charge-transfer dipole moment μ in Eq. (20) becomes large in the transition metal oxides of Perovskite type because the d-orbital e_g orients towards 2p-orbital of oxygen. Second, creations of two charge-transfer excitons by ω_p and ω_t are accompanied by the annihilation

Charge Transfer Exciton

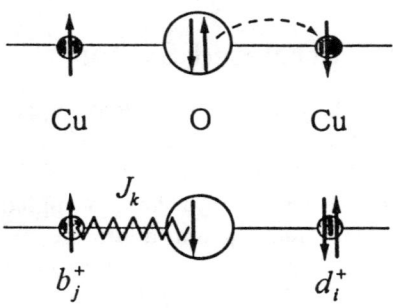

$$c_{i\sigma} = f_{i\sigma} b_i^+ + \xi_{\sigma\sigma'} f_{i\sigma'}^+ d_i ,$$

$$\sum_\sigma f_{i\sigma}^+ f_{i\sigma} + b_i^+ b_i + d_i^+ d_i = 1$$

$$P \sim \mu \sum b_j^+ d_i^+ (f_{i\uparrow} f_{j\downarrow} - f_{i\downarrow} f_{j\uparrow}) + \text{h.c.}$$

Fig. 6. Charge-transfer exciton in the system of transition-metal and oxygen ions, forming a bound state of holon and doublon associated with the annihilation of a spinon pair

of a pair of spinons as Eq. (20) describes, while annilations of two charge-transfer excitons are accompanied by the creation of a pair of spinons. These spinons interact with each other, so that creations and annihilations of four excitons shown in Fig. 5(c), become possible for many different sites within the correlation length of spinons in the ground state. When we denote by n_J the number of unit cells within their correlation region, the magnitude of $\chi^{(3)}$ increases by the factor n_J^2 in contrast to the conventional case 3.1, and by the factor n_J even in comparison to the giant two-photon absorption in 3.2. This colossal enhancement originates in long-range correlation of spinons in the electronic ground state which results in large freedom of choosing the second, third and fourth dipolar interaction at j, i, and i', relative to the original $j' = 0$, with the external fields. The freedom of choosing the sites of dipolar transitions within the correlation range brings about the colossal enhancement of $\chi^{(3)}$. This correlation range diverges at 0 K for quasi-one-dimensional systems and at Néel temperature for two- or three- dimensional crystals. Therefore, we will be able to confirm the prediction of colossal two-photon absorption by observing the temperature dependence of $\chi^{(3)}$ as well as the absolute value of $\chi^{(3)}$.

References †M. Fiebig is a CREST (JST) postdoctoral fellow.

[1] M. Fiebig, D. Fröhlich, B.B. Krichevtsov, and R.V. Pisarev. Phys. Rev. Lett. **73**, 2127 (1994).

[2] M. Fiebig, D. Fröhlich, and R.V. Pisarev, J. Appl. Phys. **81**, 4875 (1997).

[3] M. Muto, Y. Tanabe, T. Iizuka-Sakano, and E. Hanamura, Phys. Rev. **B57**, 9586 (1998).

[4] Y. Tanabe, M. Muto, M. Fiebig, and E. Hanamura, Phys. Rev. **B58**, in press.

[5] M. Fiebig, D. Fröhlich, B.B. G. Sluyterman v. L., and R.V. Pisarev. Appl. Phys. Lett. **66**, 2906 (1993).

[6] V. N. Muthukumar, R. Valenti, and C. Gros, Phys. Rev. Lett. **75**, 2766 (1995); Phys. Rev. **B54**, 433 (1996).

[7] Y. Tanabe, M. Muto, and E. Hanamura, Solid State Commun. **102**, 643 (1997).

[8] M. Inoue and Y. Toyozawa, J. Phys. Soc. Jpn. **20**, 363 (1965).

[9] E. Hanamura, Solid State Commun. **12**, 951 (1973).

[10] G.M. Gale and A. Mysyrowicz, Phys. Rev. Lett. **32**, 727 (1974).

[11] H. Ezaki, T. Tokihiro, and E. Hanamura, Phys. Rev. **B50**, 10506 (1994).

[12] S.E. Barnes, J. Phys. F: Metal Phys. **6**, 1375 (1976); **7** 2637 (1977).

[13] F.C. Zhang and T.M. Rice, Phys. Rev. **B37**, 3759 (198).

Part IV

High T$_c$ Cuprates

Fermi-Liquid Versus Pseudogap Behaviors in Filling-Control Transition-Metal Oxides

A. Fujimori[1], T. Yoshida[1], A. Ino[1], T. Mizokawa[1], C. Kim[2], Z.-X. Shen[2]
Y. Taguchi[3], T. Katsufuji[3], Y. Tokura[3], H. Eisaki[4], S. Uchida[4], K. Kishio[4]

[1] Department of Physics, University of Tokyo, Bunkyo-ku, Tokyo 113-0033, Japan
[2] Department of Applied Physics and Stanford Synchrotron Radiation Laboratory, Stanford University, Stanford, CA94305, U.S.A.
[3] Department of Applied Physics, University of Tokyo, Bunkyo-ku, Tokyo 113-0033, Japan
[4] Department of Superconductivity, University of Tokyo, Bunkyo-ku, Tokyo 113-8656, Japan

Abstract. Comparative photoemission studies of $La_{1-x}Sr_xTiO_{3+y/2}$ (LSTO) and $La_{2-x}Sr_xCuO_4$ (LSCO) have revealed contrasting behaviors in their low-energy electronic properties as the system approaches a filling-control Mott transition from the metallic side. In LSTO and overdoped LSCO, the chemical potential shift, the spectral density of states at the chemical potential μ and the quasi-particle density at μ can be consistently analyzed within Fermi-liquid theory. In underdoped LSCO, on the other hand, the Fermi-liquid picture breaks down and a relatively wide ($\sim$ 100 meV) pseudogap is opened at μ in the density of states, in addition to a smaller leading-edge shift of $\sim$ 10 meV around $\mathbf{k} \sim (\pi, 0)$ in angle-resolved photoemission spectra. The two energy scales increase with decreasing hole concentration x. The larger energy scale is probably associated with antiferromagnetic short-range order while the smaller one is attributed to the spin-gap formation or superconducting fluctuations.

1 Introduction

The nature of a second-order phase transition is characterized by its criticality in the vicinity of the transition. In particular, critical behaviors near a correlation-driven metal-insulator transition, i.e., a Mott transition, in filling-control transition-metal oxides have been a central issue in recent studies of correlated electron systems [1]. One of the most striking observations is the contrasting behaviors of the perovskite-type Ti oxides and the high-T_c Cu oxides when the Mott transition is approached from the metallic side. In $La_{1-x}Sr_xTiO_{3+y/2}$ (LSTO), the electronic specific heat coefficient γ and the Pauli-paramagnetic susceptibility are enhanced when the amount of holes $\delta = x + y$ doped into the Mott insulator $LaTiO_3$ is reduced [2, 3]. In $La_{2-x}Sr_xCuO_4$ (LSCO), on the other hand, the specific-heat γ as well as the Pauli component of the paramagnetic susceptibility *decrease* when $\delta = x$ is decreased and the system approaches the Mott insulator La_2CuO_4 [4, 5, 6]. If the metal-to-insulator transition is characterized by the disappearance of the Drude weight $D \propto \delta/m^*$, where δ is the carrier (hole) number and m^* is the effective mass of the carrier, the Mott transition in LSTO may be characterized by a divergence of the carrier effective mass "$m^* \to \infty$" whereas the transition in LSCO may be characterized by a vanishing carrier number "$\delta \to 0$" [7].

In the following, we shall describe the critical behaviors of the two systems near the Mott transitions mainly based on photoemission spectroscopic data in combination with thermodynamic data.

2 Chemical potential shift

The charge susceptibility $\chi_c \equiv \partial n/\partial \mu$, where μ is the electron chemical potential and n is the electron density, is proportional to the quasi-particle (QP) density at μ, $N^*(\mu)$, and hence to the effective mass m^* in the Fermi-liquid picture. Namely, the rate of the chemical potential shift is given by $\partial \mu/\partial n = (1+F_s^0)/N^*(\mu)$ and is therefore inversely proportional to m^*, where F_s^0 (> 0) is a Landau Fermi-liquid parameter representing the repulsive QP-QP repulsion [8]. Here, $N^*(\mu)$ is deduced from electronic specific heat measurements through $\gamma = \frac{\pi}{3} k_B^2 N^*(\mu)$. From $\partial \mu/\partial n \propto N^*(\mu)^{-1}$, it follows that, when the QP density or the effective mass is enhanced, the chemical potential shift is suppressed. Since measured photoemission spectra are referenced to the chemical potential, parallel shifts of core levels and valence levels can be attributed to a shift in the chemical potential [9].

The chemical potential shift $\Delta \mu$ in LSTO is suppressed around $\delta \sim 0$ as shown in Fig. 1 [10]. This is consistent with the mass enhancement deduced from the specific heat data. Figure 1 shows that in LSCO, too, the chemical potential shift is suppressed towards $\delta \sim 0$ as if m^* were enhanced [9]. Such a behavior of the chemical potential shift has been predicted for the two-dimensional Hubbard and $t-J$ models by numerical studies [13, 14, 15], reflecting enhanced quantum fluctuations in the paramagnetic metallic phase near the Mott transition. The result for LSCO, however, cannot be reconciled with the reduced QP density at μ deduced from the electronic specific heat if the Fermi-liquid relation $\partial \mu/\partial n \propto N^*(\mu)-1$ is assumed. This indicates a breakdown of the Fermi-liquid picture in underdoped LSCO. In overdoped LSCO, the chemical potential shift is faster than that expected from $N^*(\mu)$: $\partial \mu/\partial n > N^*(\mu)^{-1}$, indicating F_s^0 is substantial. The Fermi-liquid picture is thus applicable to overdoped LSCO.

The suppression of the chemical potential shift concomitant with the suppression of the QP density at μ in LSCO may be caused by a phase separation or spectral weight transfer between regions away from μ as we shall discuss below.

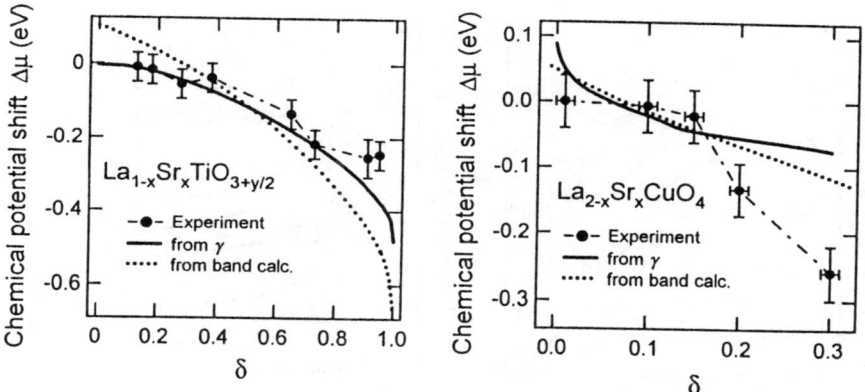

Fig. 1. Chemical potential shift $\Delta\mu$ in La$_{1-x}$Sr$_x$TiO$_{3+y/2}$ [10] and La$_{2-x}$Sr$_x$CuO$_4$ [9] as a function of hole concentration ($\delta = x + y$ and x, respectively). The solid curves show shifts deduced from the electronics specific heat [2, 3, 5] (assuming $F_s^0 = 0$) and the dotted curves show shifts deduced from local-density-approximation (LDA) band-structure calculations [11, 12].

3 Density of states at the chemical potential

In angle-integrated (k-integrated) photoemission experiments, the spectral density of states (DOS) $\rho(\omega)$ ($\equiv \int d\mathbf{k} A(\mathbf{k},\omega)$, where $A(\mathbf{k},\omega)$ is the single-particle spectral function) is measured and hence the DOS at μ, $\rho(\mu)$, can be evaluated. The spectral weight of the QP or of the coherent part of $\rho(\omega)$ is equal to the wavefunction renormalization factor $Z \equiv \langle \Psi_{N-1}|a_k|\Psi_N\rangle$ and is therefore a measure of the strength of electron correlation. One can deduce Z from the measured DOS $\rho(\mu)$ and QP density $N^*(\mu)$ through the relationship $\rho(\mu) = ZN^*(\mu)$.

Figures 2 shows the photoemission spectra [$\sim \rho(\omega)$] of LSTO and LSCO. The DOS at μ is plotted in Fig. 3. In LSTO, $\rho(\mu)$ increases to some extent with decreasing δ within the paramagnetic metallic phase ($\delta > 0.08$). Note, however, that the enhancement of $\rho(\mu)$ is not so dramatic as that of $N^*(\mu)$ deduced from the specific heat, indicating that the QP weight $Z \equiv \rho(\mu)/N^*(\mu)$ decreases toward $\delta \sim 0$ as depicted in the same figure. In fact, the increase of $\rho(\mu)$ is largely due to the increase in the band DOS $N_b(\mu)$ deduced from band-structure calculation [11]. Figures 2 and 3 also show that $\rho(\mu)$ sharply drops below $\delta = 0.08$, where the system enters the antiferromagnetic metallic phase.

In LSCO, $\rho(\mu)$ and $N^*(\mu)$ both slightly increases with decreasing δ down to $\delta \sim 0.2$ as in the Fermi-liquid system LSTO but then decreases toward $\delta \sim 0$. The latter behavior is opposite to that in LSTO. Since antiferromagnetic spin fluctuations in LSCO are known to be enhanced with decreasing δ, the similar behaviors of $\rho(\mu)$ in antiferromagnetic LSTO and underdoped LSCO imply that the antiferromagnetic fluctuations (antiferromagnetic short-range order) reduce $\rho(\mu)$ in underdoped LSCO.

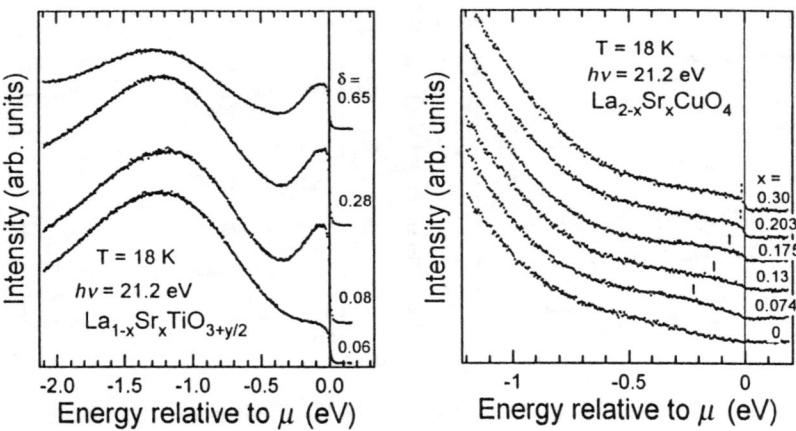

Fig. 2. Photoemission spectra $\rho(\omega)$ of $La_{1-x}Sr_xTiO_{3+y/2}$ and $La_{2-x}Sr_xCuO_4$ [10, 16]. The vertical bars denotes the minimum in the second derivative of $\rho(\omega)$. Its position measured from the chemical potential defines a "pseudogap" energy scale Δ_{PG}.

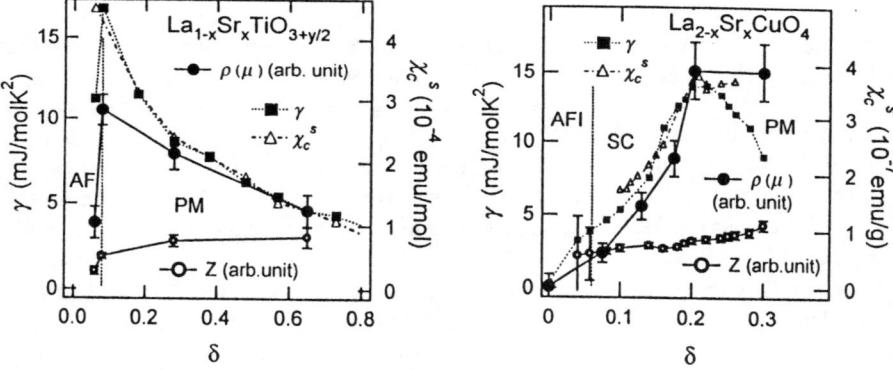

Fig. 3. Density of states at the chemical potential $\rho(\mu)$ [10, 16] compared with the electronic specific-heat coefficient $\gamma \propto N^*(\mu)$ and the Pauli susceptibility χ_s^c [2, 3, 5, 6] of $La_{1-x}Sr_xTiO_{3+y/2}$ and $La_{2-x}Sr_xCuO_4$ as a function of hole density δ ($= x + y$ or x). The quasi-particle weight $Z \equiv \rho(\mu)/N^*(\mu)$ is also plotted as a function of δ.

In spite of the contrasting behaviors of LSTO and LSCO described above, Z decreases with decreasing δ in both systems. It therefore appears that the decrease of Z, namely, the loss of coherent spectral weight at the chemical potential, is a common feature of the different types of Mott transitions.

4 Pseudogap behavior in underdoped $La_{2-x}Sr_xCuO_4$

The DOS $\rho(\omega)$ of LSTO in the paramagnetic metallic phase consists of the coherent part (the QP band) around μ and the incoherent part (a reminiscent of the lower Hubbard band) at ~ 1.5 eV below μ (Fig. 2). This spectroscopic behavior as well as the thermodynamic properties of LSTO are well described by the dynamical mean-field treatment of the Hubbard model by Kotliar and co-workers [17]. A close inspection of the DOS of the antiferromagnetic metallic phase ($\delta = 0.06$) has revealed a weak dip feature at μ, probably indicating the opening of an incomplete gap due to the antiferromagnetic long-range order.

Figure 2 shows that in LSCO the decrease of $\rho(\omega)$ with decreasing δ occurs not only at μ but in a finite energy region around $\omega = \mu$. The figure also shows that the extent of this region increases with decreasing δ. This may be rephrased that a pseudogap whose depth and width increase with decreasing δ is opened around $\omega \sim \mu$. It is interesting to note that the temperature T_χ at which the uniform magnetic susceptibility reaches a maximum increases with decreasing δ as if a pseudogap develops around μ [6]. If we denote the magnitude of the "pseudogap" by Δ_{PG} (as shown by vertical bars in Fig. 2), T_χ and Δ_{PG} show similar doping dependences as demonstrated in Fig. 4. Thus the decrease of $N^*(\mu)$ in LSCO with decreasing δ can be ascribed to the opening of a pseudogap around μ in $\rho(\omega)$.

The pseudogap discussed above has a large energy scale of ~ 100 meV, on the order of the superexchange interaction J. It is much larger than the "normal-state" gap above T_c, which has been identified as a ~ 20 meV shift of the leading edge in angle-resolved photoemission (ARPES) spectra of underdoped $Bi_2Sr_2CaCu_2O_8$ (BSCCO) compounds [18, 19]. Pseudogaps of the latter kind have also been observed in LSCO by ARPES [20]. In Fig. 5 are plotted ARPES spectra for momenta near $(\pi, 0)$ on the Fermi surface, where in the superconducting state the d-symmetry gap becomes largest. Figure 5

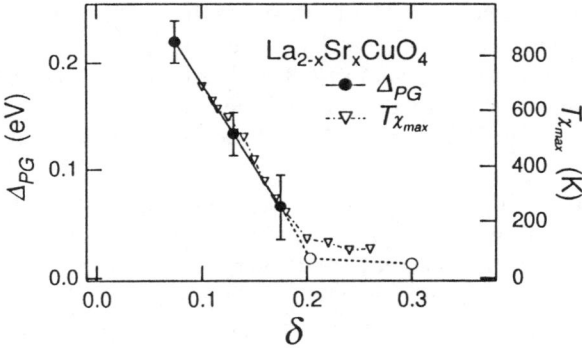

Fig. 4. Magnitude of the "pseudogap" Δ_{PG} [16] compared with the temperature T_χ at which the magnetic susceptibility takes the maximum [5]. We find $\Delta_{PG} \sim 3k_B T_\chi$.

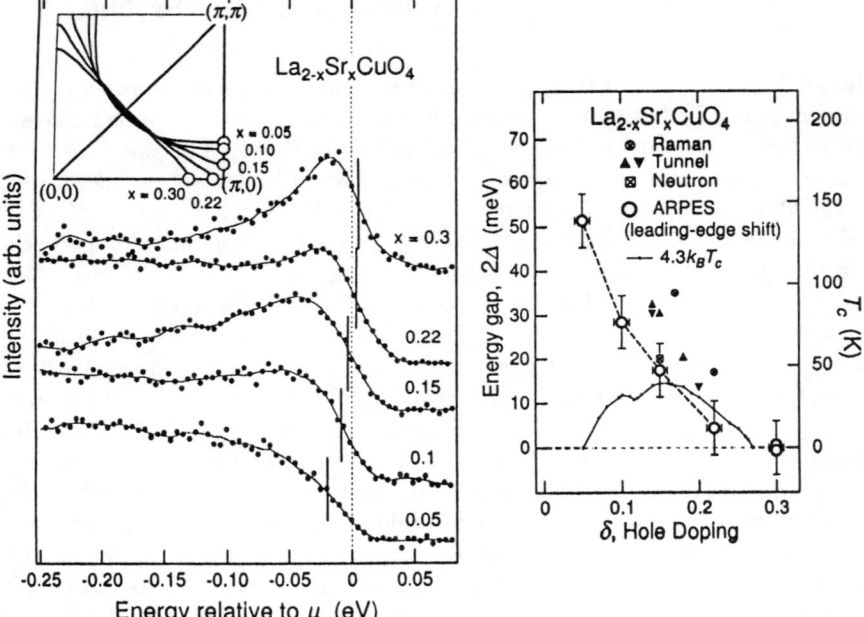

Fig. 5. Left: ARPES spectra of $La_{2-x}Sr_xCuO_4$ for $\mathbf{k} \sim (\pi, 0)$ in the Fermi surface. The leading edge is marked by vertical bars. The inset shows the Fermi surfaces and the $\mathbf{k}$ points where the ARPES spectra were taken. Right: The leading-edge position in the ARPES spectra compared with the energy gaps seen in the scanning tunneling spectroscopy [21], Raman scattering [22] and neutron scattering [23] experiments.

shows that the leading edge is shifted away from μ with decreasing δ. The edge position is plotted in Fig. 5 and are compared with the gap size obtained from tunneling spectroscopy [21], Raman scattering [22] and inelastic neutron scattering [23] experiments.

The flat band around $(\pi, 0)$ is one of the most remarkable features in the ARPES spectra of high-T_c cuprates. Since the band has a large flat portion in the $\mathbf{k}$ space, it is expected to give a peak in the QP density $N^*(\omega)$ as well as in the DOS $\rho(\omega)$. In fact, the binding energy of the flat band increases with decreasing δ as does the binding energy Δ_{PG} of the pseudogap feature in $\rho(\omega)$.

The leading edge shift is certainly more directly associated with the occurrence of the superconductivity and has a much smaller energy scale than Δ_{PG}, implying that there are two different energy scales in the underdoped cuprates. The composition-temperature diagram demonstrates the two crossover temperature (energy) scales in underdoped LSCO. The lower crossover line is due to the opening of a spin gap and to a spinon-pairing proposed by Fukuyama and co-workers [24]. The higher crossover line would be more intimately associated with antiferromagnetic fluctuations or short-

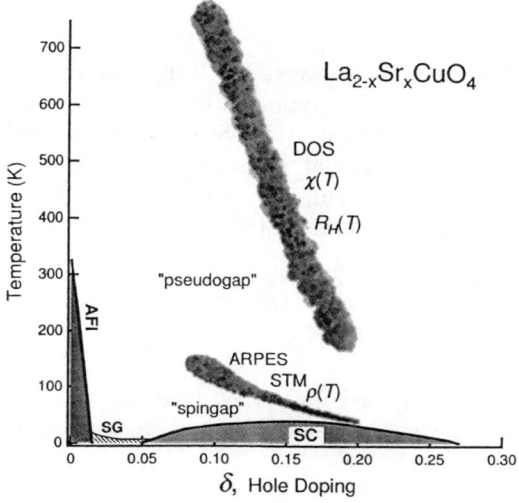

Fig. 6. Experimental "phase diagram" of $La_{2-x}Sr_xCuO_4$ deduced from the magnetic susceptibility [5], tunneling [21], Raman scattering [22] and photoemission [16, 20] studies. The crossover line for the Fermi liquid has been taken from [16].

range antiferromagnetic order within the CuO_2 plane since the spectra rather smoothly evolve from antiferromagnetic insulators to underdoped superconductor as shown in Fig. 5. Laughlin [25] also pointed out a similarly smooth evolution of ARPES spectra near $\mathbf{k} = (\pi, 0)$ from antiferromagnetic insulators to superconductors. A similar phase diagram (crossover diagram) has been proposed by Emery et al. [26] and by Pines [27]. It should be noted that the higher crossover temperature is of order J for small x, being much higher than the three-dimensional ordering temperature of La_2CuO_4. It is also worth noting that the separation between the two temperature scales is considerably larger than that found for BSCCO and YBCO systems.

Finally, we remark on possible relationship between the photoemission results and the charge stripes in underdoped LSCO [28]. If we view the charge stripes as a microscopic phase separation of doped holes into hole rich and hole-deficient regions, the suppressed chemical potential shift in the underdoped region would be a natural consequence of the phase separation. It has been proposed that charge stripes, which run along the $(0,0) - (0,\pi)$ directions, disrupt the QP peak dispersing along the $(0,0) - (\pi,\pi)$ directions in Zn-doped BSCCO [29]. In underdoped LSCO, indeed, the QP features along the $(0,0) - (\pi,\pi)$ direction are found to be almost totally destroyed [20]. More systematic studies are necessary to relate the charge stripes with the photoemission spectra and eventually with the superconductivity in the high-T_c cuprates.

Acknowledgment

Enlightening discussions with M. Imada, S. Maekawa, T. Tohyama and M. Sato are gratefully acknowledged. This work is supported by a Grant-in-Aid for Scientific Research from the Ministry of Education, Science, Sports and Culture of Japan, the New Energy and Industrial Technology Development Organization (NEDO), Special Coordination Fund from Science and Technology Agency of Japan and the U. S. DOE, Office of Basic Energy Science and Division of Material Science.

References

[1] See e.g., M. Imada, A. Fujimori and Y. Tokura: *Rev. Mod. Phys.*, in press.
[2] Y. Tokura, Y. Taguchi, Y. Okada, Y. Fujishima, T. Arima, K. Kumagai and Y. Iye: *Phys. Rev. Lett.* **70** (1993) 2126.
[3] T. Katsufuji, Y. Taguchi and Y. Tokura: *Phys. Rev.* B **56** (1997) 10145.
[4] J. W. Loram, K. A. Mirza, W. Y. Liang, and J. Osborne: *Physica* C **162-164** (1989) 498.
[5] N. Momono, M. Ido, T. Nakano, M. Oda, Y. Okajima and K. Yamaya: *Physica* C **233** (1994) 395.
[6] T. Nakano, M. Oda, C. Manabe, N. Momono, Y. Miura and M. Ido: *Phys. Rev.* B **49** (1994) 16000.
[7] M. Imada: *J. Phys. Soc. Jpn.* **62** (1993) 1105.
[8] N. Furukawa and M. Imada: *J. Phys. Soc. Jpn.* **61** (1992) 3331.
[9] A. Ino, T. Mizokawa, A. Fujimori, K. Tamasaku, S. Uchida, T. Kimura, T. Sasagawa and K. Kishio: *Phys. Rev. Lett.* **79** (1997) 2101.
[10] T. Yoshida, A. Ino, T. Mizokawa, A. Fujimori. Y. Taguchi, T. Katsufuji and Y. Tokura: submitted.
[11] K. Takegahara: *J. Electron Spectrosc. Related Phenom.* **66** (1994) 303.
[12] L. F. Mattheiss, *Phys. Rev. Lett.* **58** (1987) 1028.
[13] N. Furukawa and M. Imada: *J. Phys. Soc. Jpn.* **62** (1993) 2557.
[14] E. Dagotto, A. Moreo, F. Ortolani, J. Riera and D. J. Scalapino: *Phys. Rev. Lett.* **67** (1991) 1918.
[15] J. Jaklič and P. Prelovšek: *Phys. Rev. Lett.* **77**, (1996) 892.
[16] A. Ino, T. Mizokawa, K. Kobayashi, A. Fujimori, T. Sasagawa, T. Kimura, K. Kishio, K. Tamasaku, H. Eisaki and S. Uchida: submitted.
[17] A. George, G. Kotliar, W. Krauth and M. J. Rozenberg: *Rev. Mod. Phys.* **68** (1996) 13.
[18] D. S. Marshall, D. S. Dessau, A. G. Loeser, C.-H. Park, A. Y. Matsuura, J. N. Eckstein, I. Bozovic, P. Fournier, A. Kapitulnik, W. E. Spicer and Z.-X. Shen: *Phys. Rev. Lett.* **76** (1996) 4841.
[19] H. Ding, T. Yokoya, J. C. Campuzano, T. Takahashi, M. Randeria, M. R. Norman, T. Mochiku, K. Kadowaki and J. Giapintzakis: *Nature* **382** (1996) 51.
[20] A. Ino, C. Kim, T. Mizokawa, Z.-X. Shen, A. Fujimori, M. Takaba, K. Tamasaku, H. Eisaki and S. Uchida: submitted.
[21] N. Momono, T. Nakano, M. Oda and M. Ido: *J. Phys. Chem. Solids*, in press.
[22] X. K. Chen, J. C. Irwin, H. J. Trodahl, T. Kimura and K. Kishio: *Phys. Rev. Lett.* **73** (1994) 3290.

[23] K. Yamada, S. Wakimoto, G. Shirane, C. H. Lee, M. A. Kastner, S. Hosoya, M. Greven, Y. Endoh and R. J. Birgeneau: *Phys. Rev. Lett.* **75** (1995) 1626.
[24] T. Tanamoto, H. Kohno, and H. Fukuyama: *J. Phys. Soc. Jpn.* **61** (1992) 1886.
[25] R. B. Laughlin: *Phys. Rev. Lett.* **79** (1997) 1726.
[26] V. J. Emery, S. A. Kivelson and O. Zacher: *Phys. Rev.* B **56** (1997) 6120.
[27] D. Pines: cond-mat/9702187.
[28] J. M. Tranquada, B. J. Sterlieb, J. D. Axe, Y. Nakamura and S. Uchida: *Nature* **375** (1995) 561.
[29] P. J. White, D. L. Feng, C. Kim, H. Ikeda, R. Yoshizaki, G. D. Gu, N. Koshizuka and Z.-X. Shen: submitted.

Metal-Insulator and Superconductor-Insulator Transitions in Correlated Electron Systems

Masatoshi Imada[1], Fakher F. Assaad[2], Hirokazu Tsunetsugu[3] and Yukitoshi Motome[4]

1) Institute for Solid State Physics, University of Tokyo, Roppongi, Minato-ku, Tokyo 106-8666, Japan
e-mail address imada@issp.u-tokyo.ac.jp
2) Institut für Theoretische Physik, Teilinstitut III, Universität Stuttgart, Pfaffenwaldring 57, D-70550 Stuttgart, Germany
3) Institute of Applied Physics, University of Tsukuba, Tsukuba, Ibaraki 305-8573, Japan
4) Department of Physics, Tokyo Institute of Technology, Oh-okayama, Meguro-ku, Tokyo 152-8551, Japan

Abstract

Quantum transitions between the Mott insulator and metals by controlling filling in two-dimensional square lattice are characterized by a large dynamical exponent $z = 4$ where the origin of unusual metallic properties near the Mott insulator are ascribed to the proximity of the transition. The scaling near the transition indicates the formation of flat dispersion area due to singular momentum dependence of the single-particle renormalization. The flat dispersion controls critical properties of the Mott transition. An instability of the flat dispersion to the d-wave superconducting order is discussed. We also discuss a case of the Mott transition for a model of Mn perovskite compounds with orbital degeneracy where orbital correlation length shows critical divergence toward the metal-insulator transition.

1 Introduction

Metal-insulator transitions (MIT) driven by strong correlation effects are called Mott transitions and are a subject of recent intensive studies in two- and three-dimensional systems [1]. For one-dimensional (1D) systems, the Mott transitions are relatively well understood where rigorous results can be obtained for some nontrivial cases. In infinite-dimensional systems where the dynamical mean field theory becomes exact, the MIT can also be studied in a controlled way. However, in 1D systems, the Fermi level (and the Fermi surface) is represented by only two points while in the infinite-dimensional systems, the single-particle self-energy is site-diagonal and has no wavenumber dependence. Therefore, in both cases, charge excitations near the Fermi level are not allowed to have momentum dependence along the Fermi surface. On the contrary, as we see below, the momentum dependence of the single-particle renormalization may have singular dependence along the Fermi surface in finite-dimensional systems such as in two and three dimensions and therefore would show quite different features from one- and infinite- dimensional systems.

In two or higher dimensional systems, there exists no evidence to exclude the Fermi liquid phase as the only one stable fixed point of ordinary metals. However, this is not necessarily an isotropic metal. When charge excitations in a part of the Fermi surface have slow dynamics, they are liable to couple strongly to other degrees of freedom (such as spin fluctuations) thus yielding even slower dynamics. This synergetics may selfconsistently generate singular momentum dependence of the single-particle renormalization, where at a particular region (or points) on the Fermi surface, the Fermi liquid description breaks down. This particular region may be characterized by flat dispersion with mass enhancement due to the real part of the selfenergy as well as by strong damping due to the imaginary part.

Metallic phase near the Mott insulator is known to show various unusual properties in terms of standard metals. However, in any experimental systems clear phase boundary is not observed between the paramagnetic metallic phase near the Mott insulator and more or less standard metals observed far from the correlated insulator. This strongly suggests that the unusual properties are not due to the appearance of a new phase with adiabatic discontinuity from the weakly correlated metals, but due to a proximity of the MIT. Therefore, it is important to understand critical properties of the MIT. Metal-insulator transitions are typical examples of quantum phase transitions which takes place at zero temperature by changing a parameter to control quantum fluctuations rather than thermal fluctuations. In this paper, we discuss that the unusual properties can be understood from novelty of the quantum criticality at the transition to the Mott insulator [2, 3, 4]. In quantum phase transitions, dynamical fluctuations play roles even in high spatial dimensions. These are correctly treated in the "dynamical mean field theory" which provides exact result if spatial dimensionality is infinite and spatial fluctuations are suppressed [5]. In real materials, however, quantum fluctuations with strong interaction effects are most conspicuous in systems with low-dimensional anisotropy. In this case, fluctuations near quantum phase transition points are enhanced and appear as combined effects of strong spatial and dynamical quantum fluctuations.

The Drude weight defined from the coefficient of the δ-function in the frequency-dependent conductivity at $T = 0$ as

$$\sigma(\omega) = D\delta(\omega) + \sigma_{\text{reg}}(\omega) \tag{1}$$

and the charge compressibility κ defined below are the two important and relevant quantities to describe the Mott transition. The Mott insulator has two basic properties: one is insulating property and the other, incompressibility. The Drude weight and the charge compressibility are both zero in the Mott insulating phase while both nonzero in metallic phases. Therefore, they are totally nonanalytic at the transition point. The nonanalyticities in these two quantities are also common in the transition to the band insulator, while they do not have singularities in case of the Anderson localization transition.

We here discuss in more detail the important difference between the Anderson transition and the MIT to the Mott insulator. In the Anderson transition, it is well known that the singularity at the transition point appears in the diffusion constant or the relaxation time of the carrier τ (inverse of the imaginary part of the single-particle self-energy) but neither in the carrier effective mass m^*, nor in the density of states $N(\omega)$ etc. [6] Therefore the DC conductivity $\propto n\tau/m^*$ is a good probe to determine the character of the transition because it is directly proportional to the relaxation time. In contrast, in the transition to the Mott insulator, the singularity may not appear in the relaxation time of the carrier but appears in the ratio of the carrier density n to the effective mass m^*. The DC conductivity in the ideal case (without disorder at $T = 0$) is always infinite in the metallic phase and cannot be a good probe in the ideal condition (it can be a probe to some extent if the disorder or thermal effects play some role) while the Drude weight $D \propto n/m^*$ and the charge compressibility κ discussed here become relevant quantities with singularities.

We note that the quantities which explicitly depend on the relaxation time arising from thermal or disorder effects have to be treated carefully. As we see below it is possible that the MIT is controlled by a particular part of flat dispersion on the Fermi surface. However, the relaxation-time-dependent quantities such as DC transport properties can be "contaminated" by the relaxation time and the real criticality could be obscured because the contribution from the part of flat dispersion could be suppressed due to shorter relaxation time as compared to the other part.

We also note the difference between the Mott transition and magnetic quantum phase transitions which take place within metals. In the latter case, self-consistent renormalization theory was developed [7, 8] where, at $T = 0$, a Gaussian treatment from the paramagnetic phase and the Hartree-Fock-RPA description are justified above the upper critical dimension. There the anomalous character of metals is mainly ascribed again to the anomaly in the relaxation time although a weak effect on the renormalization factor is expected.

In §2, we review results on novel universality class of two-dimensional MIT driven by correlation effects. The scaling theory with the universality class characterized by the dynamical exponent $z = 4$ offers a unified description for the unusual properties of metal near the Mott insulator. Numerical results on the charge compressibility, spin correlations, dynamical conductivity in the metallic phase as well as the localization length in the insulator are consistently understood from this quantum criticality. The strong momentum dependence of the renormalization around $(\pi, 0)$ spot is suggested as the basis of the scaling description. Superconductor-insulator transitions have been studied motivated from instability of anomalous metallic states associated with this novel universality class of the MIT. The results are summarized in §3. MITs with orbital degeneracy is the subject of §4.

2 Metal-Insulator Transition in Two Dimensions

2.1 Numerical Results

The Mott insulator and metallic states near the MIT have been intensively studied subject by numerical approach. Here, we see that the 2D single-band Hubbard and t-J models show consistency with the scaling theory [3, 4]. It is discussed later that the single-particle excitation of the 2D Hubbard model around $(\pm\pi, 0)$ and $(0, \pm\pi)$ in the momentum space is renormalized to form a flat quartic dispersion. This singular renormalization effect determines the universality class of the transition between the Mott insulator and metals and constitutes a basis for the scaling theory. This circumstance is a unique feature of the Mott transition in finite-dimensional systems in contrast with one and infinite dimensions.

Before the clarification of the full momentum dependence, the singular momentum dependence was suggested from the "momentum integrated" quantities. The first signature was observed in the charge compressibility.

The charge compressibility κ or charge susceptibility χ_c are defined as $\chi_c = n^2\kappa = \partial n/\partial\mu$ where n is the electron density and μ is the chemical potential. The charge susceptibility vanishes in the Mott insulating phase because of the incompressibility. However, quantum Monte Carlo results on 2D systems show that doped systems become more and more compressible near the MIT in the metallic side [9, 10]. The scaling plot of χ_c shows singular dependence on the doping concentration δ in the form

$$\chi_c \propto 1/|\delta|^p \tag{2}$$

with $p = 1$ for $\delta \neq 0$ [9, 10, 11]. Similar is also observed in the 2D t-J model [12, 13]. When the charge susceptibility is singularly divergent as a power of δ for $\delta \to +0$ in 2D, the single-particle description of low-energy excitations has to show the divergence of the effective mass of relevant particle. This is in contrast with the usual MIT between metals and the band insulator, where the number of carriers vanishes.

The second signature of unusual renormalization was observed in the measurement of the localization length in the insulating phase [14]. A large numerical advantage to observe the transition from the insulator side is that we can avoid the negative sign problem known in the quantum Monte Carlo method. The insulator undergoes a transition to a metal when the chemical potential μ approaches the critical point μ_c from the region of the charge gap. In the insulating phase, the single-particle Green function defined as

$$\mathcal{G}(r,\tau) = \langle T c(r,\tau) c^\dagger(0,0)\rangle \tag{3}$$

provides the localization length ξ_l defined by

$$\mathcal{G}(r, \omega = \mu) \equiv \int_0^\infty \mathcal{G}(r,\tau,\mu)d\tau \sim e^{-r/\xi_l}. \tag{4}$$

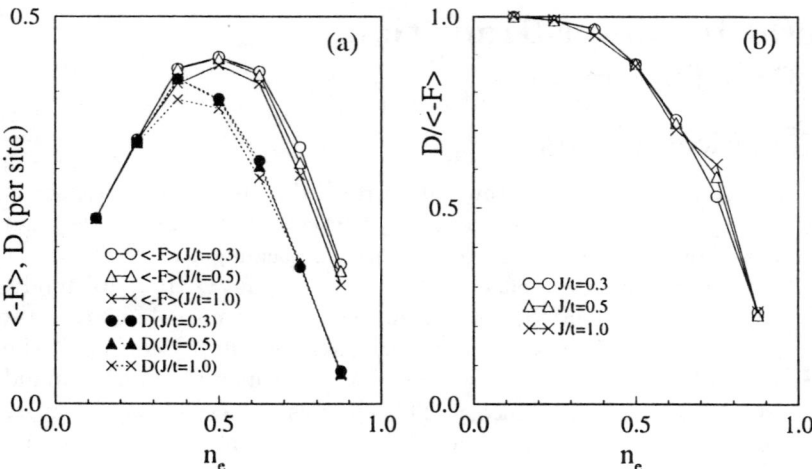

Figure 1: (a) Drude weight D and total weight $\langle -F \rangle$ in the optical conductivity of the 2D t-J model with 4×4 sites at zero temperature. (b) The ratio of the two weights $D/\langle -F \rangle$.

ξ_l may be regarded as the localization length of the wavefunction of virtually created state at the chemical potential inside the gap. The scaling of ξ_l near μ_c calculated with the above method by Assaad and Imada [14] for the two-dimenional Hubbard model shows

$$\xi_l \sim |\mu - \mu_c|^{-\nu}$$
$$\nu = 0.26 \pm 0.05. \qquad (5)$$

The obtained ν is consistent with $\nu = 1/4$.

The third signature is seen in the Drude weight. The Drude weight was calculated for the t-J model by the exact diagonalization[15]. The doping concentration dependence in Fig. 1 shows that the total kinetic energy $\langle -F \rangle$ is proportional to δ while the Drude weight D vanishes much faster consistently with $D \propto \delta^2$. It should be noted that this is in sharp contrast with a naive expectation that the carrier density is proportional to δ, where the Drude weight in the Drude theory is proportional to n/m^* and should be linear in the carrier density n divided by the carrier effective mass m^*. As we discuss in §2.3, the frequency dependence of the conductivity in numerical results also suggests that the most part of the conductivity weight in the sum rule is exhausted in a rather universal line shape of incoherent conductivity $\sigma_{\text{reg}}(\omega)$ and is consistent with a quick collapse of the Drude part with decreasing doping.

The above signatures observed in the charge compressibility, localization length and dynamical conductivity are hard to understand when isotropic renormalization of charge excitations is assumed along the large Fermi surface which satisfies the Luttinger theorem.

2.2 Scaling Theory of Mott Transition

The numerical results on the charge compressibility and spin correlations in 2D have inspired extensive studies to clarify the nature of the Mott transition in low-dimensional systems [2, 3, 4]. Because the MIT is controlled by quantum fluctuations, we may take the control parameters such as the electron chemical potential μ or the bandwidth t. Using this control parameter, the distance from the critical point is measured by Δ. The control parameter can either be the chemical potential to control the filling or the bandwidth (or the interaction). The scaling theory assumes the existence of single characteristic length scale ξ which diverges as $|\Delta| \to 0$ and a single characteristic frequency scale Ω which vanishes as $|\Delta| \to 0$. The correlation length ξ is assumed to follow $\xi \sim |\Delta|^{-\nu}$ as $\Delta \to 0$, which defines the correlation length exponent ν. The frequency scale Ω is determined from the quantum dynamics of the system independently of the length scale. The dynamical exponent z determines how Ω vanishes in relation to $1/\xi$ as $\Omega \sim \xi^{-z} \sim |\Delta|^{z\nu}$. The hyperscaling asserts the homogeneity in the singular part of the free energy density f_s in $d+1$ dimensional path integral in terms of arbitrary length-scale transformation parameter b:

$$f_s(\Delta) \sim b^{-(d+z)} f_s(b^{1/\nu}\Delta) \sim \Delta^{\nu(d+z)}. \tag{6}$$

When the inverse temperature β and the linear dimension of the system size L are both large but finite, finite-size scaling function $\mathcal{F}$ is expected to hold as

$$f_s(\Delta) \sim \Delta^{\nu(d+z)} \mathcal{F}(\xi/L, \xi^z/\beta) \tag{7}$$

in the combination of non-dimensional arguments.

Combining expressions in the path integral formalism for the Drude weight and the compressibility, respectively, with the finite-size scaling form derived from the hyperscaling, one obtains useful scaling forms for physical quantities [3, 4, 16]. The scaling of D is obtained:

$$D \propto \Delta^\zeta \tag{8}$$

with $\zeta = \nu(d + z - 2)$, while

$$\delta \propto \Delta^{-\alpha+\nu z} \tag{9}$$

and

$$\kappa \propto \Delta^{-\alpha} \tag{10}$$

are obtained with $\alpha = \nu(z - d)$, where δ denotes the carrier concentration, namely, the density measured from the Mott insulating phase.

In the insulating phase, the charge excitation gap E_g and the localization length ξ_l defined in (4) should be determined from $E_g \propto |\Delta|^{z\nu}$ and

$$\xi_l \propto |\Delta|^{-\nu}, \tag{11}$$

where we have used the fact that ν and z are identical in both sides of the transition point. In the metallic side, the scaling of the specific heat $C =$

$\beta^2 \partial^2 (\ln Z)/\partial \beta^2$ is given from (7):

$$C = \Delta^{\nu(d+z)} T \frac{\partial^2}{\partial T^2} \mathcal{F}(\xi/L = 0, \xi^z T). \tag{12}$$

At the critical point, the specific heat is given from the Δ independent term as $C \propto T^{d/z}$. If $C = \gamma T$ is satisfied at low temperatures in the metallic phase, the coefficient γ follows

$$\gamma \propto \Delta^{\nu(d-z)}. \tag{13}$$

Another interesting quantity is the coherence temperature T_F below which the electron motion becomes quantum mechanical and degenerate. It is clear that T_F has to approach zero as $\Delta \to 0$ in a continuous transition. In case of the Fermi liquid, T_F is nothing but the Fermi temperature. The existence of single characteristic energy scale with singularity at $\Delta = 0$ leads to the scaling of T_F in the form

$$T_F \propto \Delta^{\nu z}. \tag{14}$$

When the control parameter Δ is the chemical potential μ measured from the critical point μ_c, it represents the filling control MIT. In this case, it is possible to derive a useful scaling relation. Because the doping concentration is $\delta = -\frac{\partial f_s}{\partial \mu} = -\frac{\partial f_s}{\partial \Delta}$, it scales as

$$\delta \sim \Delta^{\nu(d+z)-1} \tag{15}$$

From the comparison of Eq.(15) and Eq.(9), $\nu z = 1$ is derived. The characteristic length scale is then

$$\xi \sim \delta^{-1/d}, \tag{16}$$

which is the length scale of the "mean hole distance". The above scaling description is known to be valid for the MIT to the band insulator, all the MIT in 1D [17] and the Anderson transitions [18].

2.3 Universality Class

When we assume the above scaling theory with Eq.(10), the quantum Monte Carlo result (2) implies that there exists a new universality class $z = 1/\nu = 4$ [2, 3]. As we see above in (5) and (11), the localization length also suggests that $\nu = 1/4$. These are independent estimates of the exponent and a check for the consistency of the scaling theory. On the contrary, these two results can be explained neither by the Hartree-Fock approximation nor by the $d = \infty$ results. Indeed, neither the Hartree-Fock nor the $d = \infty$ results satisfy the hyperscaling assumption (6). For example, the Hartree-Fock approximation predicts $\nu = 1/2$.

The large dynamical exponent z leads to unusual suppression of coherence in the metallic phase in various aspects. The scaling theory predicts that the Drude weight scales as $D \propto \delta^{1+\frac{z-2}{d}}$. As is well known, $D \propto \delta$ is satisfied in the usual MIT, which is consistent with the scaling theory when $z = 2$. When the universality class is characterized by $z = 1/\nu > 2$, D is suppressed stronger than δ-linear dependence at small δ. When $z = 4$ is satisfied in 2D, $D \propto \delta^2$ is predicted. This is one of the indications for the unusual suppression of coherence in this universality class. Numerical results indeed showed consistency with $D \propto \delta^2$ as we discussed in §2.1 [15].

From the sum rule, the ω-integrated conductivity gives the averaged kinetic energy

$$-\langle F \rangle = \int_0^\infty \sigma(\omega)d\omega. \tag{17}$$

In the strong coupling limit as in the t-J model, the total kinetic energy $\langle F \rangle$ is expected to be proportional to δ as observed in Fig.1a. Then the Drude part $D \propto \delta^{1+\frac{z-2}{d}}$ becomes negligibly small at small δ in the total weight $-\langle F \rangle \propto \delta$ if $z > 2$. This means most of the total weight $-\langle F \rangle$ is exhausted in the incoherent part $\sigma_{\text{reg}}(\omega)$ in (1). In the scaling theory, the form of $\sigma_{\text{reg}}(\omega)$ is not specified and indeed it may depend on details of systems including the interband transition. However, for the intraband contribution, $\sigma(\omega)$ is expected to follow $\sigma(\omega) \sim (1 - e^{-\beta\omega})/\omega$ at very high temperatures T larger than the bare bandwidth t_B. In fact the conductivity is given by

$$\sigma(\omega) = \frac{1 - e^{-\beta\omega}}{\omega}C(\omega) \tag{18}$$

with the current correlation function $C(\omega)$ defined by

$$C(\omega) \equiv \int_{-\infty}^\infty dt \, e^{i\omega t}\langle j(0)j(t)\rangle \simeq \frac{\sigma_0/\tau}{-i\omega + \frac{1}{\tau}}, \tag{19}$$

where if the carrier dynamics is incoherent, $C(\omega)$ is expected to have a broad featureless structure due to rapid decay of current correlation in time $\tau \sim 1/t_B$. When the temperature is lowered, this incoherent part $\sim (1 - e^{-\beta\omega})/\omega$ is in general transferred to the Drude part $\sim \frac{1}{-i\omega + \gamma}$ below the coherence temperature. In the usual band-insulator-metal transition, this transfer can take place for the dominant part of the weight because $-\langle F \rangle$ and D both may be proportional to δ. Therefore, the incoherent part is totally reconstructed to the Drude part with decreasing temperature below T_F. However, for $z > 2$, this transfer takes place only in a tiny part of the total weight because the relative weight of the Drude part to the whole weight is $\delta^{(z-2)/d}$ which vanishes as $\delta \to 0$ for $z > 2$. Even at $T = 0$, the major part of the conductivity weight must be exhausted in the incoherent part. Near the transition point, since we have no reason to have a dramatic change of the form of the incoherent part $\sigma_{\text{reg}}(\omega)$ at any temperatures from the scaling point of view, the optical conductivity more or less follows the form

$$\sigma_{\text{reg}}(\omega) \sim C\frac{1 - e^{-\beta\omega}}{\omega} \tag{20}$$

even for $t_B > \omega$ and $t_B > T$ where we have neglected model dependent feature such as the interband transition. This "intraband" incoherent weight may be proportional to $C \propto \delta$ in the strong coupling limit. Such dominance of the incoherent part with the scaling (20) is consistent with the numerical results of 2D systems[19], which lends further support for $z > 2$ in 2D. On the contrary, the numerical results in 1D by Stephan and Horsch [20] support that the incoherent part is small at low temperatures. This may be due to $z = 2$ in 1D because the majority of the weight seems to be exhausted in the Drude weight.

Another indication for the suppression of coherence may be seen in the coherence temperature T_F. The scaling of T_F is given by $T_F \propto \delta^{z/d}$. Because the standard MIT is characterized by $T_{F0} \propto \delta^{2/d}$, the relative suppression T_F/T_{F0} is proportional to $\delta^{\frac{z-2}{d}}$. When $z = 4$ as suggested by numerical results in 2D, we obtain $T_F \propto \delta^2$ in 2D in contrast with $T_{F0} \propto \delta$ for $z = 2$.

In the Mott insulating phase, entropy coming from the spin degrees of freedom is released essentially to zero when the antiferromagnetic order exists. The entropy is also released when the spin excitation has a gap as in the spin gap phase. Even without such clear phase change, the growth of short-ranged correlation progressively releases the entropy with decreasing temperature. When carriers are doped, an additional entropy due to the charge degrees of freedom is introduced in proportion to δ. This additional entropy is assigned not solely to the charge degrees of freedom but also to the spin entropy through their mutual coupling which destroys the antiferromagnetic long-range order. This additional entropy $\propto \delta$ has to be released below the coherence temperature T_F. Therefore, a natural consequence is that if T-linear specific heat characterizes the degenerate (coherent) temperature region, the coefficient γ should be given by $\gamma \sim \delta/T_F \sim \delta^{1-\frac{z}{d}}$. This is indeed the scaling law we obtained in (13) and (15). From this heuristic argument, it turns out that the metallic phase near the Mott insulator is characterized by large residual entropy at low temperatures which is also related with the suppression of the antiferromagnetic correlation as compared to the insulating phase. It may also be said that the anomalous suppression of the charge coherence at small δ is caused by short-ranged antiferromagnetic correlations which scatters carriers incoherently.

The strong coupling of spin and charge seems to yield the same scaling even for the spin correlation. The numerical result suggests that the antiferromagnetic transition takes place more or less simultaneously with the Mott transition [9]. The equal time spin structure factor $S(Q)$ at its maximum shows the scaling $S(Q) \propto 1/\delta$ from

which the magnetic correlation length ξ_{AFM} is scaled by $\delta^{-1/2}$. The incommensurate wavevector Q approaches (π, π) as $\delta \to 0$. This is the same scaling as (16) and implies that the criticality of the antiferromagnetic transition is also involved in the same universality.

2.4 Flat Dispersion and Strong Momentum Dependence

To clarify the origin of the large dynamical exponent, $z = 4$, it is necessary to study explicit momentum dependence of the charge excitations. The realization of the hyperscaling must be the consequence of the fact that some small number of singular points in the momentum space appear and the MIT is controlled by the charge excitations around those points. If the charge excitations near the

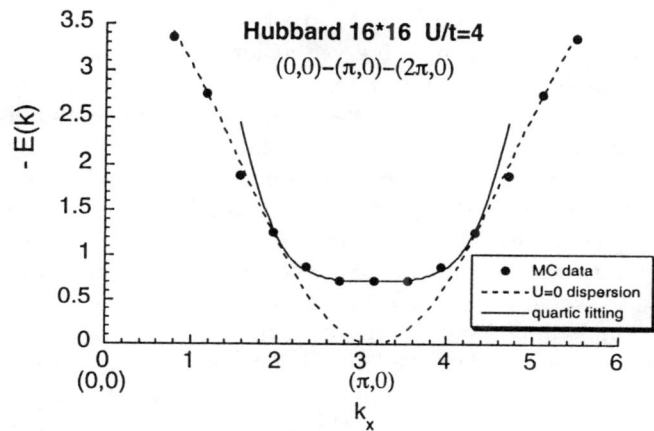

Figure 2: The dispersion of the Hubbard model along the k_x axis, from $(0,0)$ through $(\pi,0)$ till $(2\pi,0)$ obtained from the peak position of the single-particle spectral weight $A(k,\omega)$ for the 16 by 16 lattice at zero temperature. Black circles are the Monte carlo data at $U/t = 4$.

whole $d-1$ dimensional Fermi surface would contribute equally and isotropically, it would not satisfy the hyperscaling at all because it would introduce a length scale of the Fermi wavenumber which does not vanish at the MIT point. This additional length scale would invalidte the assumption (6). Furthermore, the dynamical exponent $z=4$ directly has to lead to the appearance of a k^4 dispersion around those singular points because by definition z is nothing but connecting characteristic wavenumber and energy scales. If a flat dispersion such as k^4 appears at some points on the Fermi surface, the holes will be doped predominantly to this region because of this flatness and this flat region will determine the character of the MIT.

To understand whether the above reasoning works, the single-particle spectral function $A(k,\omega)$ was calculated and analyzed in detail. $A(k,\omega)$ of a doped single hole to the Mott insulator phase of the Hubbard model has peak structures which disperses in the momentum space. Figure 2 shows this dispersion around $(\pi,0)$ and clearly indicates that the flat dispersion does appear around $(\pi,0)$ and $(0,\pi)$, while other part of the Fermi level does not show such flatness. The numerical result further indicates that the dispersion around $(\pi,0)$ and $(0,\pi)$ is well fit by k^4 dispersion. These points have originally the van-Hove singularity in the simple Hubbard model. However, the observed k^4 dispersion is far beyond the expectation from the van-Hove singularity and the consequence of strong correlation effects. We further note that this $z=4$ universality class also holds for the case with finite next-nearest-neighbor hopping t' where the van-Hove singularity does not meet the Fermi level at the MIT point [10]. This gives us the microscopic basis to understand the anomalous feature of the MIT.

The dispersion of a doped hole in the Mott insulating state does not necessarily determine the criticality of the transition because the rigid coherent band picture is not always guaranteed in the process of further doping. However, the existence of the flat dispersion is commonly observed at low doping concentration in angle resolved photoemission (ARPES) data of high-Tc cuprates [21, 22]. The presence of this flat dispersion is also consistent with numerical results of the Hubbard models at finite doping [23]. More remarkable is the following: The ARPES data for the insulating phase of the cuprates appears to show a deeper level around $(\pi, 0)$ than the level around $(\pi/2, \pi/2)$ which is relatively closer to the Fermi level as is consistent with the case with the next nearest neighbor transfer t' [24]. If the rigid band picture would be correct, this implies that the state around $(\pi/2, \pi/2)$ would be first doped with holes rather than $(\pi, 0)$ region. However, in the underdoped region, the data indicate that the state around $(\pi, 0)$ seems to quickly emerge from the incoherent tail of $A(k, \omega)$ at the Fermi level and forms a flat dispersion, which indicates the breakdown of the rigid band picture and a universal dominance of the flat dispersion excitations upon doping. This is also consistent with the numerical compressibility data with t' discussed above. The flat dispersion could be obscured by its incoherent nature and hard to observe incoherent tails in $A(k, \omega)$ by the strong damping.

The dominance of the flat dispersion around $(\pi, 0)$ on low energy excitations should be taken with care because it has some degeneracy with the excitations around $(\pi/2, \pi/2)$. It may have an effect on the DC transport properties, although the DC properties are not good probes to understand the Mott transition as we discussed in the introduction. The ARPES data of the underdoped cuprates show a gradual formation of fragmentary "Fermi surface" first around $(\pi/2, \pi/2)$ in the normal state. This may have two reasons. One is that the carrier relaxation time around $(\pi/2, \pi/2)$ is much longer and the other is that the region around $(\pi, 0)$ is under the influence of the formation of pseudogap. The low-energy excitation around $(\pi, 0)$ is dominated by the paired singlet formation due to the instability of the flat dispersion as we will discuss in the next section. The paired singlet formed in the pseudogap region seems to be still incoherent above T_c [25, 26]. Although the MIT is governed by the excitations around $(\pi, 0)$, the DC transport may have substantial contribution from the part around $(\pi/2, \pi/2)$. This is speculated from the fragmentary Fermi surface as well as from a rather large and sensitive increase of the residual resistivity upon Zn doping around 1% in the cuprates while relatively insensitive effect of Zn on the pseudogap behavior [27, 28]. This implies that the Zn doping sensitively induces the localization of carriers around $(\pi/2, \pi/2)$, while gapped singlet excitations formed from $(\pi, 0)$ excitations is not seriously influenced by such tiny amount of Zn. The DC charge transport is strongly affected from $(\pi/2, \pi/2)$ excitations while other quantities including spin excitations are dominated by the driving excitations of MIT near $(\pi, 0)$, which makes the physical properties as if spin and charge degrees of freedom were "separated". However, the seeming separation has to be understood after considering the strong momentum dependence carefully.

3 Superconductor-Insulator Transition

In the Mott insulator, because the single-particle process is suppressed due to the charge gap, we have to consider the two-particle process expressed by the superexchange interaction. This has led to the Heisenberg model for the effective Hamiltonian of the Mott insulating phase. The Hubbard model would be sufficient even for the Mott insulating phase if its low energy excitations could be precisely considered. However, the Heisenberg model offers a better and easier way to extract physics of spin excitations and properties at low temperatures in the strong coupling regime. Similarly to this circumstance in the insulating case, even in the metallic region, such suppressions of coherence discussed above for the single-particle process make it useful and helpful to consider the two-particle process explicitly. This is particularly true in the flat-dispersion part of the momentum space. In contrast with the insulating phase, the two-particle processes in metals contain an effective pair hopping term and a term described by

$$\mathcal{H}_W = -t_W \sum_i [\sum_{\delta\sigma}(c_{i\sigma}^\dagger c_{i+\delta\sigma} + c_{i+\delta\sigma}^\dagger c_{i\sigma})]^2 \qquad (21)$$

becomes relevant [29, 30, 31]. Then the Hamiltonian containing the term $\mathcal{H}_W$ added to the Hubbard model, called the t-U-W model is expected to be a useful effective Hamiltonian to understand low temperature properties near the Mott insulating phase. It was argued that $\mathcal{H}_W$ plays the same role as the term proportional to t^2/U in the strong coupling expansion of the Hubbard model which contains the superexchange term as well as the so-called three-site term [30]. It is not clear at the moment whether the bare Hubbard model implicitly contains sufficiently large effective coupling of $\mathcal{H}_W$ to make it relevant at low energy scale or some additional elements are needed. However, by including a small amount of the $\mathcal{H}_W$ term explicitly, it allows to study how the Hubbard hamiltonian becomes unstable to the two-particle process and what type of symmetry breaking or ordering are expected as its consequence. In fact the t-U-W model is the first hamiltonian which has made possible to study the d-wave superconductor-Mott insulator transition under a controlled numerical treatment.

We note that, to treat the relevance of the $\mathcal{H}_W$ term under a proper condition, it would be necessary to first extract precisely the strong momentum dependence of the single-particle renormalization in the Hubbard model. Without clarifying the formation of the flat dispersion caused by many-body effects described in the previous section, the real role and relevance of the two-particle term cannot be understood well. Mean field analyses of the Hubbard as well as the t-J models in the literature have not properly treated this important point by neglecting the singular momentum dependence of the renormalization.

When the two-particle process becomes relevant, the coherence is not suppressed because the universality class of the two-particle transfer is given by $z = 1/\nu = 2$. The universality class $z = 2$ is numerically observed in the exponent of the localization length in the insulating side of the t-U-W model [31] and the doping concentration dependence of the Drude weight of the t-J-W model [15]. Because of small z, it always becomes more relevant than the single-particle transfer for small δ. The alteration of the universality class from $z = 4$ for the MIT to $z = 2$ for the superconductor-insulator transition shows an instability of metals near the Mott insulator to the superconducting pairing or in more general an instability of $z = 4$ universality phase to some type of symmetry-broken state.

Below, we summarize how the Mott insulator to superconductor transition takes place and how the d-wave superconducting state appears in the 2D t-U-W model. This has been studied through large-scale quantum Monte Carlo calculation [29, 30, 31]. The t-U-W model appears to show a Mott insulator to superconductor transition even when the filling is fixed at half filling. With the increase of W, the antiferromagnetic long-range order seen in the pure Hubbard model continuously decreases and at the critical amplitude of $W = W_c$, the order is destroyed. At the same W, the insulating phase appears to undergo a transition to a superconductor with $d_{x^2-y^2}$ symmetry.

A remarkable property in the superconducting phase at half filling is that the antiferromagnetic correlation is extremely compatible. The equal-time antiferromagnetic correlation decays very slowly with a power law at long distance as $\sim 1/r^\alpha$ with $\alpha \sim 1$ or slightly larger. Reflecting this compatibility, the real part of the staggered susceptibility $\chi(q = (\pi,\pi), \omega = 0)$ appears to diverge with lowering temperature to the limit $T \to 0$. A clearer understanding is obtained in the dynamical structure factor $S(q,\omega)$ (or equivalently in the imaginary part of the dynamical susceptibility $\mathrm{Im}\chi(q,\omega)$). From this quantity, it turns out that the antiferromagnetic correlation is dynamical and has a strong peak at a finite frequency with presumable divergent weight in the thermodynamic limit. This peak frequency is comparable to the pairing gap amplitude. Because of the appearance of the superconducting phase, the antiferromagnetic correlation is suppressed below the pairing energy scale. However very slow power-law decay of equal-time antiferromagnetic correlation implies that $S(q = (\pi,\pi),\omega)$ has divergent weight somewhere at finite frequency. To reconcile the antiferromagnetic and superconducting correlations, the frequency of the antiferromagnetic correlation is pushed out from low-frequency to the frequency above the pairing gap. When the temperature is lowered at $W > W_c$ from high temperatures, $S(q = (\pi,\pi),\omega)$ first shows growth of a broad peak around $\omega = 0$. This appears to be an immatured state which does not differentiate the antiferromagnetic and singlet correlations. This is reminiscent of the SO(5) scenario with approximate symmetry of superconducting and antiferromagnetic state [32], although the approximate symmetry in this model has not been examined in detail yet. With further lowering of the temperature, the peak position shifts to a finite frequency and is sharpened. In this temperature range pseudogap is formed where $\mathrm{Im}\chi(q = (\pi,\pi),\omega)$ decreases at low ω. At lower temperatures, the peak at finite frequency becomes even sharper.

4 Mott Transition with Orbital Degeneracy

Orbital degeneracy plays important roles in addition to the spin degeneracy in a wide class of materials at the MIT [1]. It provides another origin of residual entropy near the transition point and may offer richer structure of the transition. Effects of orbital degeneracy was recently examined by projector quantum Monte Carlo method at zero temperature for a model of Mn perovskite compounds with degeneracy of two e_g orbitals, $d_{x^2-y^2}$, and $d_{3z^2-r^2}$, under the condition of complete spin polarization and two-dimensional configuration [33]. In Mn perovskite compounds, the Mott insulator is in the d^4 configuration where one of two e_g orbitals are occupied. Through the MIT, the spin configuration is always completely ferromagnetic within a plane mainly due to a strong Hund's rule coupling to high-spin t_{2g} electrons. However, the charge transport is strongly incoherent with very small Drude weight in the metallic region [34]. This suggests a crucial role of orbital degeneracy (and presumably also the lattice degrees of freedom coupled to it) for the incoherent charge dynamics. The spin polarized but orbitally degenerate model may be a minimal model to capture this circumstance. The Hamiltonian of this model is defined as

$$\mathcal{H} = -\sum_{\langle ij \rangle, \nu, \nu'} t_{i,j,\nu,\nu'} (c_{i\nu}^\dagger c_{j\nu'} + h.c.)$$
$$+ U \sum_{i, \nu \neq \nu'} (n_{i\nu} - \frac{1}{2})(n_{i\nu'} - \frac{1}{2}) - \mu \sum_{i\nu} n_{i\nu}, \qquad (22)$$

where the creation (annihilation) of the single-band electron at site i with orbital ν is denoted by $c_{i\nu}^\dagger$ ($c_{i\nu}$) with $n_{i\nu}$ being the number operator $n_{i\nu} \equiv c_{i\nu}^\dagger c_{i\nu}$. An important difference from the ordinary Hubbard model is that the transfer has an anisotropy with dependence on the $x^2 - y^2$ and $3z^2 - r^2$ orbitals, where we label the orbital $x^2 - y^2$ and $3z^2 - r^2$ as 1 and 2, respectively below. In 2D configuration, the nearest-neighbor transfer is scaled by a single parameter t as $t_{11} = \frac{3}{4}t$, $t_{22} = \frac{1}{4}t$, $t_{12} = t_{21} = \pm\frac{\sqrt{3}}{4}t$, where, in $\pm$, $+(-)$ is for the transfer to the $y(x)$ direction. In the absence of U, the diagonalized two bands have dispersions both with the bandwidth $4t$, where one dispersion is obtained by a parallel shift of $2t$ from the other.

The present numerical result is summarized as follows. When U is increased, the Mott gap Δ_c opens at "half filling", namely at $\langle n \rangle = 1$. However, the opening of the gap is substantially slower than the case of the ordinary single-band Hubbard model (with spin). For example, $\Delta_c \simeq 0.1$ at $U/t = 4$ is compared with $\Delta_c \simeq 0.66$ for the ordinary Hubbard model at $U/t = 4$. At $U/t = 3$, Δ_c is not distinguished from zero within the error bar. Although the opening of the Mott gap is slow within this model, the insulating state at $\langle n \rangle = 1$ is stabilized if we introduce the Jahn Teller distortion. A realistic Jahn-Teller coupling for the Mn compounds has stabilized the Jahn-Teller distorted insulating state with staggered order of $3x^2 - r^2$ and $3y^2 - r^2$ orbitals. Another important observation is that realistic amplitudes of the Jahn-Teller coupling in the absence of U is far insufficient in stabilizing the Jahn-Teller distortion with a realistic stabilization energy of several hundred K. The Mott insulating state in the experimental situation may well be resulted from synergy of U and the Jahn-Teller coupling. In the model (22) with 2D configuration, the long-ranged orbital order increases at $\langle n \rangle = 1$ with increasing U. For example the staggered orbital polarization at $U =$

$4t$ is around 0.3. However, the orbital order appears to be destroyed immediately upon doping. In the metallic region $n \neq 1$, the short-ranged correlation of the staggered orbital order is critically enhanced as $\langle n \rangle \to 1$ similarly to the ordinary 2D single-band Hubbard model. For large U, there seems to exist a critical region near $\delta \equiv 1 - n = 0$ where $T(Q) \propto 1/\delta$ as the same as the spin structure factor of the ordinary Hubbard model. Here $T(Q)$ is the equal-time structure factor at (π,π) for the orbital correlation. This critical enhancement of the orbital correlation strongly suggests that the orbital correlation in this Mn model may play a similar role to spins in the Hubbard model in realizing highly incoherent charge dynamics.

References

[1] For a review see M. Imada, A. Fujimori and Y. Tokura: to appear in Rev. Mod. Phys. **70** (1998) No.4.

[2] M. Imada: J. Phys. Soc. Jpn. **63** (1994) 4294.

[3] M. Imada: J. Phys. Soc. Jpn. **64**(1995) 2954.

[4] M. Imada: J. Low Temp.Phys. **99**(1995) 437.

[5] For a review, see A. Georges, W. Krauth and M.J. Rozenberg: Rev. Mod. Phys. **68** (1996) 13.

[6] For a review, see D. Belitz and T.R. Kirkpatrick: Rev. Mod. Phys. **66** (1994) 261.

[7] T. Moriya: *Spin Fluctuations in Itinerant Electron Magnetism*, Springer Series in Solid-State Sciences Vol.56 (Springer, Berlin, 1985).

[8] A.J. Millis: Phys. Rev. B **48** (1993) 7183.

[9] N. Furukawa and M. Imada: J. Phys. Soc. Jpn. **61** (1992) 3331.

[10] N. Furukawa and M. Imada: J. Phys. Soc. Jpn. **62** (1993) 2557.

[11] N. Furukawa, F.F. Assaad and M. Imada: J. Phys. Soc. Jpn. **65** (1996) 2339.

[12] J. Jaklič and P. Prelovšek: to appear in Adv. Phys.; Cond-mat/9603081.

[13] M. Kohno: Phys. Rev., **55** (1997) 1435.

[14] F.F. Assaad and M. Imada: Phys. Rev. Lett., **76** (1996) 3176.

[15] H. Tsunetsugu and M. Imada: J. Phys. Soc. Jpn. **67** (1998) 1864.

[16] M. A. Continentino: Phys. Rep. **239** (1994) 179.

[17] M. Imada: J. Phys. Soc. Jpn. **63** (1994) 3059.

[18] F. Wegner: Z. Phys. B **25** (1976) 327.

[19] J. Jaklič and P. Prelovšek: Phys. Rev. B **52** (1995) 6903.

[20] W. Stephan and P. Horsch: Phys. Rev. B **42** (1990) 8736.

[21] Z. -X. Shen and D. S. Dessau: Phys. Rep. **253** (1995) 1.

[22] K. Gofron et al.: Phys. Rev. Lett. **73** (1994) 3302.

[23] R. Preuss, W. Hanke, C. Grober, and H.G. Evertz: Phys. Rev. Lett. **79** (1997) 1122.

[24] B.O. Wells et al: Phys. Rev. Lett. **74** (1995) 964.

[25] D.S. Marshall et al.: Phys. Rev. Lett. **76** (1996) 4841.

[26] H. Ding et al.: Nature **382** (1996) 51.

[27] K. Mizuhashi et al.: Phys. Rev. B **52** (1995) R3884.

[28] Y. Fukuzumi et al.: Phys. Rev. Lett. **76** (1996) 684.

[29] F. F. Assaad, M. Imada and D. J. Scalapino: Phys. Rev. Lett. **76** (1996) 4592 .

[30] F. F. Assaad, M. Imada and D. J. Scalapino: Phys. Rev. B **56** (1997)15001.

[31] F. F. Assaad and M. Imada: Phys. Rev. B **57** (1998) to appear.

[32] S.C. Zhang: Science **275** (1997) 4126.

[33] Y. Motome and M. Imada: unpublished.

[34] Y. Okimoto, T. Katsufuji, T. Ishikawa, T. Arima, and Y. Tokura: Phys. Rev. B **55** (1997)4206.

High Temperature Superconductors as a Member of Transition Metal Oxides

Sadamichi Maekawa

Institute for Materials Research, Tohoku University, Sendai, 980-8577 Japan

Abstract. Many transition metal oxides have perovskite-type crystal structure. In manganites with this structure, the orbital degeneracy of $3d$ electrons in Mn ions brings about the characteristic properties including ferromagnetism. Relation between crystal structure and electronic structure in the cuprates is examined in the ionic model. Then, the relation between electronic structure and high temperature suoerconductivity is examined in the cluster model approach.

1 Introduction

Since the discovery of high temperature superconducting cuprates, extensive study of transition metal oxides with perovskite-type crystal structure has been done from the new theoretical and experimental viewpoints. A variety of dramatic phenomena in the oxides are caused by the systematic variations of $3d$ electrons in transition metal ions. In particular, the orbital degeneracy of $3d$ electrons in perovskite manganites and related compounds has recently attracted much attention.

In this paper, we would like to examine how unique the electronic structure in high temperature superconducting cuprates is as compared with the other oxides. Many high temperature superconducting cuprates have been discovered so far. Among the variety of their layered structure, the hole-carrier superconductors contain the CuO_2 plane with apex oxygen atoms as a common structural unit.

The parent compounds of high temperature superconductors are ionic crystals, for example, La_2CuO_4. In this compound, the CuO_2 plane is alternately stacked with an $(LaO)_2$ layer (called the block layer) containing La^{3+} and O^{2-} ions. Evidently, the CuO_2 plane has a negative charge and $(LaO)_2$ layer has a positive charge. Therefore, the positive and negative charges are stacked alternately in a direction perpendicular to the layers (called the c direction). If a Sr^{2+} ion is substituted for part of the La^{3+} ion, a surplus of one negative charge per Sr^{2+} ion occurs. Accordingly, the electrical charge of the CuO_2 plane decreases by one in order to keep the entire crystal electrically neutral. We note that the carrier concentration required for the emergence of high temperature superconductivity is 0.15-0.2 per copper ion. Within this range of the concentration, the CuO_2 plane exhibits the metallic properties, but the ionic character is still kept in the c direction. This duality of the ionic crystal structure and the metallic property of the electronic structure is important for carrier doping and effectively gives the electronic structure its two-dimensionality.

In Sec. II, we examine the electronic structure of the cuprates based on the ionic model and discuss the relation between crystal structure and electronic structure. Then, introducing the covalent character into the model, the relation between electronic structure and the superconductivity is discussed in Sec. III. The results are summarized in Sec. IV.

1.1 Crystal Structure and High Temperature Superconductivity

In this section, we examine the relation between the crystal structure and high temperature superconducting transition temperature (T_c).[1,2] As shown in Fig. 1, there exist three kinds of CuO_2 planes in high temperature superconductors. It is empirically known that the presence of apex oxygens makes the CuO_2 plane easier to dope with holes; the planes with the square coordination (without apex oxygen) have not yet been doped with holes but those with the octahedral coordination have been made overdoped.

The bond length is the most direct measure of the magnitude of covalency since the hopping integrals are directly related to the bond lengths. In Figs. 2(a) and 2(b) we plot the T_c's as a function of d_p [one-half of the nearest $Cu(P)$-$Cu(P)$ distance, which is roughly equal to $Cu(P)$-$O(P)$ bond length] and d_A [the $Cu(P)$-$O(A)$ bond length]. Here, $Cu(P)$ and $O(P)$ denote the in-plane copper and oxygen, respectively, and $O(A)$ does the apex oxygen. No correlation may be seen in both figures. For example, both d_A and d_P are large in the Tl family which shows very high T_c, whereas the Bi family, which has rather small bond lengths, also shows very high T_c. This result suggests that the magnitude of covalency itself is not a major factor gov-

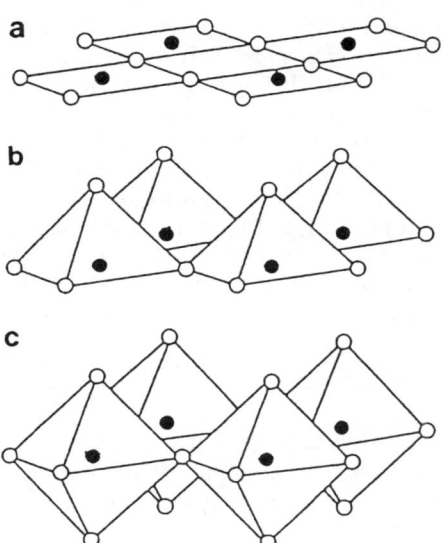

Fig. 1. The CuO_2 plane structures. : oxygen ion, : copper ion.

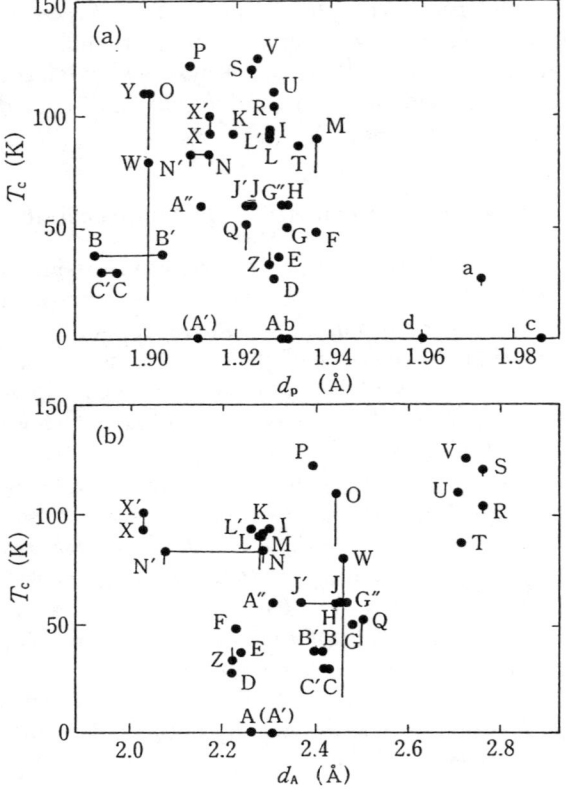

Fig. 2. (a) The T_c vs. d_P and (b) the T_c vs. d_A correlations. d_P is one-half of the nearest Cu(P)-Cu(P) bond length and d_A is the Cu(P)-O(A) bond length. The average distance is shown for orthorhombic compounds. A: $La_2SrCu_2O_{6.2}$, A': $La_{1.9}Ca_{1.1}Cu_2O_6$, A": $La_{1.6}Sr_{0.4}CaCu_2O_6$, B and B': $La_{1.85}Sr_{0.15}CuO_4$, C and C': $La_{1.85}Ba_{0.15}CuO_4$, D: $Nd_{0.66}Sr_{0.205}Ce_{0.135}CuO_4$, E: $Sm_1La_{0.75}Sr_{0.25}CuO_{3.95}$, F: $(Ba_{0.67}Eu_{0.33})_2(Eu_{0.67}Ce_{0.33})_2Cu_3O_{8+z}$, G, G' and G": $Y_{0.8}Ca_{0.2}Ba_2Cu_3O_{6.11}$, H: $YBa_2Cu_3O_{6.5}$, I: $YBa_2Cu_3O_7$, J and J': $ErBa_2Cu_3O_{6.53}$, K: $ErBa_2Cu_3O_7$, L and L': $Y_{0.9}Ca_{0.1}Ba_2Cu_4O_8$, M: $(Ca_{0.5}La_{0.5})(Ba_{1.25}La_{0.75})Cu_3O_{6+\delta}$, N and N': $Pb_2Sr_2Y_{0.5}Ca_{0.5}Cu_3O_8$, O: $Pb_{0.5}Tl_{0.5}Sr_2CaCu_2O_7$, P: $Pb_{0.5}Tl_{0.5}Sr_2Ca_2Cu_3O_9$, Q: $Tl(Ba_{0.6}La_{0.4})_2CuO_{5-\delta}$, R: $TlBa_2CaCu_2O_{7-\delta}$, S: $TlBa_2Ca_2Cu_3O_{9-\delta}$, T: $Tl_2Ba_2CuO_{6+\delta}$, U: $Tl_2Ba_2CaCu_2O_{8+\delta}$, V: $Tl_2Ba_2Ca_2Cu_3O_{10+\delta}$, W: $Bi_2Sr_2CuO_{6+\delta}$ X: $Bi_2Sr_2Ca_{0.9}Y_{0.1}Cu_2O_{8.24}$, X': $Bi_{2+x}Sr_{2+y}Ca_{1+z}Cu_2O_{8-\delta}$(x+y+z=0.1), Y: $Bi_2Sr_2Ca_2Cu_3O_{10+\delta}$, Z: $Bi_2Sr_2(Gd_{0.82}Ce_{0.18})_2Cu_2O_{10.24}$, a: $Nd_{1.85}Ce_{0.15}CuO_4$, b: $Ca_{0.86}Sr_{0.14}CuO_2$, c: $Sr_2CuO_2Cl_2$, and d: Sr_2CuO_3.

erning the T_c. However, we find that d_A is much larger than d_P. This is in strong contrast with the other transition metal oxides with perovskite-type crystal structure.[3] Such difference between d_A and d_P may provide a unique property in the electronic structure.

Let us next examine the electronic structure in the parent compounds. By neglecting the covalency for a moment, the compounds are assumed to be ionic with O^{2-} and Cu^{2+}. Even though the covalency is strong in the CuO_2 plane, the ionicity between the layers is retained predominantly.

In the ionic model, the central role is played by the electrostatic potential, which is defined for a hole at the i-site as

$$V_i = \sum_{j \neq i} \frac{Z_j e^2}{|\mathbf{r}_j - \mathbf{r}_i|}, \tag{1}$$

where $\mathbf{r}_j$ is the position of the j-th ion with electric charge $Z_j e$. V_i is the electrostatic energy to bring a hole from infinity to the i-site and is called the Madelung energy. Assuming the lattice periodicity and total charge neutrality, we calculate V_i by the standard Ewald method for the lattice summation.

The crystal-field splitting of the d-orbitals of $Cu(P)$ may be obtained from the spatial variation of the potentials around the $Cu(P)$ ion. Here, we take six points along the $Cu(P)$-$O(P)$ and $Cu(P)$-$O(A)$ bond directions at which the atomic wave function of the copper $3d$ orbital is maximum (or at the points 0.32 apart from the $Cu(P)$ site), and evaluate the Madelung potentials there. We define ΔV_d to be an average of the potential differences, those along the $Cu(P)$-$O(A)$ bond direction minus those along the $Cu(P)$-$O(P)$ bond directions. This is the crystal-field splitting of the e_g orbitals of $3d$ holes due to the Madelung potential as shown in Fig. 3. In Fig. 4, the T_c's are plotted as a function of ΔV_d. A fair correlation may be seen between T_c and ΔV_d: the compounds with a higher T_c have a larger value of ΔV_d. This tells us that the more stable a hole is in the plane, the higher the T_c is. In other words, the degeneracy of e_g orbitals is destructive of the superconductivity.

In the real situation, doped holes extend to the neighboring oxygen sites because of the covalency. Therefore, let us calculate the Madelung potentials for a hole on the apex and in-plane oxygen atoms defined as $V_{O(A)}$ and $V_{O(P)}$, respectively, and obtain the difference $\Delta V_A = V_{O(A)} - V_{O(P)}$. In Fig. 5, the maximum T_c's of hole-carrier superconductors, which have achieved by optimum doping, are plotted as a function of ΔV_A. We find that T_c scales very well with ΔV_A; all the compounds (except Z)[4] are located in the area which exhibits a characteristic curve. The curve indicates that (i) compounds with a larger ΔV_A have a higher T_c, (ii) T_c appears at $\Delta V_A \cong -2\text{eV}$ and increases rapidly with increasing ΔV_A, (iii) below this threshold the system is metallic but not superconducting, and (iv) the curve tends to have a smaller gradient $dT_c/d\Delta V_A$ for larger ΔV_A. We consider that this is not

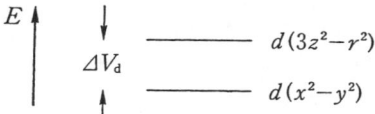

Fig. 3. The crystal-field splitting of e_g orbitals of $3d$ holes due to the Madelung potential.

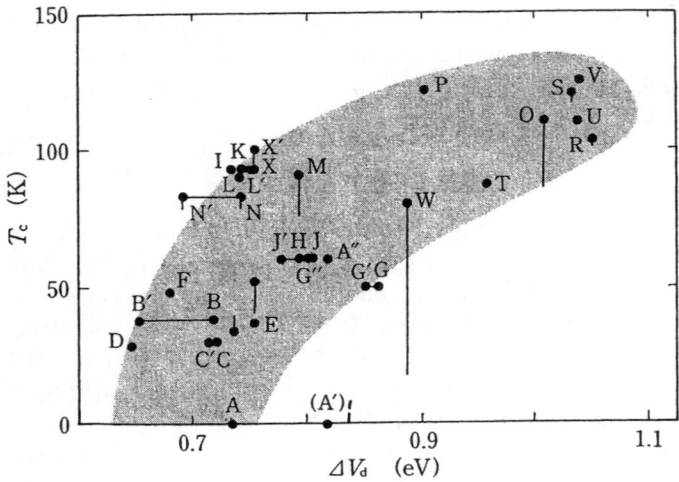

Fig. 4. The T_c vs. ΔV_d correlation. For compounds, see Fig. 2.

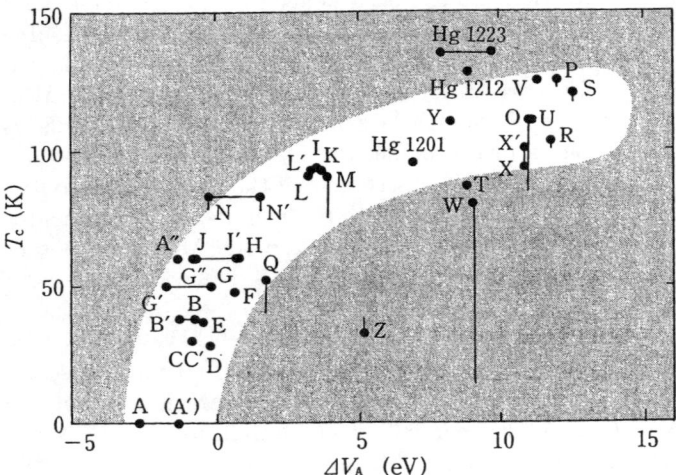

Fig. 5. The T_c vs. ΔV_A correlation. Compounds labeled Hg1223, Hg1212 and Hg1201 are $HgBa_2Ca_2Cu_3O_{8+\delta}$, $HgBa_2CaCu_2O_{6+\delta}$ and $HgBa_2CuO_{4+\delta}$, respectively. For other conpounds, see Fig.2.

just an empirical correlation but the energy level of the apex oxygen atom plays an essential role in the electronic states of the doped holes and thereby determines the maximum T_c's. In Sec. 3, we will examine the origin of this correlation on the basis of the cluster-model calculation and show that the correlation comes from the stability of Zhang-Rice local singlet.[5]

2 Electronic Structure and High Temperature Superconductivity

The correlations extracted through the ionic model strongly suggest that the maximum T_c's of hole-carrier superconductors are essentially governed by the one-body energy levels of the relevant orbitals, especially by that of the apex oxygen atoms. In this section, we introduce the covalency and examine the relation between electronic structure and high temperature superconductivity.

We take the cluster model approach and calculate the electronic structure. Figure 6 shows Cu_2O_9 and Cu_2O_{11} clusters where $2p_\sigma$ orbitals in $O(P)$, $2p_z$ in $O(A)$ and $3d_{x^2-y^2}$ and $3d_{3z^2-r^2}$ orbitals in $Cu(P)$ are explicitly given. When two holes are introduced into the clusters, we obtain insulating systems. When

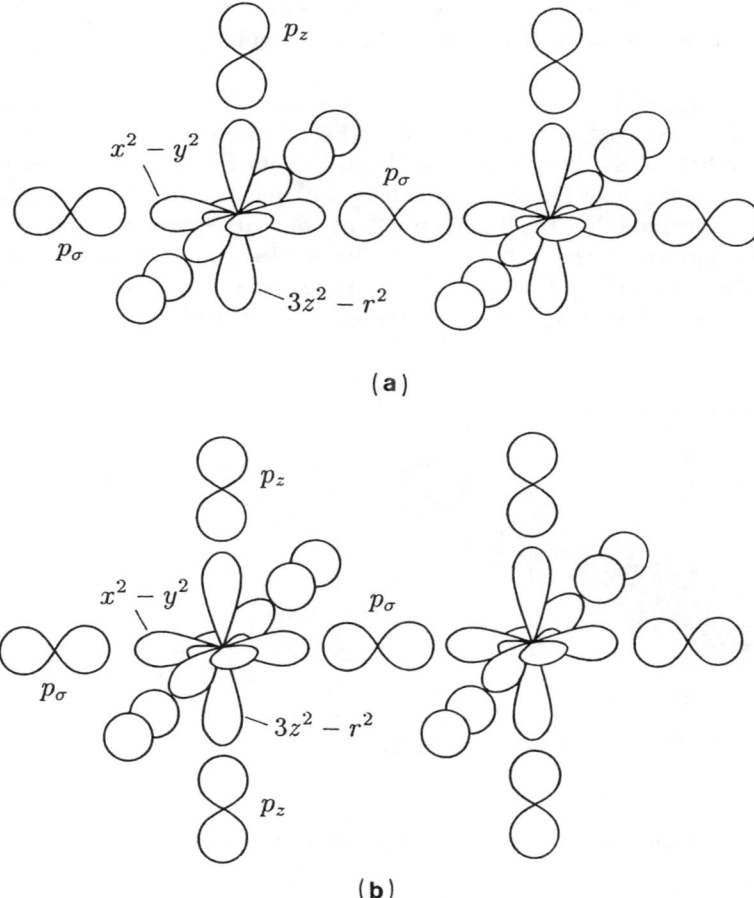

Fig. 6. (a) The Cu_2O_9 and (b) the Cu_2O_{11} clusters with their assigned orbitals.

three holes are introduced, the ground state in these clusters is represented very well by the "bonding" state of local singlets (Zhang-Rice singlet), each of the two being formed between the doped spin and each of the two Cu(P) spins. We also find[6,7] that the first-excited state is mainly represented by the "antibonding" state of the two local singlets. The difference between these lowest two energy eigenvalues may thus be a measure of the transfer energy of the local singlet. We define a parameter t to be one-half of this energy difference. The parameter t of the $t - J$ model has thus been extracted.[6,7] Note that these arguments are appropriate, in particular, when there are no effects of O(A)'s. Let us now consider the role of O(A) explicitly. We note[8] that the stability of the singlet propagation is influenced strongly by the other (mainly triplet) components involved in the first-excited states; a ratio of the singlet component to all the other components involved in the first-excited state decreases associating with the decrease in the energy level of O(A). The decrease in this ratio is nearly proportional to the decrease in the energy difference $2t$. There occurs, therefore, a crossover to a state where the first-excited states are not well written by the antibonding states of the two local singlets.

We have calculated the difference $2t$ in the clusters. The details of the calculation are presented in refs. 1 and 2. In Fig. 7, the maximum T_c's are plotted as a function of t. It is found that the maximum T_c correlates linearly with t; the T_c appears for $t \geq 0.22$ eV and increases linearly with increasing t. We may thus conclude that the maximum T_c of hole-carrier superconductors scales universally with t, the stability of the Zhang-Rice local singlet. Because t directly reflects the energy level of O(A) as shown above, the T_c versus ΔV_A correlation given in Sec. II is a direct consequence of this effect.

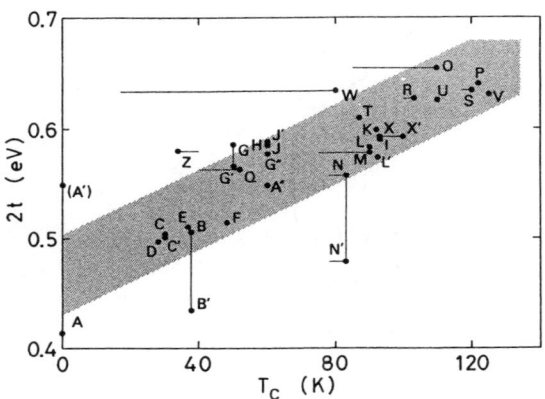

Fig. 7. The correlation between $2t$ and T_c. For compounds, see Fig. 2.

3 Conclusion

We have found that in the ionic model, the electronic energy-level structure has the correlation with the maximum T_c's in high temperature superconductors.

The cluster-model calculations have been employed to examine the electronic structure. The covalency neglected in the ionic model has been fully taken into account in this model. We have shown by taking the picture that the Zhang-Rice local singlet plays an essential role for high-T_c superconductivity, that the maximum T_c's are determined basically by the parameter t, a measure of the stability of the local singlet. The T_c appears for $t \geq 0.22$eV and increases linearly with t. We have obtained that the value of t is primarily determined by the energy level of the apex oxygen atoms and argued that the T_c versus ΔV_A correlation is a direct consequence of this effect.

The correlations between T_c and ΔV_A and thus between T_c and t show that doped holes should be stable in the CuO_2 plane for high temperature superconductivity. This is in strong contrast with the perovskite manganites where the degeneracy of orbitals of $3d$ electrons provides their unique properties.

ACKNOWLEDEGMENTS

This work has been done in collaboration with Y. Ohta and T. Tohyama.

This work was supported by a Grant-in-Aid of the Ministry of Education, Science and Culture in Japan and CREST.

References

1. Y. Ohta, T. Tohyama and S. Maekawa, Phys. Rev. B **43**, 2968 (1991).
2. S. Maekawa, Y. Ohta and T. Tohyama, in *Physics of High Temperature Superconductors*, eds. S. Maekawa and et M. Sato(Springer Series in Solid State Sciences, 106, 1992) p.29.
3. S. Ishihara, S. Okamoto and S. Maekawa, J. Phys. Soc. Jpn. **66**, 2965(1997).
4. Compound Z, $Bi_2Sr_2(Gd_{1-x}Ce_x)Cu_2O_{10+y}$, shows a very low T_c.[9] The number of holes so far doped in experiment[10] is rather small($p \geq 0.1$); we expect a higher T_c at optimum doping, which can be consistent with our correlation. The experiment also suggests a valley in T_c versus p curve, which may be another reason of the low T_c. Further experimental works are desired.
5. F. C. Zhang and T. M. Rice, Phys. Rev. B **37**, 3759 (1988).
6. S. Maekawa, J. Inoue and T. Tohyama, in *Strong Correlations and Superconductivity*, eds. H. Fukuyama, S. Maekawa and A. P. Malozemoff (Springer Series in Physics, 89, 1989), p.66.
7. H. Eskes, G. A. Sawatzky and L. F. Feiner, Physca C **160**, 424(1989).
8. T. Tohyama and S. Maekawa, J. Phys. Soc. Jpn. **59**, 1760 (1990); Physica B **165-166**, 1019(1990).
9. T. Tokura, T. Arima, H. Takagi, S. Uchida, T. Ishigaki, H. Asano, R. Beyers, A. I. Nazzal, P. Lacorre and J. B. Torrance, Nature **342**, 980(1989).
10. T. Arima, Y. Tokura, H. Takagi, S. Uchida, R. Beyers and J. B. Torrance, Physica C **168**,79(1990).

Electronic Structure of Underdoped Superconductors and Related Mott Insulators

Zhi-Xun Shen

Departments of Physics, Applied Physics, and Stanford Synchrotron Radiation Laboratory, Stanford University, Stanford, CA 94305, USA

Abstract. In this short paper, we give a brief review of our experimental effort to investigate the issue of normal state pseudogap in the underdoped regime, and the connection of the pseudogap of the electronic structure of the insulator.

1 Introduction

Over the past five years, we have been focusing on the electronic structure of underdoped cuprate superconductors, and related Mott isulators. In this short paper, we give a brief overview of the results we obtained at the Stanford Synchrotron Radiation Laboratory.

2 Discussion on Electronic Structures

Our effort is motivated by the desire to understand the evolution of the electronic structure as a function of doping. In the optimally and overdoped samples, ARPES experiments by many groups have provided a good evidence for a large Fermi surface that is consistent with predictions by band theory [1]. On the other hand, it was soon discovered that these results from the overdoped metal cannot be trivially connected to that of the insulators. This realization stemmed from the experiment on an insulator, $Sr_2CuO_2Cl_2$, which also has the crucial CuO_2 plane [2]. Along the (0, 0) to (π, π) direction, the data from the insulator looks very similar to that of the metal. On the other hand, the dispersive quasiparticle feature near (π, 0) is pushed down to much lower energy relative to maximum energy along the (0,0) to (π, π) line. Figure 1 compares the results from a typical metal, $Bi_2Sr_2CaCu_2O_{8+\delta}$, to that of a $Sr_2CuO_2Cl_2$. Thus, the electronic structure near (π, 0) region is the key to the non-trivial electronic structure evolution from the undoped isulator to overdoped metal.

To further address this issue, we started the doping dependant study with undoped $Bi_2Sr_2CaCu_2O_{8+\delta}$ [3]. Figure 2 compares the data from optimally and underdoped $Bi_2Sr_2CaCu_2O_{8+\delta}$ along the (0, 0)–(π, 0)–(π, π) high symmetry directions. Along (0, 0) to (π, π) direction, the results from the two samples are very similar, with a dispersive feature moving up in energy and eventually crossing the Fermi level near (0.4π, 0.4π). In other words, there is a well defined Fermi surface crossing along (0, 0) to (π, π) direction in

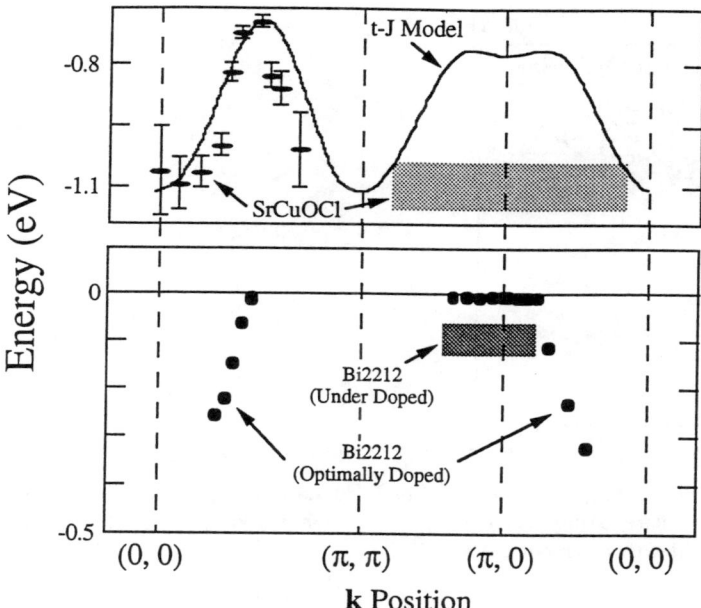

Fig. 1. The E versus k relationship of the experimental data from $Bi_2Sr_2CaCu_2O_{8+\delta}$, plotted against a calculated dispersion for the t–J model. Note that the dispersion along $(0,0)$ to (π,π) direction in the underdoped sample is very similar to that of the optimally doped case and has been omitted from the figure for clarity

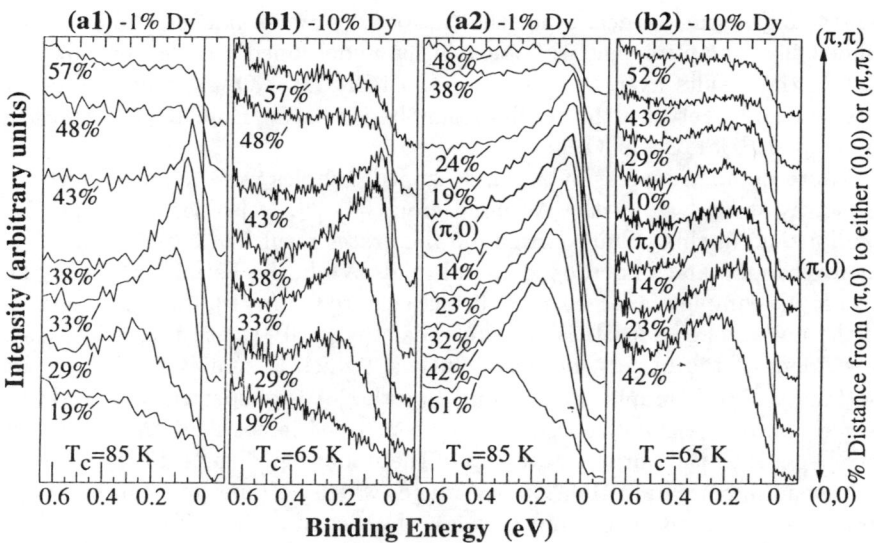

Fig. 2. ARPES spectra from $Bi_2Sr_2Ca_{1-x}Dy_xCu_2O_{8+\delta}$ single crystal thin films with 1% and 10% Dy. The symmetry points identified correspond to those indicated in the two-dimensional Brillouin zone of Fig. 1. The data were collected along the cuts (1) and (2) of Fig. 1 at 110 K.

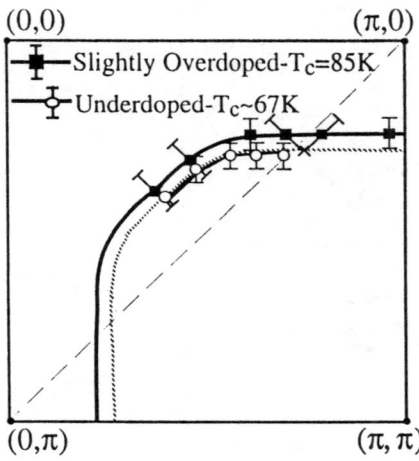

Fig. 3. Fermi level crossings from two $Bi_2Sr_2CaCu_2O_{8+\delta}$ samples of differing oxygen content. The entire Brillouin zone can be reconstructed by four-fold rotation about (0, 0). The data were recorded at 100 K

both samples. Near $(\pi, 0)$ region, the situation is different in these samples. For the overdoped sample ($T_c \approx 85$ K), there is a clear Fermi surface crossing along $(\pi, 0)$ to (π, π), with the feature right at the Fermi level. For the underdoped sample ($T_c \approx 67$ K), on the other hand, the feature never reaches the Fermi level. We attribute this to the opening of a gap along the underlying Fermi surface. Since the (0, 0) to (π, π) is not gapped, we concluded that an anisotropic gap opens in the underdoped sample. As shown in Fig. 1, the results from the underdoped $Bi_2Sr_2CaCu_2O_{8+\delta}$ sample interpolate smoothly between the results from the insulating $Sr_2CuO_2Cl_2$ and the overdoped $Bi_2Sr_2CaCu_2O_{8+\delta}$.

More detailed studies of the underdoped samples were followed both by ourselves as well as by the Argonne group [4]–[6]. (Also see the paper by Campuzano in this volume.) Figure 3 illustrates results of Fermi surface for the overdoped and underdoped samples [4]. While the Fermi surface of the overdoped sample is a large Fermi surface centered at (π, π), the Fermi surface of the underdoped sample is an "arc" near the (0, 0)–(π, π) line. Obviously, a segment of the underlying Fermi surface (which is similar to that of the optimally doped sample as determined by the locus of spectral weight drop) is gapped. In these and subsequent studies [7], it is recognized that there are two aspects to the normal state gap of the underdoped samples. The first is the broad maximum in 100–200 mV range (which becomes more pronounced in the more underdoped symples), while the second is the shift of the leading edge. Figure 4 shows underdoped sample at $(\pi, 0)$ in the normal as well as the superconducting state, clearly illustrating the presence of two energy scales. Here, we refer the leading edge gap and the high-energy hump as high and low energy pseudogaps, respectively. These features have qualitatively the

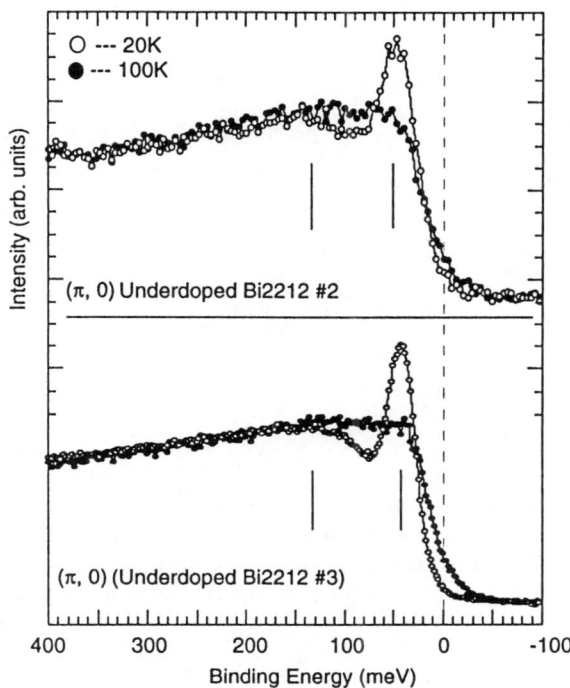

Fig. 4. Normal and superconducting state spectra for two underdoped $Bi_2Sr_2CaCu_2O_{8+\delta}$ samples at 20 K and 100 K. The two vertical bars indicate the possible two energy scales

similar angular dependence to that of the d-wave gap function. In particular, angular dependence of the low energy pseudogap is very similar to that of the d-wave superconducting gap, as shown in Fig. 5 [8]. Given the similarity between the low energy pseudogap and the superconducting gap, it is likely that the former is also related to the superconductivity. This fact is correlated with the fact that the gap increases as the samples become more underdoped, although T_c decreases at the same time, as illustrated in Fig. 6. This suggests that the T_c reduction in the underdoped regime is not related to the decrease of pairing strength. The low energy pseudogap is likely related to the pairing fluctuations in the normal state, and it becomes very poorly defined in the more underdoped regime where the carrier density is very low.

Unlike the low energy pseudogap, the high energy pseudogap is clearly a feature that can be connected to the insulator, as depicted in Fig. 1 in that it interpolates between the property of the metal and the property of the insulator. The doping dependence of these two aspects of the gaps are strongly correlated. As the sample becomes overdoped, both the low and high energy pseudogaps collapse rapidly [7]. The correlated behavior of the two "gaps" are correlated in Fig. 7. Figure 8 shows the $(\pi, 0)$ spectra for five samples, insulating $Ca_2CuO_2Cl_2$, and Dy doped $Bi_2Sr_2CaCu_2O_{8+\delta}$ with

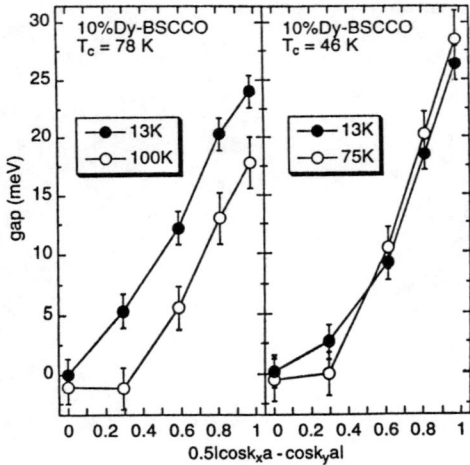

Fig. 5. Leading edge gaps of underdoped single crystal thin films ($T_c \approx 78$ K and 46 K, respectively) in the superconducting and normal states. k-space positions were selected at the underlying Fermi surface to facilitate measuring the energy gap. The similarity of the gaps measuered at two temperatures suggests a common origin for them

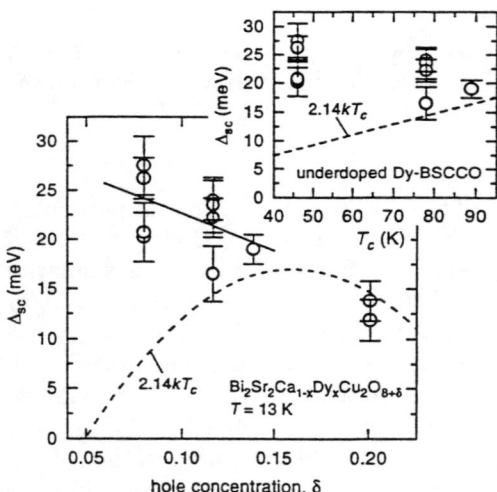

Fig. 6. Inset: The superconductiong state gap Δ_{sc} measured at 13 K on $Bi_2Sr_2Ca_{1-x}Dy_xCu_2O_{8+\delta}$ plotted vs. T_c. Δ_{sc} is the gap maximum minus the gap minimum in k-space. (In fact, the gap minimum is close to zero in all cases.) The nine samples came from three growth and annealing runs, and therefore fall into three T_c groups. The dashed line is the standard BCS mean-field d-wave prediction with $\Delta_{sc} = 2.14kT_c$, shown to highlight the non-mean-field trend of the data. Main panel: Δ_{sc} vs. doping δ, with δ inferred from $T_c/T_{c,max}$ (see [8]). The energy scale from T_c ($2.14kT_c$, dashed line) shows very different behavior from the linear fit to the Δ_{sc} data points (straight line)

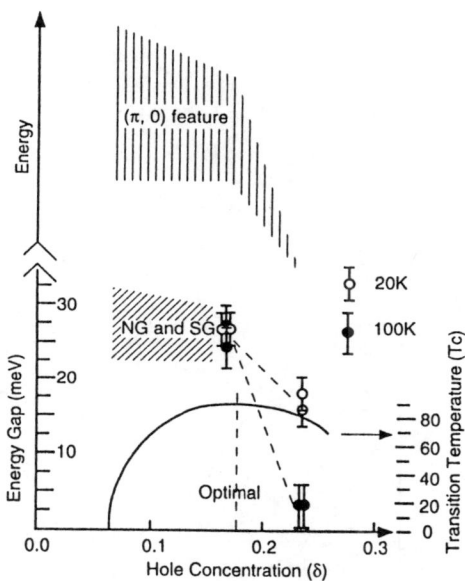

Fig. 7. The doping dependence of the maximum of the low and high energy pseudogaps, as well as the superconducting gap. The data suggest the two gaps are correlated, as they disappear together

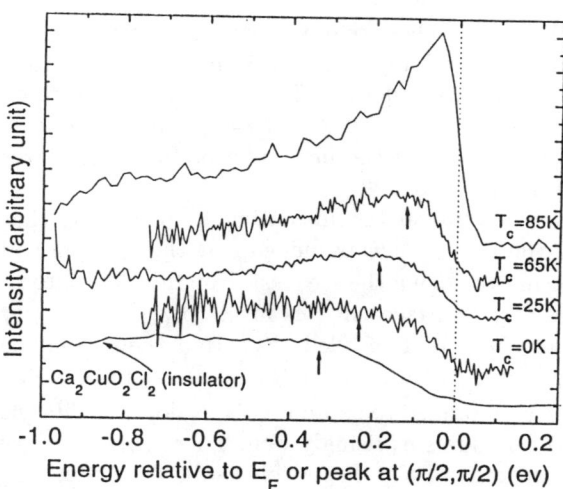

Fig. 8. $(\pi, 0)$ spectra as a function of doping, for $Bi_2Sr_2CaCu_2O_{8+\delta}$ samples with T_c from $\approx$ 0 K–85 K, as well as for the insulator $Sr_2CuO_2Cl_2$. The spectra at this point in k-space evolve smoothly with doping

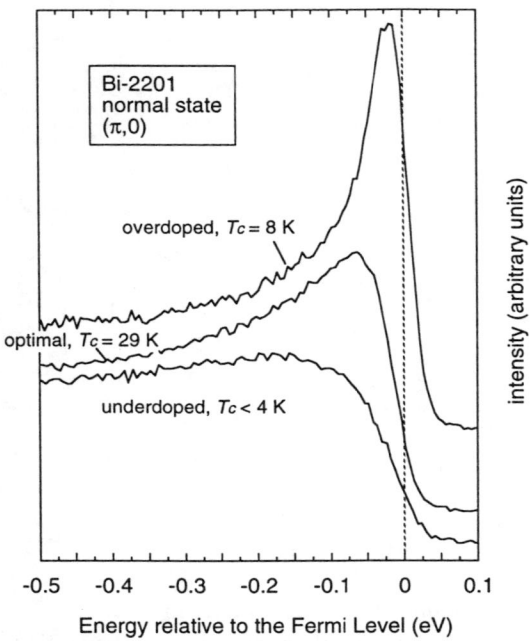

Fig. 9. $(\pi, 0)$ spectra as a function of doping for $Bi_2Sr_2CaCu_2O_{8+\delta}$ samples.

$T_c \approx 0, 25$ K, 65 K and 85 K. It is clear that the $(\pi, 0)$ spectra evolves continuously from the insulator to metal. In particular, it means that the 100–200 meV high energy pseudogap in the underdoped sample can be connected to the property of an insulator as first pointed out by Laughlin [9]. We have found a similar behavior of the electronic structure evolution in a single layer cuprate of Bi2201, as shown in Fig. 9 [10].

Most recently, we found a feature in the insulating $Ca_2CuO_2Cl_2$ that more convincingly suggests the d-wave-like pseudogap in the underdoped regime is a property of the insulator. Using the steepest drop in n(**k**) contour, we identified an "underlying Fermi surface" in this insulator, which is very close to what band structure gives for half-filled case. Had there been no interaction, this should have been a constant energy contour. Instead, we found that this contour has been modified to exhibit a dispersion that fits the d-wave ($|\cos k_x a - \cos k_y a |$) function surprisingly well. Therefore, somehow the interaction gives a d-wave form to the dispersion along the underlying Fermi surface contour. Once doped, one expects that the chemical potential drops to the "node point" near $(\pi/2, \pi/2)$, and the d-wave dispersion turns into the d-wave like pseudogap. This unexpected finding in the insulator strongly suggests that d-wave superconductivity is a derivative of a property existing already in the insulator.

Acknowledgements

I would like to thank D.S. Dessau, B.O. Wells, A.G. Loeser, D.S. Marshall, J.M. Harris, C. Kim, P.J. White, F. Ronning, D.L. Feng, A. Kapitulnik, J. Eckstein, I. Bosovic, J.R. Birgeneau, M.A. Kastner and L.L. Miller for collaborations. Experiments were performed at SSRL which is operated by the Division of Chemical Sciences, Office of Basic Energy Sciences, Department of Energy.

References

1. Shen, Z.-X. and Dessau, D.S. (1995) Physics Report **253**, 1–162
2. Wells, B.O. et al. (1995) Phys. Rev. Lett. **74**, 964
3. King, D.M. et al. (1995) J. of Phys. Chem. Solids **56**, 1865
4. Marshall, D.S. et al. (1996) Phys. Rev. Lett. **76**, 4841
5. Loeser, A.G. et al. (1996) Science **273**, 325
6. Ding, H. et al. (1996) Nature **382**, 51
7. White, P.J. et al. (1996) Phys. Rev. B **54**, R15669
8. Harris, J.M. et al. (1996) Phys. Rev. B **54**, R15665
9. Laughlin, R.B. et al. (1997) Phys. Rev. Lett. **79**, 1726
10. Harris, J.M. et al. (1997) Phys. Rev. Lett. **79**, 143

Destruction of the Fermi Surface in Underdoped Cuprates

Juan Carlos Campuzano[1,2], H. Ding[1,2], M.R. Norman, and M. Randeria[3]

[1]Physics, University of Illinois at Chicago, 845 W. Taylor St., Chicago, IL 60608
[2]Argonne National Laboratory, 9700 S. Cass Ave., Argonne, IL 60439
[3]Tata Institute for Fundamental Research, Mumbai, India

Abstract. We review some of our recent angle-resolved photoemission results on $Bi_2Sr_2CaCu_2O_8$ as a function of doping. We find that the Fermi surface is progressively destroyed as the temperature is lowered below the pseudogap temperature T^*. Furthermore, the superconducting gap behaves quite differently at different **k**-points. All of these results point to a non-mean-field behavior of the superconducting transition.

Angle-resolved photoemission (ARPES) of the high temperature superconductors is quite unusual, in that the lineshape is strongly temperature and doping dependent. We begin by discussing the overdoped samples of $Bi_2Sr_2CaCu_2O_8$, which have a simpler electronic structure. In Fig. 1a we show the temperature dependence of the spectra near the $(\pi,0)$ point, indicated in the insert of Fig. 1b.

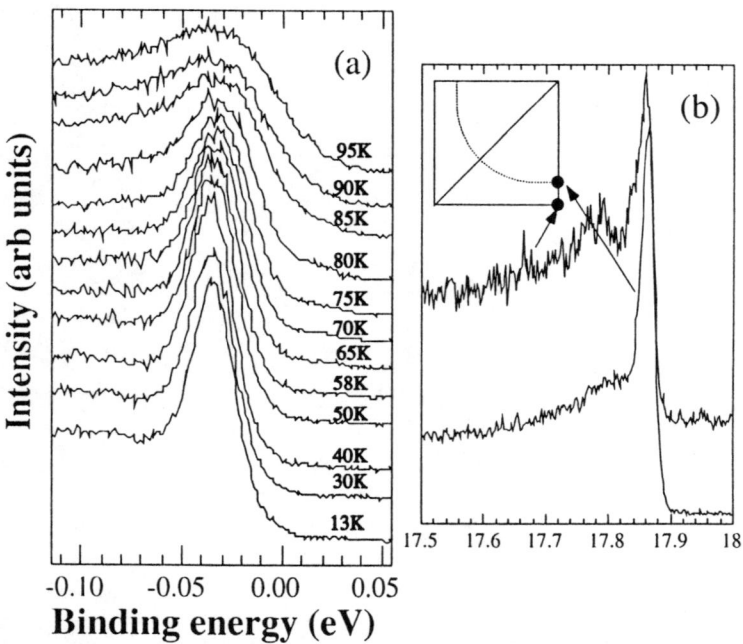

Fig. 1: a) Temperature dependence of the ARPES lineshape a at a resolution of ~20 meV; b) ARPES lineshape at 5K at a resolution of 6 meV. (Note that there is a tail on the leading edge due to impurity scattering)

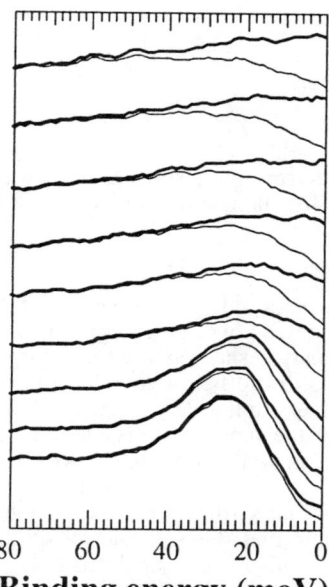

80   60   40   20   0
Binding energy (meV)

Fig. 2: Selected data in Fig. 1 (thin lines) divided by the effective Fermi function (thick lines)

We have shown [1] that the ARPES photoemission intensity is given by a product of the Fermi function $f(\omega)$ and the spectral function $A(\mathbf{k},\omega)$, scaled by the matrix element for ARPES, $I(\mathbf{k})$, which depends on the photon energy and its polarization, as well as the final states

$$I(\mathbf{k},\omega) = I_0(\mathbf{k}) f(\omega) A(\mathbf{k},\omega) \tag{1}$$

Since $I(\mathbf{k})$ only determines the overall amplitude of the spectrum, we will ignore it in the present discussion. The other terms are of interest to us, since they determine the lineshape. $A(\mathbf{k},\omega)$ is the probability of adding or removing a particle from a many-body interacting system, and in turn is given by the imaginary part of the self energy

$$A(\mathbf{k},\omega) = -\frac{1}{\pi} \Im m G(\mathbf{k},\omega + i0^+) \tag{2}$$

The data in Fig. 1 can then be most easily understood if we divide each curve by the effective Fermi function, as shown in Fig. 2:

The top curve, at $T=95$K, above T_c, shows a broad spectrum centered at $\omega = 0$. This broad spectrum has a full width at half-maximum of 400 meV, and shows no evidence for a quasiparticle peak, exhibiting the unusual nature of the states in this material. As the temperature is lowered, two changes occur in the spectrum: 1) the peak moves to higher binding energy as the superconducting gap opens, and 2) the peak gets much sharper as the lifetime of the state increases. We have shown [2] that the spectrum moving to higher binding energy is in fact due to the superconducting energy gap, as one can in addition observe the clear signature of particle-hole mixing in the energy dispersion. The sharpening of the peak is a consequence of the reduced scattering rate in the superconducting state, which is strongly temperature-dependent because it is dominated by electron-electron scattering [3]. The leading term in electron-electron scattering is given by the bubble diagram in Fig. 3a:

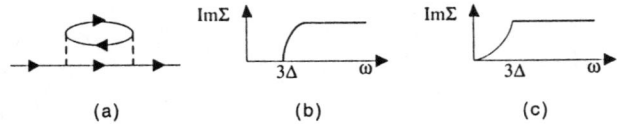

Fig. 3: a) Leading-order diagram for electron-electron scattering; b) expected behavior of the imaginary part of the self-energy for an *s*-wave and c) *d*-wave superconductor.

Since each line in this diagram represents a quasiparticle which has a gap Δ, scattering in an *s*-wave superconductor would be suppressed over an energy range of 3Δ, as shown in Fig. 3b. This causes the spectrum to start to sharpen at a much larger binding energy than Δ, giving rise to the formation of a dip below the sharp peak, followed by a hump, where the spectral function recovers to that of the normal state. On the other hand, in a simple *d*-wave superconductor the suppression is expected to have a much softer energy dependence of ω^2 [4], as shown in Fig. 3c. However, a closer look at Fig. 1 reveals a more interesting story than this simple picture. First, the fact that SC and normal state data match beyond 90 meV (they continue to match for energies beyond those in the figure). This means that the self-energy of the electrons in the normal and superconducting state are equivalent beyond this energy. Second, note that the drop from the hump to the dip and the rise to the sharp peak occurs within the experimental resolution, and therefore the dip must be intrinsically quite deep. These simple observations has non-trivial consequences. They imply that the decay of $\Im m\Sigma$ is much sharper, decaying at least as ω^6, closer to the case in Fig. 3b. We arrive at this conclusion from a simulation [5] of the spectral function using the form

$$A(\mathbf{k},\omega) = \frac{1}{\pi} \frac{\Im m\Sigma}{(\omega - \varepsilon_\mathbf{k} - \Re e\Sigma)^2 + \Im m\Sigma^2} \qquad (3)$$

taking into account the observed momentum and energy resolution as shown in Fig. 4. For details, see Ref [5].

Basically, there is a step in $\Im m\Sigma$, unlike the standard analysis of a *d*-wave pairing state [4], which doesn't give a dip at all. In fact, the data are not only consistent with a step in $\Im m\Sigma$, but the depth of the dip is such that it is best fit by a peak in $\Im m\Sigma$ at the dip energy, followed by a rapid drop to a small value. What are the consequences of this behavior in $\Im m\Sigma$? If $\Im m\Sigma$ has a sharp drop, then by Kramers-Kronig transformation, $Re\Sigma$ will have a sharp peak at the dip energy. This peak can very simply explain the unusual SC state dispersion shown in Fig. 5, as it will cause a low energy quasiparticle pole to appear even if the normal state binding energy is large.

The most transparent way to appreciate this result is to note that a sharp step in $\Im m\Sigma$ is equivalent to the problem of an electron interacting with a sharp (dispersionless) mode, since in that case, the mode makes no contribution to $\Im m\Sigma$ for energies below the mode energy, and then makes a constant contribution for energies above. This problem has been treated by Engelsberg and Schrieffer, and extended to the superconducting state by Scalapino and coworkers [6]. The difference in our case is that since the effect only occurs *below* T_c, it is a consequence of the opening of the superconducting gap in the electronic energy spectrum, and thus of a collective origin, rather than a phonon as in Ref. [6]. To facilitate comparison to this classic work, in Fig. 5c we plot the position of the low energy peak and higher binding energy hump as a function of the energy of the single broad peak in the normal state.

This plot has a striking resemblance to that predicted for electrons interacting with a sharp mode in the superconducting state [7], and one clearly sees the low

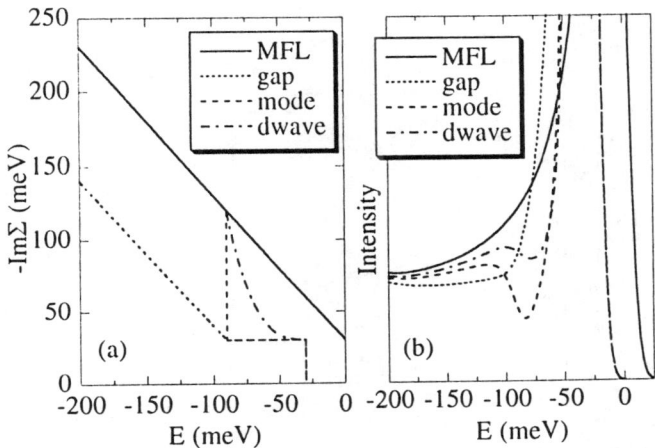

Fig. 4: a) $\Im m\Sigma$ for MFL (solid line), gapped MFL (dotted line), gapped MFL plus mode (dashed line), and simple d-wave model (dashed-dotted line). Parameters are $\alpha=1$, $\omega_c=200$ meV, $\Delta=30$ meV (0 for MFL), $\Omega_0=2\Delta$, and $\Gamma_0=30$ meV. b) Spectral functions (times a Fermi function with $T=14$ K convolved with a resolution Gaussian of $\sigma=7.5$ meV for these four cases ($\varepsilon=-34$ meV)

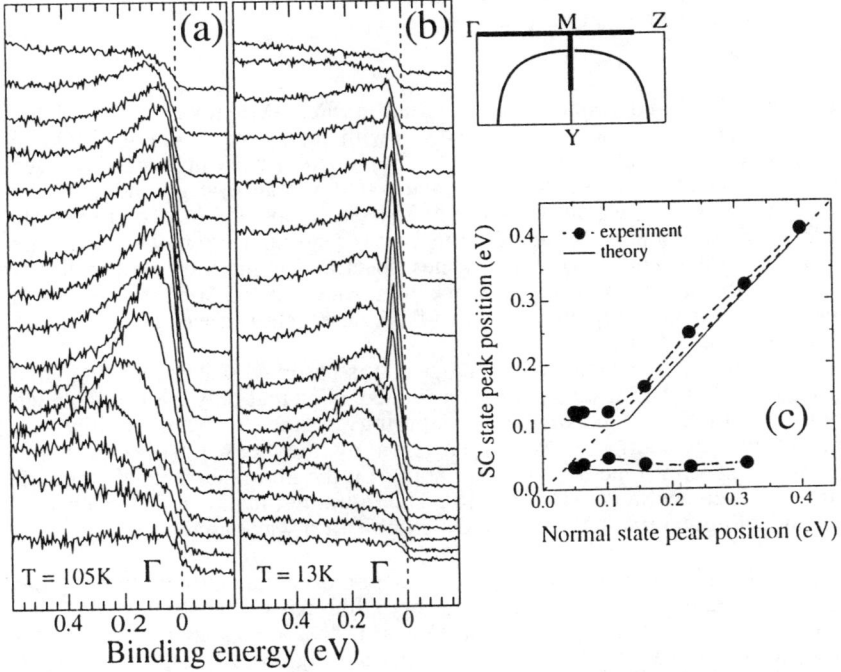

Fig. 5: EDCs in (a) the normal state (105 K) and (b) the superconducting state (13 K) along the line (0,0) to (π,0). The zone is shown as an inset with the curved line representing the observed Fermi surface. (c) Positions (eV) of the sharp peak and the broad hump in the SC state versus normal state peak position obtained from (a) and (b). Solid points connected by a dashed line are the data; the dotted line represents the normal state dispersion, and the solid lines are from a model discussed in Ref. 5.

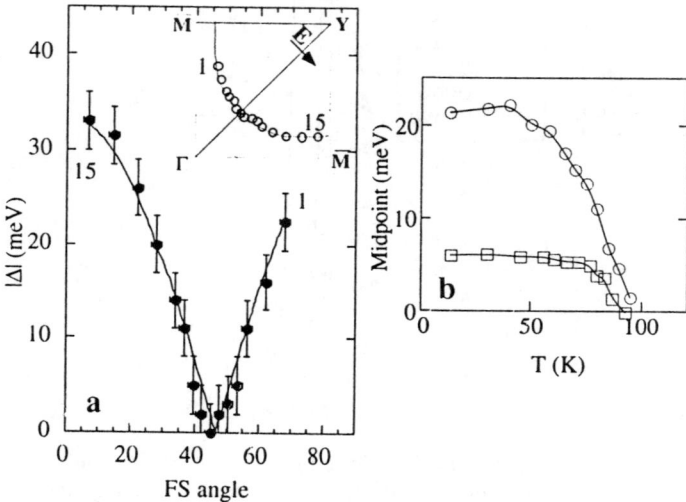

Fig. 6: a) The superconducting gap, extracted from fits, versus angle on the Fermi surface (filled circles) compare to a d-wave gap (solid curve). Locations of measured points and the Fermi surface are shown in the inset. b) Plot of the midpoint of the leading edge of the spectra as a function of T, showing that the gap at all k's close simultaneously in an overdoped sample.

energy pole which is a consequence of the peak in $\Re e\Sigma$. Moreover, the predicted spectral functions of this model, when convolved with energy resolution, give a good representation of the data shown in Fig. 1 (with the probable peak in $\Im m\Sigma$ discussed above due to the peak in the SC density of states) [5]. On general grounds, the flat dispersion of the low energy peak seen in Fig. 5c is a combination of two effects: (1) the peak in $\Re e\Sigma$, which provides an additional mass renormalization of the SC state relative to the normal state, and thus pushes spectral weight towards the Fermi energy, and (2) the superconducting gap, which pushes spectral weight away. This also explains the strong drop in intensity of the low energy peak as the higher binding energy hump disperses.

We now return to the superconducting gap. It is observed that the leading edge of the spectrum in Fig. 1 is always resolution-limited. Note that only the leading edge is resolution-limited, as quasiparticle damping begins to occur below the superconducting gap energy Δ, which is given by the peak position, i.e. the maximum in the spectral function. At other points on the Fermi surface, we must fit the spectra, as discussed in detail in Ref. [8]. The resulting functional dependence of the superconducting gap in $Bi_2Sr_2CaCu_2O_8$ exhibits a d-wave character, as shown in Fig. 6a.

We also plot an estimate of the gap as a function of temperature by plotting the mid-point of the leading edge, shown in Fig. 6b. Although we must stress that this quantity really does not have any physical meaning, it does at least give an indication of the behavior of the gap vs. T. As can be seen from Fig. 6b, the gap closes at the same temperature at all points along the Fermi surface, i.e. the transition exhibits the classic mean-field behavior with the gap closing at T_c. We will show below a better estimate of the gap using a phenomenological analysis.

As the doping is reduced, the behavior drastically changes. The most notable difference is, that even at optimal doping, as the temperature is reduced, a pseudogap appears in the excitation spectrum, as shown in Fig. 7 [9,10].

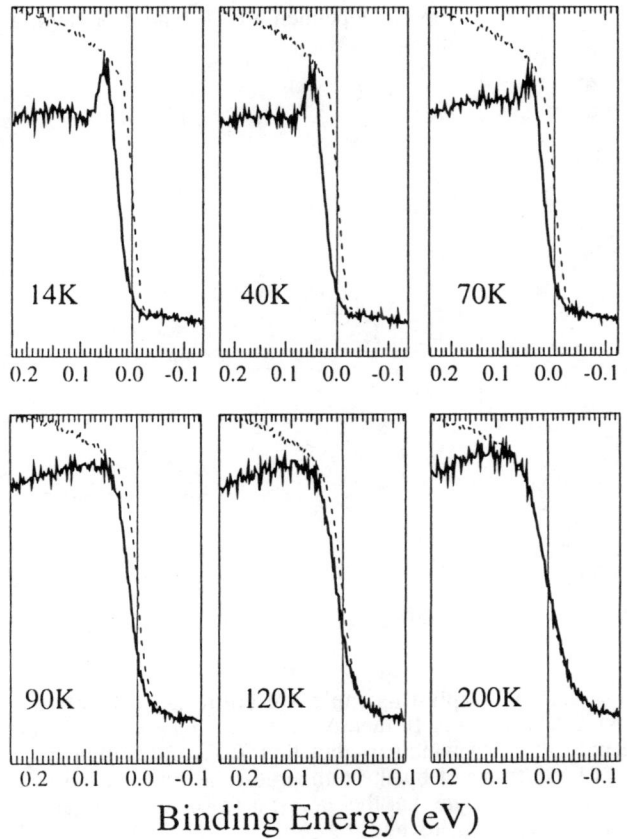

Fig. 7: ARPES spectra at the Fermi surface along the *M-Y* direction for an 85K underdoped Bi2212 sample at various temperatures (solid curves). The dotted curves are reference spectra from polycrystalline Pt.

In Fig. 7 we plot spectra at the Fermi surface along the $(\pi,0)$-(π,π) direction of an underdoped (T_C = 83K) sample at six different temperatures. Note that above T_c, i.e. at 90K, there is a sizeable (16 meV) shift between the leading edge of the sample (solid line) and that of polycrystalline Pt (dotted line) which is used as a chemical potential reference. This pseudogap eventually disappears at a much higher temperature T^* (~200K in this case).

It is significant to note that there are always two features in the spectra, one that is related to the quasiparticle peak in the superconducting state, and gives rise to the sharp leading edge in the pseudogap state, and another feature at higher binding energy, the "hump" described above, which remains in the pseudogap state and only disappears at a higher temperature. The pseudogap that we describe here is associated with the feature at low binding energy, the leading edge gap [11].

We find that T^* increases with decreasing doping in the underdoped region, and merges with T_c in the overdoped region [10], as shown in Fig. 8. In Fig. 8 we also plot the position of the sharp coherent peak near $(\pi,0)$ (see first panel of Fig. 7) as a function of doping, or carrier concentration x. Since this sharp peak is essentially resolution limited, one can regard the position of its maximum as the value of the

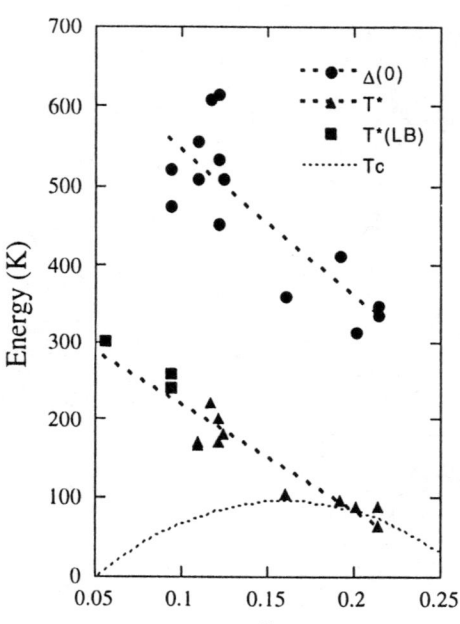

Fig. 8: Dependence of T^* and Δ on doping (in °K)

gap, Δ_0. Despite some considerable sample-to-sample variation, Δ_0 follows the general trend of increasing with decreasing x. In fact, Δ_0 seems to scale with T^*, not with T_C. This is consistent with theories which predict that T_C is controlled by a phase stiffness temperature [12,13], and not by the temperature at which a pairing gap opens. On the other hand, one may argue that the gap near $(\pi,0)$ is no longer the superconducting gap, since it has no relationship to T_C.

Let us address this problem by looking at some experimental evidence. Temperature dependent measurements in underdoped samples, described in more detail below, reveal a gap that smoothly evolves through T_C, suggesting that the gaps below and above T_C have the same origin, i.e. the pseudogap is closely related to the superconducting gap. We have also found that the low temperature (T = 14K) gap of an underdoped (T_C = 83K) sample has a very similar momentum dependence as the gap of the overdoped 87K sample which has the d-wave gap shown in Fig. 6 [10]. This is a strong indication that in underdoped samples the gap below T_C near $(\pi,0)$ is still the superconducting gap. It is interesting to note that, although having little effect on the gap size near $(\pi,0)$, T_C has a strong effect on the lineshape.

It is easy to understand how in an overdoped sample they system evolves from one with a full Fermi surface to one with the point-nodes of the d-wave state by spontaneous symmetry breaking at T_C. But the question is, how does the underdoped system evolve? At high temperatures, the system exhibits the full Fermi surface, if we define the Fermi surface as the locus of gapless excitations, since we are discussing high temperatures. We find [14] the unusual result that as the temperature is lowered, different regions of the Fermi surface begin to gap at different temperatures, as shown in Fig. 9.

In Fig. 9 we show the temperature evolution of the spectra at three points along the Fermi surface, marked a,b, and c in panel d. At point a, the Fermi surface crossing point with the largest gap, the pseudogap fills in at ~180 K. The surprise is that at points (b) and (c) closer to the d-wave node, the pseudogap fills in at progressively lower temperatures, of 120 and 95 K, respectively. This data implies

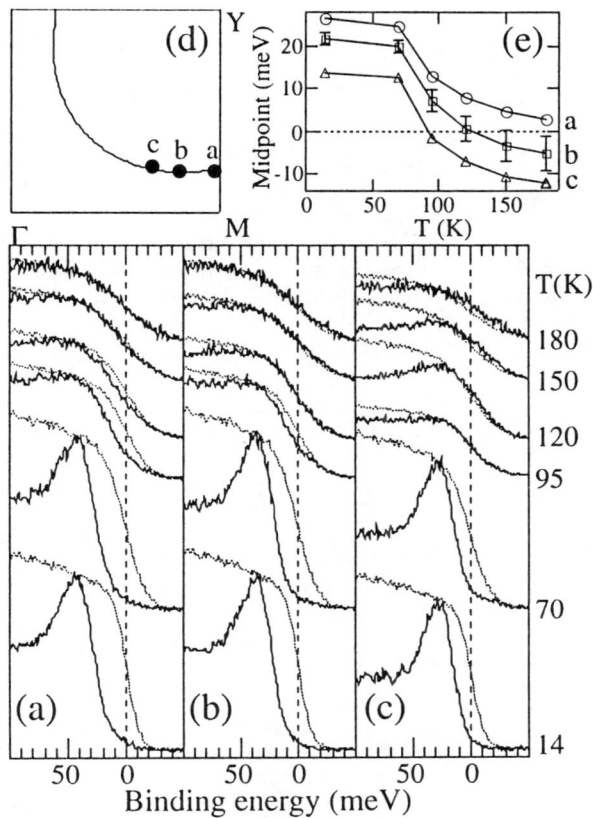

Fig. 9: a,b,c) Data obtained at three **k** points in the Y quadrant (d) of the Brillouin zone for an 85K underdoped Bi2212 sample at various temperatures (solid curves). The dotted curves are reference spectra from polycrystalline Pt used to determine the chemical potential. Note the closing of the spectral gap at different T for different **k**. This feature is also apparent in the plot (e) of the midpoint of the leading edge of the spectra as a function of T.

that, as the temperature is lowered, the Fermi surface is progressively destroyed, as shown in Fig. 10

This is quite a novel and unusual situation, where the Fermi surface changes topology without a change in symmetry. The pseudogap suppression first opens up near $(\pi,0)$ and progressively gaps out larger portions of the Fermi contour, leading to gapless arcs which shrink with decreasing T. Contrast this behavior to that of the more conventional T-dependence of an overdoped 87K sample shown in Fig. 6b, where the gap for different **k** values closes at the same T. It is important to note that the arcs found here are not the hole pockets of a lightly doped antiferromagnet [11]. We find that although the d-wave superconducting gap below T_c smoothly evolves into the pseudogap above T_c, the gaps at different **k** points are not related to one another above T_c the same way as they are below, implying an intimate, but nontrivial relation, between the two. This can more easily be seen by the use of the symmetrization method [14], described in Fig. 11, which effectively eliminates the Fermi function from ARPES data and permits us to focus directly on the spectral function A.

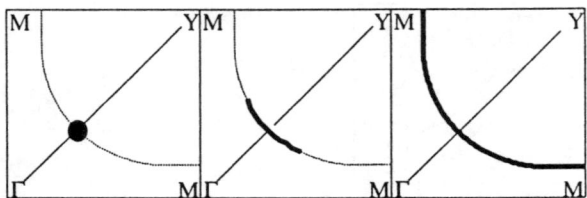

Fig. 10: Schematic illustration of the T-evolution of the Fermi surface in underdoped cuprates. The d-wave node below T_c (left panel) becomes a gapless arc above T_c (middle panel) which expands with increasing T to form the full Fermi surface at T^* (right panel).

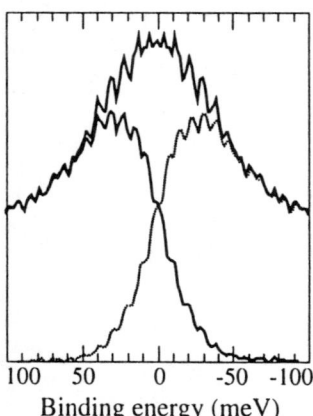

100   50   0   -50   -100
Binding energy (meV)

Fig. 11: Given ARPES data described by $I(\omega) = I_0 A(\mathbf{k},\omega) f(\omega)$ (with the sum over a small momentum window about the Fermi momentum $\mathbf{k}_F$), we can generate the symmetrized spectrum $I(\omega) + I(-\omega)$. Assuming particle-hole (p-h) symmetry for a small range of ω and $\varepsilon_\mathbf{k}$, we have $A(\varepsilon_\mathbf{k},\omega)=A(-\varepsilon_\mathbf{k},-\omega)$ for $|\omega|,|\varepsilon|$ less than few tens of meV. It follows, using the identity $f(-\omega) = 1-f(\omega)$, that $I(\omega) + I(-\omega) = \Sigma_\mathbf{k} I_0 A(\mathbf{k},\omega)$

We now discuss why the T dependence of the Fermi arc is not simply due to inelastic scattering above T_c broadening the d-wave node. This is most clearly illustrated in Fig. 12, which shows symmetryzed data at two **k**-points on the Fermi surface. It is apparent that the gap "fills in" for **k**-point 1 (at the maximum gap) as T is raised, whereas it "closes" for **k**- point 2. Fig. 12b shows the gap values extracted at these two points using a phenomenological model [17]

$$\Sigma(\mathbf{k},\omega) = -i\Gamma_1 + \Delta^2/[\omega + \varepsilon(\mathbf{k}) + i\Gamma_0] \qquad (4)$$

This purely phenomenological self-energy is simply obtained by adding a damping term Γ_0 to the Nambu-Gorkov propagator to remove the singularity at $\omega=0$ and take into account the filling-in of the pseudogap above T_c. Γ_0 can be thought of as a "pair lifetime". The term Γ_1 describes the quasiparticle lifetime.

We believe that the unusual T-dependence of the pseudogap anisotropy will be a very important input in reconciling the different crossovers seen in the pseudogap regime by different probes. The point here is that each experiment is measuring a **k**-sum weighted with a different set of **k**-dependent matrix elements or kinematical factors (e.g., Fermi velocity). For instance, quantities which involve the Fermi velocity, like dc resistivity above T_c and the penetration depth below T_c (superfluid

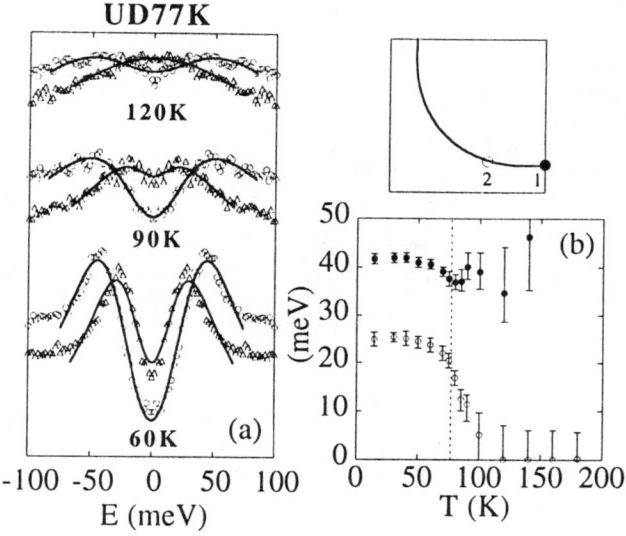

Fig. 12: a) Symmetryzed data for a $T_C=77K$ underdoped sample for three temperatures at (open circles) $\mathbf{k}_F$ point 1 in the inset, and at (open triangles) $\mathbf{k}_F$ point 2, compared to the model fits. b) $\Delta(T)$ for these two points (filled and open circles), with T_C marked by the dashed line.

density), should be sensitive to the region near the Γ-Y direction, and would thus be affected by the behavior we see at **k** point c. Other types of measurements (e.g. specific heat and tunneling) are more "zone-averaged" and will have significant contributions from **k** points a and b as well, thus they should see a more pronounced pseudogap effect. Interestingly, other data we have indicate that the region in the Brillouin zone where behavior like **k** point c is seen shrinks as the doping is reduced, and thus appears to be correlated with the loss of superfluid density [15]. Further, we speculate that the disconnected Fermi arcs should have a profound influence on magnetotransport given the lack of a continuous Fermi contour in momentum space.

We thank the National Science Foundation for Grants DMR 9624048, DMR 91-2000 through the Science and Technology Center for Superconductivity, DMR 9726550,and the U.S. Dept. of Energy, Basic Energy Sciences, contract W-31-109-ENG-38.

References

1. M. Randeria, *et al.*, Phys. Rev. Lett. 74, 4951-4954 (1995).
2. J.C. Campuzano, *et at.*, Phys. Rev. **B53** (1996) 14737-14740.
3. M. C. Nuss *et al.*, Phys. Rev. Lett. **66**, 3305 (1991); D. A. Bonn *et al.*, Phys. Rev. Lett. **68**, 2390 (1992).
4. S. M. Quinlan, P. J. Hirschfeld, and D. J. Scalapino, Phys. Rev. B. **53**, 8575 (1996).
5. M. Norman and H. Ding, Phys. Rev. B **57** (1988) R11089.
6. S. Engelsberg and J. R. Schrieffer, Phys. Rev. **131**, 993 (1963); J. R. Schrieffer, *Theory of Superconductivity* (W. A. Benjamin, New York, 1964); D. J. Scalapino, in *Superconductivity*, ed. R. D. Parks (Marcel Decker, New York, 1969), Vol 1, p. 449.

7. See Fig. 50 of Scalapino, Ref. [6].
8. H. Ding, *et at.*, Phys. Rev. B **54**, (1996) R9678.
9. Loeser, A.G., *et al.*, Science **273**, (1996) 325.
10. H. Ding, *et al.* Nature. **382** (1996) 51.
11. H. Ding *et al.*, Phys. Rev. Lett. **78**, (1997) 2628.
12. M. Randeria *et al.*, Phys. Rev. Lett. **69**, (1992) 2001.
13. V. Emery and S. Kivelson, Nature **374**, (1995) 434.
14. M. Norman, *et al.*, Nature **392** (1998) 157.
15. Y. J. Uemura, *et al.*, Phys. Rev. Lett. **62** (1989) 2317.
16. For a review, see: P.A. Lee, Normal state properties of the oxide superconductors: a review, in *High Temperature Superconductivity* (ed. K.S. Bedell, *et al.* 96-116 Addison-Wesley, New York, 1990).
17. M.R. Norman, *et al.*, Phys. Rev. B**57** (1998) R11093.

Effects of Stripe Order on Charge Dynamics in La-Cuprates

S. Uchida, N. Ichikawa, T. Noda, and H. Eisaki

Department of Superconductivity, The University of Tokyo, Bunkyo-ku, Tokyo 113-8656, Japan

S. Tajima and N. L. Wang

Superconductivity Research Laboratory, ISTEC, Koto-ku, Tokyo 135-0062, Japan

Abstract. Charge transport and optical conductivity spectra have been investigated for $La_{2-x-y}Nd_xSr_xCuO_4$ single crystals in the stripe ordered phase. The charge dynamics turns out to be basically metallic despite that nearly entire volume of sample is in the antiferromagnetically ordered, suggesting a spatially separation of the conduction paths from the spin domains. The charge dynamics also show a critical change upon increasing x across 1/8, and in the highly doped materials the superconducting order seems to coexist with the static stripe order. We argue that the dynamical stripe fluctuations may affect the charge dynamics in the Nd-free $La_{2-x}Sr_xCuO_4$.

1. Introduction

Among the high-T_c cuprates $La_{2-x}Sr_xCuO_4$ (LSCO) is known to have unique spin excitation spectrum. In the whole range of the superconducting phase ($0.05 < x < 0.25$) the incommensurate peak splitting around the position of the antiferromagnetic (AF) peak of the undoped La_2CuO_4 is observed in the inelastic neutron scattering [1, 2]. It was exlained in terms of dynamical incommensurate spin fluctuations tied to a nesting Fermi surface of LSCO [3]. However, recent inelastic neutron scattering experiments have offered evidence for similar incommensurate splitting in $YBa_2Cu_3O_{7-y}$ (YBCO) and $Bi_2Sr_2CaCu_2O_{8+y}$ (BSCCO) [4, 5]. Many researchers now believe that the incommensurate peaks are indicative of dynamical modulation of the AF spin structure in the hole-doped CuO_2 planes.

One atttractive explanation for the modulated spin structure is the possibility of stripe fluctuations. The stripe model postulates that not only spin density but also the charge density is spatially modulated. The static stripes are seen in nickelates [6] and manganites [7] in which the doped holes segregate and line up in rows separating AF antiphase spin domains. A neutron diffraction study on Nd doped $La_{1.6-x}Nd_{0.4}Sr_xCuO_4$ (LNSCO) with $x = 0.12$ revealed elastic magnetic superlattice peaks at $(0.5 \pm \epsilon, 0.5, 0)$ where $\epsilon = 0.12$ [8]. For this compound static charge modulation is also seen in neutron and x-ray scatteinr studies [8, 9]. The static charge order shows up as a splitting around the Bragg lattice peak, $(2 \pm 2\epsilon, 0, 0)$, due to the small atomic displacements induced by the charge density modulation. This leads to a model of spin and charge stripes pictured in Fig. 1.

One criticism to the stripe model in cuprates is that so far there is no evidence for higher harmonics of the magnetic diffraction peaks, indicating a sharp spin/charge density modulation as depicted in Fig. 1. Also, there is no evidence for dynamical charge stripes which should be associated with the incommensurate inelastic neutron peaks, except for a circumstantial evidence given by the recent photoemission (ARPES) result on BSCCO [10].

In this article we show the experimental results of charge transport and optical conductivity spectrum of LNSCO. We argue how the stritic and/or dynamical "stripe" order affects the charge dynamics and whether the one-dimensional charge stripe model is compatible with the observed charge dynamics. Arguments are also on the possible effects of dynamical "stripe" fluctuations in LSCO where the fluctuations would be slower (stronger) than those in YBCO and BSCCO.

2. Charge Dynamics in the Static Stripe Phase

2.1 Neutron Diffraction Studies

The static spin order (modulation) has been observed for a series of Nd doped crystals $La_{1.6-x}Nd_{0.4}Sr_xCuO_4$ with x=0.10, 0.12, 0.15, and 0.20 [11]. In the elastic neutron scattering they all show very sharp (almost resolution limited) incommensurate peaks at $(0.5 \pm \epsilon, 0.5, 0)$. The variation of the incommensurability ϵ with x is basically the same as that obtained from inelastic neutron scattering on LSCO [12] as illustrated in Fig. 2. ϵ increases linearly with x up to x=0.12 and then levels off for $x > 0.12$. The addition of Nd changes the tilt pattern of the CuO_6 octahedra so as to have the tetragonal (LTT) symmetry. The tilt pattern in the LTT phase are favorable for the charge stripes in Fig. 1, so that the coupling between the tilt modulation and the charge density

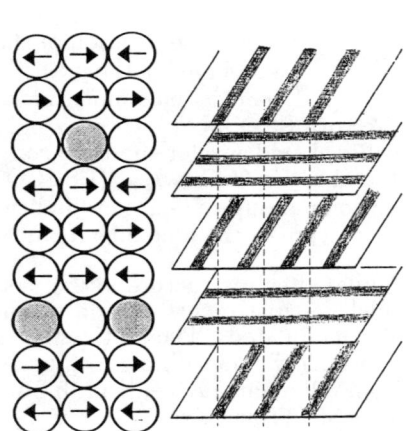

Fig.1 A model of spin and charge stripes in the CuO_2 planes corresponding to $\epsilon = 1/8$. The Cu ions are represented by circles (not shown are the oxygen ions). The holes occupy every other Cu site on the charge stripe which separates regions of the AF ordered spin domains. A possible stacking of the stripe ordered CuO_2 planes is also shown.

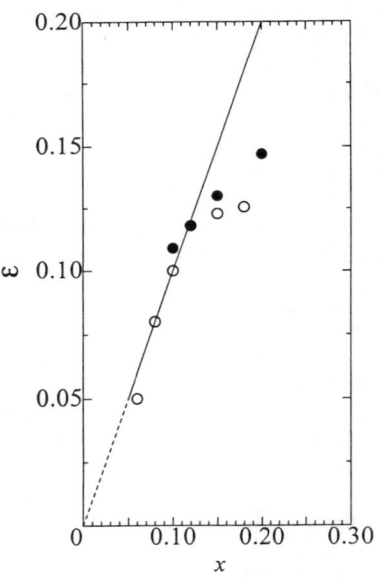

Fig.2 The incommensurability ϵ vs Sr composition x for LNSCO (closed circles, determined by the elastic neutron scattering) and for LSCO (open circles, the data from the inelastic neutron scattering [12]).

modulation would likely pin the charge tripes and hence stabilize the stripe spin domains. In fact, for LNSCO with $x=0.12$, after cooling the crystal through the orthorhambic (LTO) to LTT phase transition near 70K, the charge-order peaks appear first at ~60K and then the magnetic peaks follow below 50K [8].

Recently, the static spin modulation has also been observed for LSCO with doping level $x=0.12$ [13] known to produce a sharp dip in T_c vs x as well as for oxygen-doped $La_2CuO_{4+\delta}$ [14] with a relatively high value of $T_c(42K)$. Curiously, both compounds do not show any indication of the LTT tilt modulation and the magnetic peaks develop upon the onset of superconductivity.

In contrast to the observation of static and/or dynamical spin modulation for a wide class of cuprate materials, so far there is no evidence for dynamical charge modulation in the neutron scattering, and the static charge modulation is observed only for LNSCO with $x=0.10$ and 0.12. The charge superlattice peaks are observed at $(2 \pm 2\epsilon, 0, 0)$ where ϵ is the same as that in the magnetic peaks and identical to the doping level x. For $x=0.15$ and 0.20 the charge peaks have not been detected, although they show the structural phase transition to the LTT phase.

2.2 Charge Transport

The temperature dependences of the in-plane resistivity (ρ_{ab}) are shown in Fig. 3 for four x's of LNSCO. A small resistivity jump seen at $T_d \sim 70K$ for all the samples signals the LTO-LTT phase transition [15]. At low temperatures in

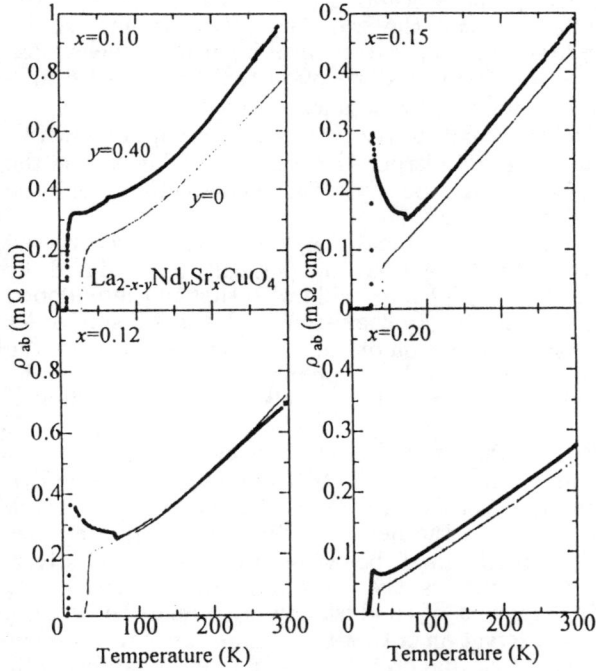

Fig.3 Temperature dependence of the in-plane resistivity for $La_{1.6-x}Nd_{0.4}Sr_xCuO_4$ with $x=0.10$, 0.12, 0.15 and 0.20 indicated by dark lines. For comparison, the data for $La_{2-x}Sr_xCuO_4$ with the same x is shown by light line in each frame.

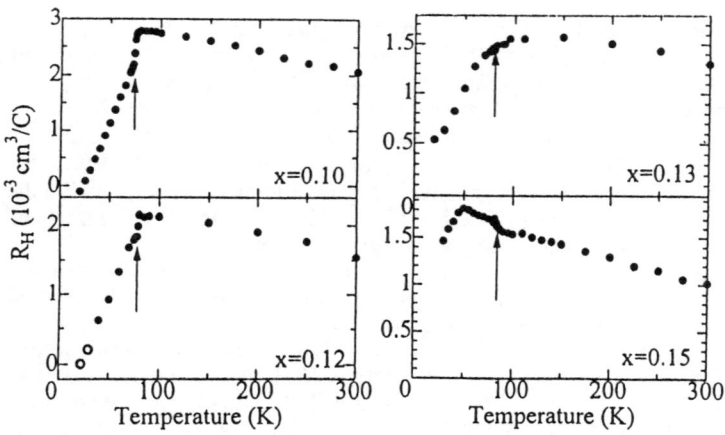

Fig.4 Temperature dependence of the Hall coefficient for $La_{1.6-x}Nd_{0.4}Sr_xCuO_4$ with x in the range between 0.10 and 0.15.

the LTT phase $d\rho_{ab}/dT$ is negative except for $x=0.10$. However, this does not imply that the LTT or the static stripe phase is an insulator. The resistivity has been measured for $x=0.12$ under high magnetic fields which kill the spurious superconductivity below 10K [16] perhaps occuring in a minor franction of the sample volume due to spatial fluctuation of the Sr content. It turns out that ρ_{ab} logarithmically increases with lowering temperature until it saturates below 2K. The saturated value of ρ_{ab} is well below the Mott-Ioffe-Regel limit (~ 1.6 mΩ· cm) and near or slightly below the universal resistance in two dimensions, $h/4e^2$ (~ 400 $\mu\Omega$· cm), near which a superconductor-insulator transition is known to take place in Zn doped cuprates in the underdoped regime [17]. This is also the case with other compositions, indicating a poorly metallic conduction along the planes in the statically ordered state.

The μSR experiments, on the other hand, suggest that, in the ordered phase, more than 95% of sample volume is antiferromagnetically ordered and the size of Cu magnetic moment is 0.2-0.3 μ_B comparable with that in the undoped AF insulators [18]. The metallic conductivity coexisting with the AF spin order is apparently prerequisite to spatial separation of the conduction paths from the magnetic domains and thus consistent with the stripe model. Note that metallic stripes are in contrast to the charge stripes in the nickelates and manganites in which charges are strongly localized on the stripes.

The Hall coefficient (R_H) show a more dramatic behavior in the ordered phase (Fig. 4). R_H also exhibits a jump at the LTO-LTT phase transition (T_d). Below T_d a rapid decrease in R_H is observed for $x=0.10$ and 0.12 for which the static charge modulation is seen in the neutron diffraction. R_H continues to decrease till it becomes vanishingly small below ~ 20K. As the resistivity remains finite and low at low temperatures, the vanishingly small values of R_H implies that the off-diagonal component of conductivity, σ_{xy}, tends to be zero. A possible explanation for this observation is that the charge carriers can move only in one direction in the CuO_2 plane and do not respond to external magnetic fields. This is just what is expected from the stripe model in Fig. 1 if the charges are confined within the charge stripe.

For x larger than 1/8 the behavior of R_H in the LTT phase is no more dramatic. The magnitude of R_H decreases at low temperatures, but it remains finite and fairly large. The charge transport in the CuO_2 plane retains two-dimensionality in the stripe ordered phase which apperars to be linked with the fact that the evidence for the charge modulation is difficult to detect in the neutron scattering.

2.3 Optical Conductivity Spectrum

The in-plane optical spectra have been measured for $La_{1.6-x}Nd_{0.4}Sr_xCuO_4$ with x=0.10, 0.12 and 0.15 [19]. For all the compositions no dramatic change has been observed upon cooling across the LTO-LTT phase transition temperature T_d. The optical conductivity spectra $\sigma_{ab}(\omega)$ at various temperatures are shown in Fig. 5 for x=0.12 which is the extreme case in that the static charge modulation is most cleary seen and superconductivity is almost completely suppressed. In contrast to the remarkable changes seen in the dc ($\omega = 0$) transport, the temperature dependence of the spectrum is very weak, and the specral features are typical of metals with conductivity slowly increasing with lowering temperature in the infrared (IR) region. The negative T coefficient of the dc (and microwave [20]) conductivity below T_d is indicative of enhanced disorder/impurity scattering in the static ordered state due possibly to reduced dimensionality.

Although the onset of the static order does not appreciably affect the optical conductivity spectrum, we find that the optical conductivity of LNSCO is largely different from that of LSCO even at room temperature for $x \leq 1/8$. As displayed in Fig. 6, the room-temperature conductivity of LNSCO is significantly reduced over a wide frequency range, up to 7000cm^{-1} for x=0.12 and to 3500cm^{-1} for x=0.10. By contrast, the conductivity does not change by the Nd addition in the case of x=0.15 except in the lowest frequency region below 50cm^{-1}. The spectral weight, the integated conductivity up to $\sim$ 9000cm^{-1} ($\sim$1eV) is reduced by 26% for x=0.12 and 7% for x=0.10, while the change is less than 1% for x=0.15. [21] The reduced spectral weight is shifted to the charge-transfer gap region around 2eV. The results of optical conductivity in-

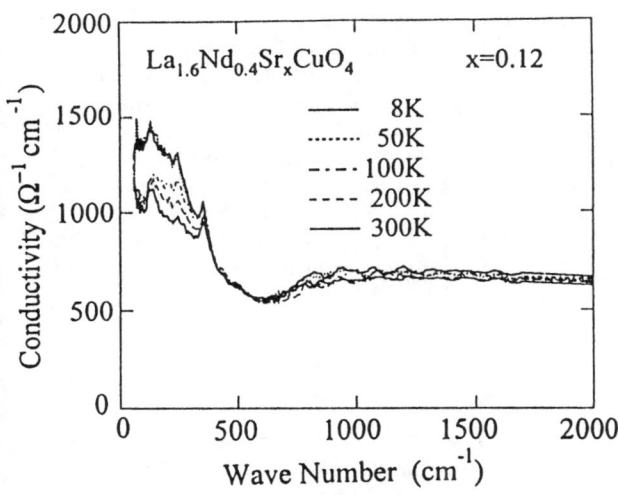

Fig.5 In-plane optical conductivity spectra of $La_{1.48}Nd_{0.4}Sr_{0.12}CuO_4$ at various temperatures.

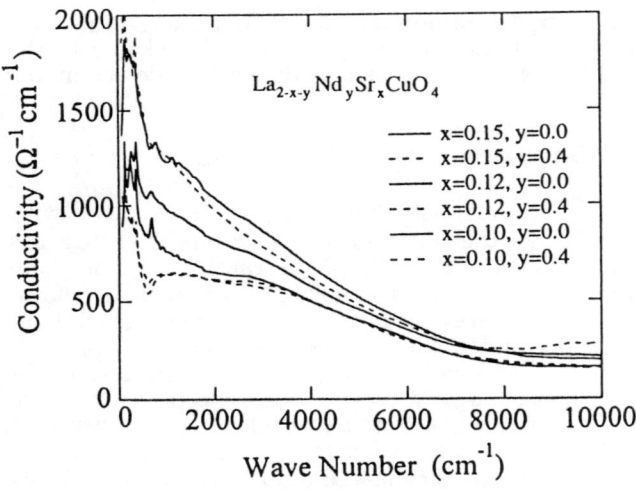

Fig.6 In-plane optical conductivity spectra of LSCO (dashed curves) and LNSCO (solid curves) at room temperature for x=0.10, 0.12 and 0.15.

dicate that there exist strong stripe fluctuations in LNSCO at temperatures well above T_d. Moreover, they demonstrate that such fluctuations influence the charge dynamics up to unexpectedly high energies.

2.4 Summary of the LaNdSrCuO System

We have shown that the properties, particularly the charge dynamics in the static stripe phase are distinctive between the Sr compositional region $x \leq 1/8$ and $x > 1/8$. The characteristic differences between the two regions are summarized below : (i) The static charge modulation is observable for $x \leq 1/8$, while it is hard to detect for $x > 1/8$. (ii) The incommensurability increases with x as $\epsilon \sim x$ up to $x=1/8$ and then saturates when x exceeds $1/8$. (iii) The Hall coefficient becomes vanishingly small for $x \leq 1/8$, suggestive of one-dimensional charge transport but the effect of static order on R_H is much weaker for $x > 1/8$, indicating that the charge dynamics restore two-dimensionality. (iv) The in-plane optical conductivity is strongly suppressed for $x \leq 1/8$ over a wide frequency range even at temperature well above T_d, while the Nd doping has only a little effect on the spectrum for $x > 1/8$. (v) Superconductivity is almost completely suppressed for $x \leq 1/8$, but the superconducting order appears to coexist with the static stripe order for $x > 1/8$ [18, 22].

3. Stripe Fluctuations

3.1 LaNdSrCuO System

The T dependence of $\sigma_{ab}(\omega)$ gives evidence for strong fluctuations even at temperatures well above T_d. As a consequence $\sigma_{ab}(\omega)$ shows only a small change on cooling down the sample across T_d. In the resistivity above T_d, the Nd doping induces a residual component as seen in Fig. 3 by a parallel shift of $\rho_{ab}(T)$ with respect to that for the Nd free crystals. This does not necessarily imply that Nd works as an impurity that elastically scatters carriers in the CuO_2 planes. The induced residual resistivity is smallest when $x = 0.12$ in which the stripes would be nearly static at $T > T_d$ and also in the crystal without Nd [13] due possibly to the commensurability pinning effect. So, the additional pinning effect due to Nd doping is expected to be relatively weak at this particular composition. In this respect, the residual resistivity likely arises from the stripe fluctuations slowed down by the Nd doping.

3.2 Fluctuations in LaSrCuO

We further speculate that the stripe fluctuations are present even in LSCO without Nd and affect the charge dynamics of this system. There are several circumstantial evidences for that :

(1) The in-plane resistivity has a residual component even without Zn or Nd [23], and does not show the characteristic T-linear dependence observed in many other high-T_c cuprates at high temperatures. ρ_{ab} of LSCO tends to saturate at elevated temperatures which is displayed in Fig. 7 where the measured isobaric (constant pressure) resistivity is transformed to the isochoric (constant volume) values [24]. Only for highly doped LSCO ($x = 0.20$) the T-linear resistivity is observed up to 1000K or higher. This result suggests that additional or different scattering mechanisms, both elastic and inelastic, are operating in the CuO_2 planes of LSCO.

(2) In connection with this, the optical conductivity spectrum of LSCO is significantly different from the spectra of other cuprates. Figure 8 shows $\sigma_{ab}(\omega)$ of LSCO with $x = 0.15$ together with the spectrum of YBCO having hole density near 0.15 (in the underdoped regime). In addition to the broader Drude peak in LSCO due to larger elastic scattering rate, the conductivity is higher in the mid-IR region, indicative of stronger inelastic scattering in LSCO.

(3) Weaker pseudogap in the c-axis optical conductivity spectrum of LSCO and corresponding slower increase in the c-axis resistivity with lowering temperature [25] as compared with those for the underdoped YBCO [26] (Fig. 9).

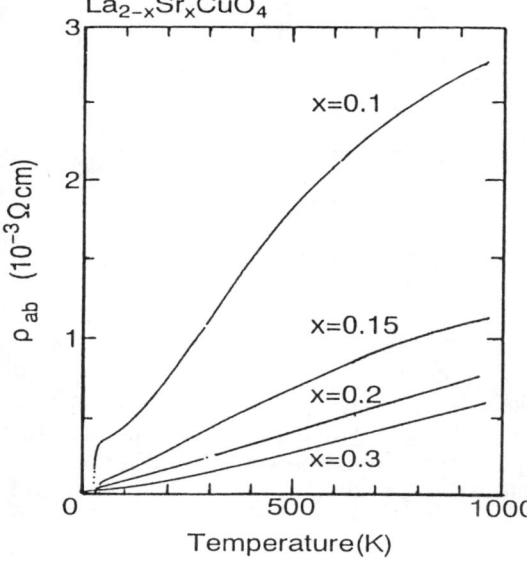

Fig.7 Temperature dependence of the in-plane resistivity for $La_{2-x}Sr_xCuO_4$. The resistivity at constant volume is plotted by transforming the measured resistivity at constant pressure using the available data of the thermal expansion coefficient and the pressure dependence of resistivity.

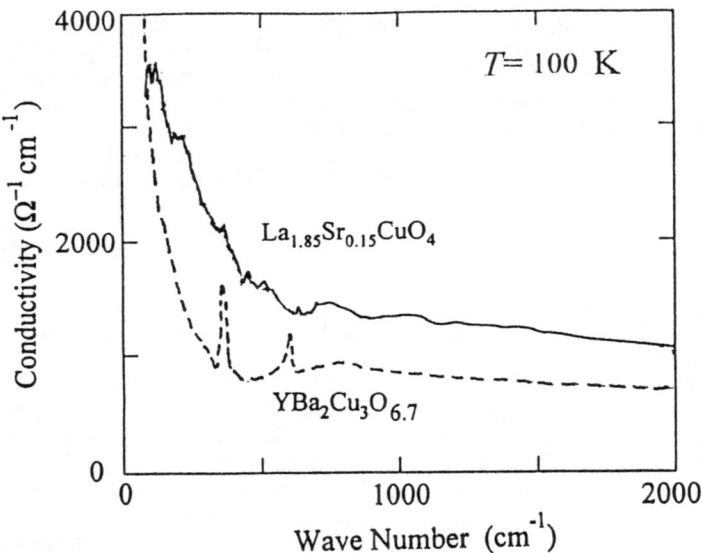

Fig.8 Optical conductivity spectrum of $La_{1.85}Sr_{0.15}CuO_4$ at $T=100K$ compared with the spectrum of an underdoped $YBa_2Cu_3O_{6.7}$ at 100K with nearly the same hole density.

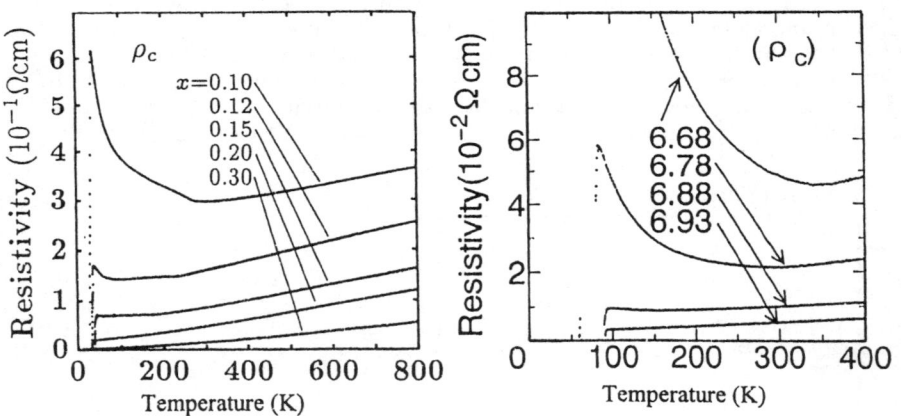

Fig.9 Temperature dependence of the c-axis resistivity for LSCO and YBCO with various hole densities.

(4) Pseudogap as well as spin gap in LSCO is hard to see in ARPES [27] and NMR [28]. These three gaps have close connection with each other. The difficulty to observed a clear spin gap in LSCO is understandable as a consequence of stripe fluctuations, since most of the sample volume would be virtually AF ordered.

(5) PDF analysis of neutron diffraction of LSCO suggests the presence of local tilt disorder of the CuO_6 octahedra in the form of a mixture of LTO and LTT tilt directions [29].

(6) The relatively low T_c values in LSCO might be due to slower stripe fluctuations as we have seen that the T_c suppression with Nd doping is apparently connected with the slowing down of the fluctuations.

(7) The stripe order or stripe fluctuations may be viewed as a manifestation of intrinsic phase separation into the AF insulator with lower hole density and the metallic phase with higher hole density. In this regard, the very small shift of the chemical potential with doping into LSCO [30] is a natural consequence of phase separation.

At present, it is not clear what role is played by the dynamical spin modulation, observed as incommensurate neutron peaks, in YBCO and BSCCO and whether such modulation is intimately tied to high-T_c superconductivity [31, 32]. A tentative conclusion on superconductivity is that the CuO_2 planes can support superconductivity when only spin modulation, either static or dynamics, is observed but the charge modulation is too weak to be detected. Obviously, efforts to search for evidence of dynamical charge modulation in the superconducting samples are required to give a firm support for the stripe model.

We thank J. M. Tranquada, J. D. Axe, J. E. Ostenson, D. K. Finnemore, B. Nachumi, Y. J. Uemura, H. Kitano, and A. Maeda for collaborations. The present work was supported by NEDO for the International Joint Research Program and the Research and Development of Industrial Science and Technology Frontier Program and also by a COE Grant from the Ministry of Education, Science and Culture, Japan.

References

[1] S -W. Cheong et al., Phys. Rev. Lett. **67**, 1791 (1991).
[2] T. R. Thurston et al., Phys. Rev. B**46**, 9128 (1992).
[3] T. Tanamoto, H. Kohno, and H. Fukuyama, J. Phys. Soc. Jpn. **61**, 1886 (1992).
[4] P. Dai, H. A. Mook, and F. Dogan, Phys. Rev. Lett. **80**, 7738 (1998).
[5] H. A. Mook et al., unpublished.
[6] J. M. Tranquada, D. J. Butlrey, V. Sachan, and J. E. Lorenzo, Phys. Rev. Lett. **73**, 1003 (1994).
[7] S. Mori, C. H. Chen, and S. -W. Cheong, Nature **392**, 473 (1998).
[8] J. M. Tranquada, B. J. Sternlieb, J. D. Axe, Y. Nakamura, and S. Uchida, Nature **375**, 561 (1995).
[9] M. V. Zimmerman et al., Europhys. Lett. **41**, 629 (1998).
[10] Z. -X. Shen et al., Science **280**, 259 (1998).
[11] J. M. Tranquada, J. D. Axe, N. Ichikawa, A. R. Moodenbaugh, Y. Nakamura, and S. Uchida, Phys. Rev. Lett. **78**, 338 (1997).
[12] K. Yamada et al., Phys. Rev. B**57**, 6165 (1998).
[13] T. Suzuki et al., Phys. Rev. B**57**, R3229 (1998).
[14] Y. S. Lee et al., unpublished.
[15] Y. Nakamura and S. Uchida, Phys. Rev. B**46**, 5841 (1992).
[16] Y. Ando, G. Boebinger, N. Ichikawa, and S. Uchida, unpublished.
[17] Y. Fukuzumi, K. Mizuhashi, K. Takenaka, and S. Uchida, Phys. Rev. Lett. **76**, 684 (1996).
[18] B. Nachumi et al., unpublished.

[19] S. Tajima, N. L. Wang, M. Takaba, N. Ichikawa, and S. Uchida, to appear in J. Phys. Chem. Solids.
[20] H. Kitano, A. Maeda, N. Ichikawa, and S. Uchida, unpublished.
[21] S. Tajima *et al.*, unpublished.
[22] J. E. Ostenson, S. Bud'ko, M. Breitwisch, D. K. Finnemore, N. Ichikawa, and S. Uchida, Phys. Rev. B**56**, 2826 (1997).
[23] T. R. Chien, Z. Z. Wang, and N. P. Ong, Phys. Rev. Lett. **67**, 2088 (1991).
[24] Y. Ito, S. Uchida, C. Murayama, and N. Mori, unpublished.
[25] Y. Nakamura and S. Uchida, Phys. Rev. B**47**, 8369 (1993).
[26] K. Takenaka, K. Mizuhashi, H. Takagi, and S. Uchida, Phys. Rev. B**50**, 6534 (1994).
[27] A. Ino *et al.*, Phys. Rev. Lett. **79**, 2101 (1997).
[28] H. Yasuoka, Hyperfine Interactions **20**, 1 (1996).
[29] E. S. Bozin, S. J. L. Billinge, G. H. Kwei, and H. Takagi, cond-mat / 9807151.
[30] A. Ino *et al.*, to appear in Phys. Rev. Lett.
[31] J. Zaanen and O. Gunnarson, Phys. Rev. B**40**, 7391 (1989).
[32] V. J. Emery and S. A. Kivelson, Physica C**263**, 44 (1996).

Spin Excitations in Pure and Zn-Substituted $YBa_2Cu_3O_{6+x}$

B. Keimer

Department of Physics, Princeton University, Princeton, NJ 08544, USA

Abstract. The effects of varying the hole concentration and the concentration of nonmagnetic (Zn) impurities on the spin excitation spectra of $YBa_2Cu_3O_{6+x}$ are compared and contrasted. With decreasing hole concentration, the magnetic resonance peak observed below T_c for $x = 1$ is shifted to lower energies, but remains sharp and continues to set in at T_c. By contrast, a minute concentration (0.5%) of nonmagnetic impurities broadens the spectrum dramatically. The broadened spectral distribution persists up to temperatures significantly higher than T_c.

1. Introduction

The relationship between antiferromagnetism and superconductivity in the cuprates remains at the center of lively debate. The parent compounds of the cuprate high temperature superconductors are antiferromagnetic insulators, and in the metallic state magnetic correlations peaked at or near the antiferromagnetic wave vector $q_0 = (\pi, \pi)$ become more and more prominent as the hole concentration is reduced. The spin excitation spectra in fact seem to evolve smoothly from the underdoped metallic to the insulating antiferromagnetic state. On the other hand, near the hole concentration corresponding to the optimal superconducting transition temperature, the spin susceptibility is much reduced. In particular, while spin excitations in underdoped $YBa_2Cu_3O_{6+x}$ are readily observable by neutron scattering, near $x = 1$ ($T_c = 93K$) only an upper bound on the dynamical susceptibility in the normal state could be established [1].

This observation has cast some degree of doubt on magnetic pairing pairing models that rely on spin fluctuations to drive superconductivity. However, even in near-optimally doped $YBa_2Cu_3O_{6+x}$ a prominent, sharp spin excitation appears in the superconducting state [1]-[5]. This "magnetic resonance peak" is characterized by the wave vector $q_0 = (\pi, \pi)$, energy 40 meV, and spectral weight comparable to that of antiferromagnetic spin waves in the same energy range. This mode has hence been linked directly to the antiferromagnetic state. One theory views it as a Goldstone boson of an SO(5) symmetry group that encompasses both antiferromagnetism and superconductivity [6]. In the framework of this model, a quantitative prediction of the resonance spectral weight has been made that is in good accord with the experimental observation. The resonance energy is directly related to the hole concentration and independent of the superconducting gap. Several other models also associate the resonance peak directly with the spin excitations in the antiferromagnetic insulator [7, 8]. Another category of theories takes the band susceptibility calculated for $YBa_2Cu_3O_7$ as a starting point. Since the dynamical susceptibility at the resonance energy is one to two orders of magnitude larger than the Lindhard susceptibility of a noninteracting band metal, large renormalization factors have to be introduced. Interlayer tunneling [9], electron-electron interactions [10], band structure singularities [11] or combinations thereof are among the possible physical origins of the enhanced dynamical susceptibility. All of these models associate the resonance energy with the superconducting gap.

Here we review the evolution of the spin excitations of $YBa_2Cu_3O_{6+x}$, especially the magnetic resonance peak, with both hole concentration and disorder. It is important to understand the effect of disorder because in a real material the hole concentration cannot be changed without introducing disorder at the same time. At least for some hole concentrations in $YBa_2Cu_3O_{6+x}$, charge transfer from the "chain layer" to the CuO_2 sheets can be accomplished without breaking the integrity of the chains, and hence with minimal disorder. In other high-T_c compounds such as $La_{2-x}Sr_xCuO_4$, disorder effects are believed to be more important. Here we monitor the effect of a controlled number of nonmagnetic (Zn) impurities in the CuO_2 layers on the spin excitations in $YBa_2Cu_3O_7$. Zn does not change the hole concentration and is particularly interesting because it has been known for some time that it induces a local staggered on the neighboring copper sites [12]. One may thus expect that Zn impurities, not unlike reduction of the hole concentration, nudge the system closer to the antiferromagnetic state, albeit in a spatially highly nonuniform manner. The local magnetic clusters induced by the Zn impurities scatter conduction electrons near the unitary limit and lead to a rapid suppression of the superconducting transition temperature T_c [13].

2. Hole Concentration

Our samples are large ($\sim 2-10 cm^3$) $YBa_2Cu_3O_{6+x}$ single crystals which contain voids and microcracks through which oxygen can migrate, so that they can be uniformly oxygenated. (The same microstructural features render the crystals unsuitable for transport measurements.) The oxygen content was adjusted by post-annealing in sealed quartz tubes for about two weeks. Four oxygenation states (x=0.2, 0.5, 0.7, and 1) were investigated. The neutron scattering measurements were carried out on triple axis spectrometers at the High Flux Beam Reactor at the Brookhaven National Laboratory, the Orphée reactor at the Laboratoire Léon Brillouin, Saclay, France, and at the Institut Laue Langevin, Grenoble, France. As described in detail elsewhere [1, 5], the imaginary part of the dynamical magnetic susceptibility, $\chi''(\mathbf{q}, \omega)$, can be separated from phonon scattering with the aid of lattice vibrational calculations, and by studying the momentum, temperature and doping dependence of the neutron scattering cross section. In pure (Zn-free) $YBa_2Cu_3O_{6+x}$, this procedure was verified by measurements with polarized neutron beams [1, 18]. Absolute units for $\chi''(\mathbf{q}, \omega)$ were extracted by calibrating to the phonon spectrum [1].

Because of the bilayer structure, the spin wave spectrum of the antiferromagnetic insulator (x=0.2) consists of two branches: an acoustic branch in which local spins in directly adjacent CuO_2 layers precess in phase, and an optical branch in which they precess out of phase. Since the layers are antiferromagnetically coupled, the acoustic branch is odd, the optical branch even under permutation of the layers. While the acoustic branch has only a very small gap at $\mathbf{q_0} = (\pi, \pi)$, the optical branch has a gap of ~ 70 meV [14, 15]. It can be shown that for each branch the **q**-integrated susceptibility, $\chi''_{2D}(\omega) = \int d^2q \, \chi''(\mathbf{q}, \omega)$, is energy independent at low energies (*i.e.*, far from the zone boundary).

The doping induced changes in the spin excitation spectra can be summarized as follows. In the underdoped regime, the general shape of the spin excitation spectrum is quite similar to the antiferromagnetic insulator. The spectrum is split into a gapless odd and a gapful even branch and shows a remnant of the spin wave dispersion away from $\mathbf{q_0} = (\pi, \pi)$ at high energies [16]. However, the **q**-integrated susceptibilities in both channels are no longer energy-independent below about 200K, but develop broad peaks with decreasing temperature that go along with the opening of a spin pseudo-gap (Fig. 1). On top of this gradual buildup of spectral weight, an additional enhancement is observed in the odd channel (but not in the even channel) below the superconducting transition temperature (Figs. 1 and 2). As this superconductivity-induced anomaly is very sharp (resolution limited) in energy, it has been termed "magnetic resonance peak". (Presumably due to disorder, the resonance peak is slightly broadened for x=0.5, see below.) As the hole concentration increases, the resonance energy increases, from 25 meV for x=0.5 to 33 meV for x=0.7 and 40 meV for x=1 [18]. (Similar data were reported by Dai *et al.* [17] and Bourges *et al.* [19].) At the same time, the normal

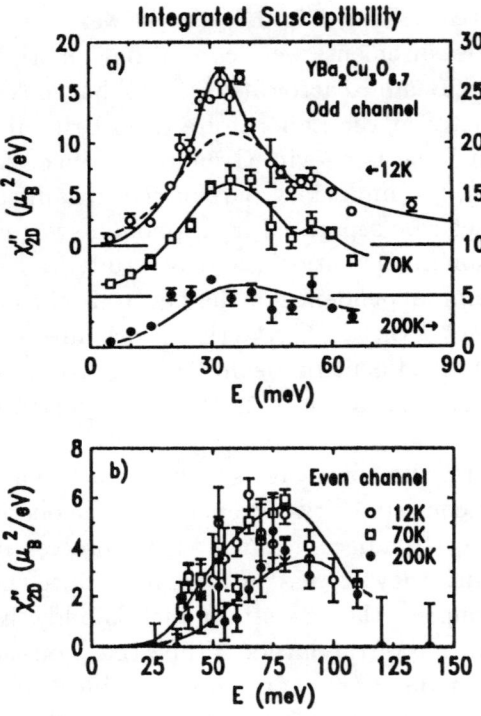

Figure 1: **q**-integrated susceptibility of YBa$_2$Cu$_3$O$_{6.7}$ in odd and even channels. The dashed line in the upper panel is the susceptibility immediately above T$_c$ = 67K. The solids line are guides to the eye.

state susceptibility is rapidly reduced, and the superconductivity-induced anomaly involves a progressively larger fraction of the normal state spectral weight (Fig. 2). For x=1, the normal state susceptibility is too weak to be separated from the experimental background, and the resonance peak in the superconducting state is the only clearly identifiable magnetic excitation [1, 5].

We can turn this around and start with the optimally doped material. The two main effects of reducing the hole concentration are a reduction of the resonance energy and an enhancement of the normal-state susceptibility. The resonance remains sharp in energy (at least down to x=0.7) and its temperature dependence is clearly correlated with the superconducting transition. Although one may regard the broad peak in the **q**-integrated susceptibility above T$_c$ (Fig. 1) as a precursor to the resonance peak, superconductivity thus has a clearly discernible effect on the spin excitation spectrum at all hole concentrations.

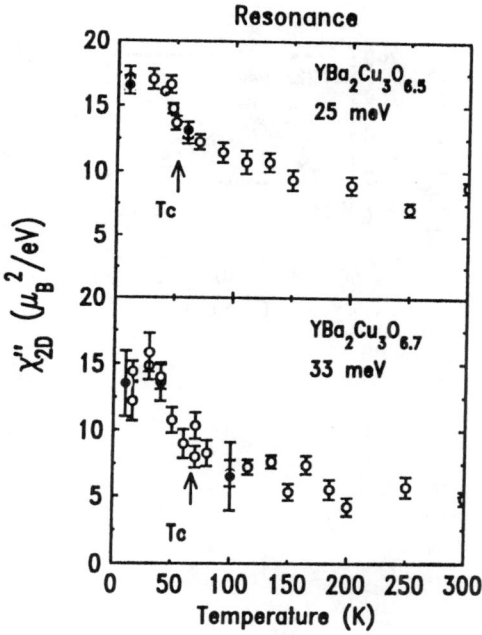

Figure 2: Temperature dependence of the **q**-integrated susceptibility at the resonance energy of $YBa_2Cu_3O_{6.5}$ ($T_c = 52K$) and $YBa_2Cu_3O_{6.7}$ ($T_c = 67K$). The closed circle were taken with polarized, the open circles with unpolarized neutrons.

3. Nonmagnetic Impurities

As mentioned above, the introduction of Zn impurities also reduces T_c and leads to enhanced antiferromagnetic correlations, at least locally. In order to elucidate the effect of nonmagnetic impurities on the spin excitations, a single crystal of composition $YBa_2(Cu_{0.995}Zn_{0.005})_3O_7$ was synthesized and subjected to the same annealing procedure as the pure compounds of Section 2. After the heat treatment, the crystal showed $T_c = 87K$, consistent with earlier reports on Zn-substituted, fully oxygenated $YBa_2Cu_3O_7$ [13].

As shown in Fig. 3, the magnetic response of the Zn-substituted crystal is confined to an energy window around 40 meV, similar to the pure (Zn-free) compound [20]. However, while the magnetic resonance peak in pure $YBa_2Cu_3O_7$ is resolution limited and only an upper bound of about 3-4 meV has been established [1], the intrinsic width of the spectral distribution in $YBa_2(Cu_{0.995}Zn_{0.005})_3O_7$ is ~ 8 meV, significantly exceeding the instrumental resolution. Fig. 3 also indicates a more dramatic differ-

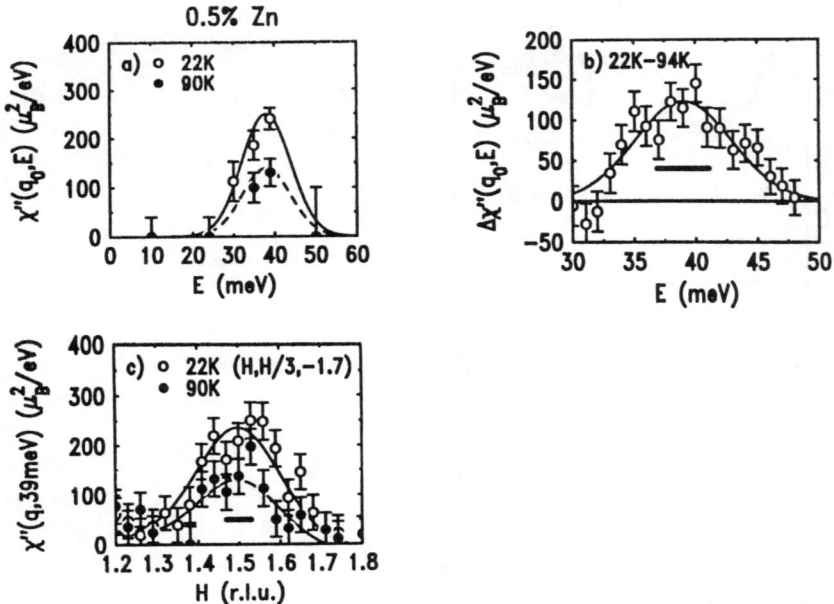

Figure 3: Absolute dynamical susceptibility of YBa$_2$(Cu$_{0.995}$Zn$_{0.005}$)$_3$O$_7$ as a function of (a), (b) energy and (c) wave vector. Panel (b) shows the difference of the susceptibility at low temperature and immediately above T$_c$. The bars indicate the instrumental resolution.

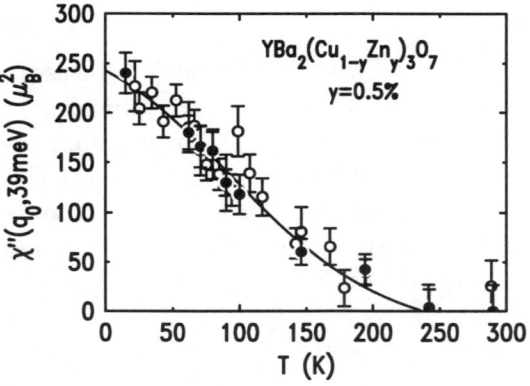

Figure 4: Peak susceptibility of YBa$_2$(Cu$_{0.995}$Zn$_{0.005}$)$_3$O$_7$ at 39 meV and $q_0 = (\pi, \pi)$ as a function of temperature. The solid circles were determined from q-scans, the open circles are peak count rates.

ence: Whereas the resonance peak in the pure system vanishes above T_c, the spectral weight in the Zn-substituted system is reduced by only about half at T_c. The complete temperature dependence of the peak intensity is shown in Fig. 4. In the pure systems (both underdoped and optimally doped), the resonance intensity follows a sharp, order parameter-like curve below T_c. By contrast, in the Zn-substituted compound the intensity shows no anomaly at T_c and goes smoothly to zero at a much higher temperature, $T \sim 200K$.

Finally, the absolute energy-integrated spectral weights of the pure and Zn-substituted compounds were normalized to each other. This quantity is *larger* in the Zn-substituted system. Heuristically, additional spectral weight around $\mathbf{q_0} = (\pi, \pi)$ might be expected if a local staggered magnetization is induced around each Zn impurity. Indeed, previous neutron scattering work on Zn-substituted cuprate superconductors has revealed additional low energy ($\lesssim 10$ meV) magnetic fluctuations centered around (π, π) [21, 22, 23]. At these low energies, no magnetic scattering is observed above background in our sample, presumably because the Zn concentration is significantly lower than in all previous studies. Rather, the additional spectral weight we observe is peaked around approximately the *same* wave vector and energy as the resonance peak in the pure system. Preliminary data on a 1% Zn-substituted sample are very similar to those taken on the $YBa_2(Cu_{0.995}Zn_{0.005})_3O_7$ material and also connect smoothly to the $YBa_2(Cu_{0.98}Zn_{0.02})_3O_7$ material investigated previously, where a broad peak centered around $\mathbf{q_0} = (\pi, \pi)$ and 35 meV was observed [21].

As the spectral distribution in the $YBa_2(Cu_{0.995}Zn_{0.005})_3O_7$ sample shares the same signature in wave vector and energy with the resonance peak in $YBa_2Cu_3O_7$, the two must be intimately related. Since the resonance peak is clearly correlated with T_c in the pure systems (both underdoped and optimally doped), the persistence of the spectral weight to higher temperatures in the Zn-substituted system may indicate some remnant of the superconductiviting state above the actual superconducting transition. A similar phenomenon appears to occur in the "spin gap" regime of the underdoped compounds.

4. Conclusions

The qualitative difference in the magnetic excitation spectra and their temperature dependences (Fig. 4) of the pure $YBa_2Cu_3O_7$ material and a material in which one in 200 copper atoms is replaced by nonmagnetic zinc is astounding. The sharpness of the resonance peak and its correlation with superconductivity hints at a delicate quantum coherence that is extremely susceptible to disorder. This coherence is not strongly affected by changing the oxygen content (and hence the hole concentration) in $YBa_2Cu_3O_{6+x}$, but it may well be disrupted by Sr donor ions and/or static "stripe" correlations in $La_{2-x}Sr_xCuO_4$ where the resonance peak is not observed. None of the theories proposed for the resonance peak [6]-[11] has anticipated its extreme sensitivity to nonmagnetic impurities. However, Fukuyama and coworkers [24] have suggested a microscopic analogy between the effect of nonmagnetic impurities on the metallic state in the quasi-two dimensional cuprate superconductors (which they describe as a resonating valence bond state) and on collective singlet ground states in quasi-one dimensional systems. A prominent member of the latter category is the spin-Peierls compound $CuGeO_3$ whose ground state is indeed qualitatively altered by a minute amount of nonmagnetic impurities [25]. It remains to be seen whether quantitative predictions for the neutron scattering cross section can be derived from this model. In any case, all models of the magnetic resonance peak should be reassessed in the light of these new observations.

Acknowledgments. This work was done in collaboration with I.A. Aksay, J. Bossy, P. Bourges, H.F. Fong, A. Ivanov, D.L. Milius, and L.P Regnault. It was supported by the National Science Foundation under grant No. DMR-9400362, and by the Packard and Sloan Foundations.

References

[1] H.F. Fong et al., Phys. Rev. B. **54**, 6708 (1996).

[2] J. Rossat-Mignod et al., Physica C **185-189**, 86 (1991).

[3] H.A. Mook et al., Phys. Rev. Lett. **70**, 3490 (1993).

[4] H.F. Fong et al., Phys. Rev. Lett. **75**, 316 (1995).

[5] P. Bourges, L.P. Regnault, Y. Sidis and C. Vettier, Phys. Rev. B **53**, 876 (1996).

[6] E. Demler and S.C. Zhang, Phys. Rev. Lett. **75**, 4126 (1995); S.C. Zhang, Science **275**, 1089 (1997).

[7] F.F. Assaad and M. Imada, Report No. cond-mat/9711172.

[8] Y. Zha, V. Barzykin and D. Pines, Phys. Rev. B **54**, 7561 (1996); D.K. Morr and D. Pines, Report No. cond-mat/9805107.

[9] L. Yin, S. Chakravarty and P.W. Anderson, Phys. Rev. Lett. **78**, 3559 (1997).

[10] D.Z. Liu, Y. Zha and K. Levin, Phys. Rev. Lett. **75**, 4130 (1995); I.I. Mazin and V.M. Yakovenko, Phys. Rev. Lett. **75**, 4134 (1995); F. Onufrieva, Physica C **251**, 348 (1995); A.J. Millis and H. Monien, Phys. Rev. B **54**, 16172 (1996).

[11] N. Bulut and D.J. Scalapino, Phys. Rev. B **53**, 5149 (1996); G. Blumberg, B.P. Stojkovic and M.V. Klein, *ibid.* **52**, 15741 (1995); A.A. Abrikosov, *ibid.* **57**, 8656 (1998).

[12] H. Alloul *et al.*, Phys. Rev. Lett. **67**, 3140 (1991).

[13] T.R. Chien, Z.Z. Wang and N.P. Ong, Phys. Rev. Lett. **67**, 2088 (1991); D.A. Bonn *et al.*, Phys. Rev. B **50**, 4051 (1994); Fukuzumi *et al.*, Phys. Rev. Lett.**76**, 684 (1996).

[14] D. Reznik *et al.*, **53**, R14741 (1996).

[15] S.M. Hayden *et al.*, Phys. Rev. B **54**, R6905 (1996).

[16] P. Bourges *et al.*, *ibid.* **56**, R11439 (1997).

[17] P. Dai *et al.*, Phys. Rev. Lett **77**, 5425 (1996).

[18] H.F. Fong, B. Keimer, D.L. Milius and I.A. Aksay, Phys. Rev. Lett. **78**, 713 (1997).

[19] P. Bourges *et al.*, Europhys. Lett. **38**, 313 (1997).

[20] H.F. Fong, P. Bourges, Y. Sidis, L.P. Regnault, J. Bossy, A. Ivanov, D.L. Milius, I.A. Aksay, and B. Keimer, to be published.

[21] Y. Sidis *et al.*, Phys. Rev. B **53**, 6811 (1996).

[22] K. Kakurai *et al.*, Phys. Rev. B **48**, 3485 (1993).

[23] M. Matsuda *et al.*, J. Phys. Soc. Jpn. **62**, 443 (1993).

[24] N. Nagaosa *et al.*, J. Phys. Soc. Jpn. **65**, 3724 (1996); H. Fukuyama, T. Tanimoto, and M. Saito, *ibid.* **65**, 1182 (1996).

[25] L.P. Regnault, J.P. Renard, G. Dhalenne, A. Revcolevschi, Europhys. Lett. **32**, 579 (1995); V. Kiryukhin *et al.* Phys. Rev. B **54**, 7269 (1996).

Static and Dynamical Incommensurate Spin Correlations in High-T_C Cuprates of $La_{2-x}Sr_xCuO_{4+y}$

K. Yamada[1,3], R. J. Birgeneau[2], Y. Endoh[3], T. Fukase[4], M. Greven[2], M. Fujita[1], K. Hirota[3], S. Hosoya[5], M. A. Kastner[2], Y.M. Kim[2], H. Kimura[3], K. Kurahashi[3], C. H. Lee[3], S. H. Lee[6], Y. S. Lee[2], H. Matsushita[3], G. Shirane[7], T. Suzuki[4], S. Ueki[3] and S. Wakimoto[3]

[1] Institute for Chemical Research, Kyoto University, Uji 611-0011, Japan
[2] Department of Physics, Massachusetts Institute of Technology, Cambridge, Massachusetts 02139, USA
[3] Department of Physics Tohoku University, Aramaki Aoba, Sendai 980-77, Japan and CREST, Japan Science and Technology Corporation
[4] Institute for Material Research, Tohoku University, Sendai 980, Japan
[5] Institute of Inorganic Synthesis, Yamanashi University, Kofu 400, Japan
[6] Center for Neutron Research, NIST, Gaithersberg, Washington, USA
[7] Department of Physics, Brookhaven National Laboratory, Upton, NY 11973, USA

Abstract. Static and dynamical spin correlations in high-T_C superconductor studied by neutron scattering are reviewed based on our recent work on hole-doped La_2CuO_4. Spin correlation in the superconducting phase exhibits a long period spatial modulation. The doping, temperature and energy dependences of the spin fluctuations revealed a concordant relation between the incommensurate dynamical spin correlation and superconductivity. No-well defined spin correlation exists beyond the upper critical doping of the superconducting phase. Below T_C, a well-defined energy-gap of ~7meV opens in the dynamical spin susceptibility for the optimally doped or slightly overdoped phases. In contrast, no well-defined energy-gap opens for the underdoped samples. In addition to the dynamical spin fluctuations, a long range incommensurate static magnetic order is observed around 1/8-doping in the superconducting states. For x=0.12, the onset temperature of the magnetic order is highest and corresponds to T_C.

1 Introduction

Doping of carriers into the Mott-type antiferromagnet of nondoped high-T_C cuprates induces remarkable changes in the magnetic as well as transport properties. The doping tranforms the insulating antiferromagnetic phase into the normal or conventional metallic one with increasing the doping rate; the high-T_C superconducting phase appears in between the two. Although the mechanism of pairing interaction of high-T_C superconductivity is not fully understood at present, the interplay between the magnetic fluctuations and high-T_C superconductivity is one of the central issues in the basic physics of the lamellar CuO_2 materials.

Neutron scattering is one of the unique as well as most direct experimental technique for studying this interplay. Compared with other local probes for magnetism such as muon spin relaxation (μSR) or nuclear magnetic resonance (NMR), neutron scattering is susceptible to both the spatial and time coherence of spin correlation. These three techniques provide information of so-called time-averaged magnetic correlation in a sample within a time-span Δt. The difference of these technique is the difference in Δt. A conventional thermal (cold) neutron scattering on 3-axis spectrometer measures the spin correlation with an energy resolution $\Delta E \sim 1(0.1)$ meV corresponding to $\Delta t \sim 10^{-12}(10^{-11})$ sec. which is much shorter than $10^{-6 \sim 8}$ sec. for μSR and $\sim 10^{-9}$ sec. for NMR.

The first stage of neutron scattering work on high-T_C cuprates started in 1987. By 1990, both neutron scattering and NMR studies already revealed the existence of the dynamical antiferromagnetic spin correlation and the energy-gap in the superconducting states [1]. Particularly, the momentum-sensitive former technique discovered a long period spatial spin modulation in the superconducting La2-1-4 cuprate [2]. In the second stage of neutron studies we have proved that these incommensurate spin fluctuations are unique in the superconducting phases [3]. In addition, the doping, temperature and energy dependences of the energy-gap in the spin fluctuations were systematically studied. These results naturally evoked following questions; what is the origin of the long period spatial modulation?; what is the role of the spin fluctuations for the superconductivity? Very recent observation of a static incommensurate magnetic long range order in the superconducting $La_{1.88}Sr_{0.12}CuO_4$ by Suzuki et al. [4] and Kimura et al. [5] triggered the third stage of neutron scattering study on $La_{2-x}Sr_xCuO_{4+y}$. At present, comprehensive neutron scattering measurements are revealing new details on the magnetic order. Then, another questions appear; does the incommensurate static magnetic order microscopically coexist with the high-T_C superconductivity and if it does, are they concordant or competing each other?

The purpose of this paper is to review the second stage of neutron scattering, to introduce a part of current data in the third stage of neutron scattering and to summarize the resolved and unresolved problems at this stage of 1998.

We note here, such development of high-T_C works in the second and third stage of neutron scattering particularly owes to the pioneering work in both the single crystal growth and the neutron scattering studies in the first stage performed dominantly in Brookhaven National Laboratory. For the former, the application of the traveling solvent floating zone (TSFZ) method is indispensable; by using this method Kojima and coworkers in Yamanashi University [6] have firstly succeeded in growing sizable superconducting single crystals of $La_{2-x}Sr_xCuO_4$. For the latter, the systematic studies have been shedding lights on the remaining problems on the spin fluctuations. We also note, besides Sr-doping, superconducting samples can be prepared by oxygen-doping using a post-growth electrochemical treatment of the insulating phase. This technique was successfully applied to single crystal of La_2CuO_4 by Wells et al. in MIT for the first time [7]. By using both types of single crystals, we have conducted the second and third stages of neutron scattering studies over a wide doping region mainly between the lower (x_{LC}) and upper (x_{UC}) critical doping of the superconducting phase.

The neutron experiments have been dominantly performed by using thermal and cold neutron 3-axis spectrometers of both reactors in JAERI and NIST due to the temporal shutdown of Brookhaven reactor HFBR. All the single crystals were grown in Tohoku University by the TSFZ method using a lamp-image furnace.

2 Fundamental phase diagram of $La_{2-x}Sr_xCuO_4$

As shown in Fig.1, holes introduced into $La_{2-x}Sr_xCuO_4$ destroy the three dimensional long range Néel order with x_{NC}~0.02. This critical hole concentration is nearly the same for the oxygen-doped sample [8]. In between x_{NC}~0.02 and x_{LC}~0.05, a so-called spin-glass phase appears; there, the zero-field-cooled magnetic susceptibility shows a cusp at around the spin freezing temperature. Due to the macroscopic phase separation into the superconducting and antiferromagnetic phases, no spin-glass phase is observed for the electrochemically oxygen-doped La_2CuO_{4+y} but for $La_{2-x}Bi_xCuO_{4+y}$ annealed under oxygen high pressures [9]. Superconducting phase develops with Sr-doping beyond x_{LC}. At the same time, incommensurate spatial spin modulation appears with magnetic inelastic scattering peaks on the two dimensional reciprocal lines at $(1/2,1/2(1\pm\delta))$ and $(1/2(1\pm\delta),1/2)$. T_C reaches a maximum value of about 38K around the optimal doping x_{OP}~0.15. The same incommensurate spin fluctuations are observed for the oxygen-doped superconductor irrespective of type of dopant [10]. In between x_{LC} and x_{OP}, there exists an anomalous doping near x_S=1/8 where the superconductivity is locally suppressed down to ~25K compared to T_C~30K for the neighboring doping. Note that the suppression of T_C is weaker compared to $La_{1.6-x}Nd_{0.4}Sr_xCuO_4$ in LTT structure. Due to the difficulty in fine tuning of doping rate by the electrochemical treatment the existence of anomalous doping as in the case of Sr-doping is not reported. Overdoping beyond x_{OP} degrades the superconductivity and beyond x_{UC}~0.27, the superconductivity disappears. No one has succeeded in preparing highly overdoped sample by oxygen-doping.

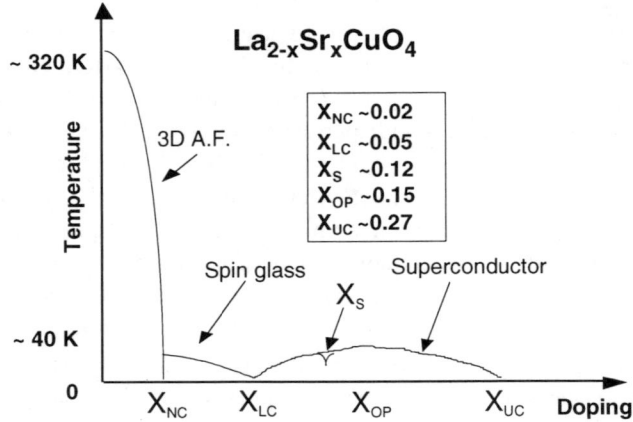

Fig. 1. Schematic phase diagram of $La_{2-x}Sr_xCuO_4$ and characteristic concentrations in the diagram.

Concerning to the spin correlation, the remaining major problems are as follows; [i] How does the commensurate spin correlation in the spin-glass phase transform into the incommensurate one in the superconducting one? [ii] What is the doping dependence of the energy-gap in the superconducting state? [iii] Is there any specific spin correlation around x_S?

3 Dynamical spin correlation

We determined the concentration range for the incommensurate spin correlation. The double-peaked spectra become noticeable with doping x=0.06, close to x_{LC}. The incommensurability δ increases with increasing the doping rate in the underdoped region. Beyond x~1/8 or x_{OP}, the δ starts to saturate with the maximum δ of around 1/8. The spatial spin modulation is robust against the pair-breaking such as Zn-substitution of Cu-site or oxygen-reduction [3]. Except the doping region close to x_{LC}, the δ is simply determined by the effective doping rate x_{eff} on CuO_2 plane calculated by x_{eff} =x+2y for $La_{2-x}Sr_xCuO_{4+y}$. Similar doping dependence of δ in the elastic peaks is also observed by Tranquada et al. [11] for the superconducting $La_{1.6-x}Nd_{0.4}Sr_xCuO_4$, there, the crystal structure is the so-called low temperature tetragonal (LTT) phase different from the low temperature orthorhombic (LTO) one for the Sr- or oxygen-doped La_2CuO_4. Furthermore, quite recently, Mook et al. [12] observed incommensurate spin fluctuations in the superconducting Y1-2-3 and Bi2-2-1-2 systems. For the former, they found the same direction of spatial spin modulation as well as the similar doping dependence of δ compared to La2-1-4. On the other hand, for the Sr- or oxygen-doped La_2NiO_4 a spin density modulation or a striped order of doped holes is formed along the diagonal direction in the square lattice of NiO_2 plane which is different from the direction for cuprates by 45°. In addition, the incommensurability in $La_{2-x}Sr_xNiO_4$ does not show any saturation with doping at least up to around x=0.5 [13]. Therefore, the incommensurate spin correlation in high-T_C cuprates is considered to be a unique nature of the hole-doped superconducting phases.

We systematically studied the peak-width of incommensurate peaks [3, 18]. The spatial coherency of the spin correlation at low energies as well as temperatures is degraded beyond x_{OP} by keeping the δ constant [3]. The coherence length shows a maximum around x=0.12. Such an elongated coherence length may correspond to the static magnetic order appearing around T_C which will be described in the next section. Beyond x_{UC}, no well-defined spin correlation exists [14]. These results suggest a concordant relation between the dynamical incommensurate spin correlation and superconductivity. We note that around x_{OP}, many transport properties exhibit a crossover from hole-like to electron-like behavior [15]. Furthermore, a change in the topology of Fermi surface is observed by angle-resolved photoemission [16]. Therefore, we expect a change in the orbital or site character of the doped holes around x_{OP}.

A linear relation between T_C and δ was obtained for the Sr-doped samples [3]. This relation is completely free from any uncertainties connected with the doping level in each sample. This empirical result directly demonstrates a concordant relation between the incommensurate dynamical spin correlation and superconductivity. Since pair-breaking does not change δ to first order, the linear relation is described between the maximum T_C and δ at a given doping in a system. However, as the maximum T_C is sensitive to the structural factors, one must carefully define the system to discuss the T_C-δ relation strictly. For example, can we expect the same T_C-δ relation between systems under high pressure and at ambient one? If the pressure induces a considerable change in the structure, the T_C-δ relation will be modified by pressure.

We studied the doping dependence of energy spectrum of the incommensurate spin fluctuations. A well-defined energy-gap of ~7meV opens for the optimally doped (x=0.15) or slightly overdoped (x=0.18) phases in the superconducting states. In contrast, no well-defined energy-gap can be seen for the underdoped (x=0.10) and heavily overdoped (x=0.25) superconducting phases. An analysis by Lee et al. [18] using a phenomenological gap-function proposed a possible reason for the missing of the energy gap; a smearing of the gap-edge corresponding to a shortening of lifetime of quasiparticles for the underdoped sample and a substantial decrease in the gap-size for the heavily doped sample. For x=0.18, gap like behavior of $\chi''(\omega)$ still remains even at $T=T_C$ in contrast to the conventional superconducting gap.

4 Static spin correlation

The opening of an energy-gap in the superconducting state suggests that the low-energy spin fluctuations within the gap degrade the superconductivity. However quite recently, besides such dynamical fluctuations, a static magnetic ordering was observed in the superconducting state of La2-1-4 system around the 1/8-doping. Before neutron scattering studies, the magnetic ordering was already indicated by µSR [19, 20] and NMR [21] measurements for both Ba-doped and Sr-doped La2-1-4 systems. However, details of the character and the relation with superconductivity were first presented by Tranquada et al. [11] by neutron scattering study on $La_{1.6-x}Nd_{0.4}Sr_xCuO_4$. In the x=1/8 sample, they found a charge segregation which preceded the magnetic order with incommensurate spin density modulation. Below the LTO-to-LTT transition, the striped domain is formed parallel to the orthorhombic distortions in the CuO_2 planes. Furthermore, they argued the competitive coexistence of both the static magnetic order and superconductivity; the onset temperature of the magnetic order is lower for samples with higher T_C.

Neutron scattering study to search for magnetic order in LTO system has conducted by Suzuki et al. [4] on $La_{1.88}Sr_{0.12}CuO_4$. They found sharp incommensurate peaks at the same positions around (π,π) as those of the dynamical spin fluctuations. The onset temperature of the peak was similar to T_C. After this measurement, Kimura et al. [5] have performed more comprehensive neutron scattering measurement by using cold neutron 3-axis spectrometers for both $La_{1.88}Sr_{0.12}CuO_4$ and nonsuperconducting $La_{1.88}Sr_{0.12}Cu_{0.97}Zn_{0.03}O_4$. Then they confirmed the static nature of the order. The resolution limited peak-width for the former indicates the static magnetic correlation length exceeds 200Å isotropically on the CuO_2 planes. The 3% Zn-substitution, on the other hand, degrades the long-range magnetic order; the peak appears at lower temperature (17 K) and the correlation length is shorter (77Å) compared to the Zn-free sample. Therefore, the long range magnetic order seems to be an intrinsic phenomenon around x=0.12. Although preliminary, such well-defined incommensurate elastic peaks were observed up to x=0.135 and down to x=0.10. The onset temperatures of these peaks are lower than that for x=0.12 and than the corresponding T_C.

Before Kimura's measurement, Hirota et al. [22] already observed incommensurate elastic peaks for Zn-doped x=0.14 sample, $La_{1.88}Sr_{0.12}Cu_{0.97}Zn_{0.03}O_4$ at similar temperature range as x=0.12. The static magnetic correlation length of about 80Å is also similar to the Zn-doped sample at x=0.12. It should be noted that the average nearest neighbor distance between the doped Zn in CuO_2 planes is calculated to be 35Å for y=0.012 and 22Å for y=0.03 which is smaller than the static correlation lengths of around 80Å determined by the q-width of the elastic peaks. Therefore, the magnetic correlation length of the static order is not directly confined by the average distance of doped Zn. Taking into the results of both x=0.12 and x=0.14, we conclude that the Zn-substitution not only degrades an existing long range magnetic order but also newly induces a static magnetic order in the dynamically fluctuating incommensurate phase. The latter function may correspond to the enhancement of low energy spin fluctuations by pair-breaking with Zn-doping which hides and/or destroys the energy-gap [17]. Therefore, phenomenologically, the reason for the missing of energy-gap in the underdoped superconducting state can be connected with the Zn-doping effect.

Stimulated by the elastic peak in the Sr-doped La_2CuO_4, Lee et al. searched for the peak in the electrochemically oxygen-doped superconducting La_2CuO_4. As a result, they found similar incommensurate peaks [23]. In this case, the onset of the magnetic order almost coincides with T_C=42K which is higher than that for the optimally Sr-doped sample. Surprisingly, the doping rate of the oxygenated sample was estimated to be about 0.12.

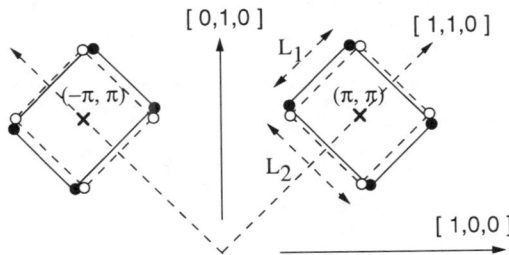

Fig. 2. Distribution of elastic incommensurate peaks around (π, π). Closed circles denote the observed positions which are deviated from the squared positions as depicted by open circles. The observed rectangulability L_1/L_2 is about 0.90 for the hole doping at 0.12.

Moreover, due to the sharpness of the elastic peak, another interesting detail on the peak-position was obtained on the oxygen-doped La_2CuO_{4+y} [23]. The four peaks around (π,π) locate at corners of a rectangle not square as shown in Fig.2. Although the reason for such "deformed" peak distribution is not understood, this rectangular geometry was also observed in the Sr-doped sample with much smaller orthorhombicity.

Therefore, the origin may not relate with the orthorhombic distortion or twinned structure.

In order to elucidate the interplay of such long range magnetic order with the superconductivity we need to extend the doping region, particularly toward the critical point x_{LC} to study the details on the transition or crossover between commensurate and incommensurate spin correlation. Quite recently, Wakimoto et al. [24] observed a mixture of commensurate and incommensurate elastic spin correlation at x=0.05 which exhibits a spin-glass behavior in the uniform magnetic susceptibility without superconducting diamagnetism. We note that a recent µSR measurement suggests a coexistence of the spin-glass and superconducting phases [25]. However, the momentum-sensitive neutron scattering revealed a more surprising fact that the observed direction of the spin density modulation is diagonal one in the square lattice of CuO_2 plane which is different from the direction of the superconducting cuprates by 45° [24].

5 Concluding remarks

We summarize in Fig.3 the overall magnetic properties of hole-doped $La_{2-x}Sr_xCuO_4$ elucidated in the second and third stages of neutron scattering studies. Based on these

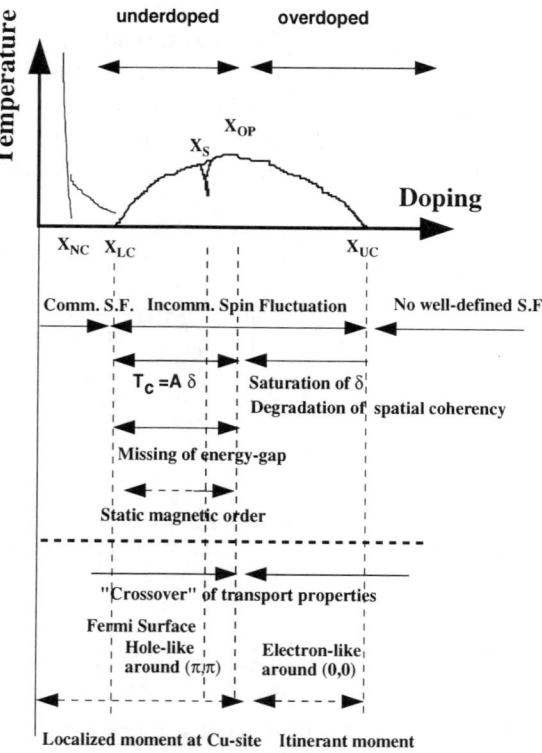

Fig.3. Overall doping dependence of magnetic properties around the ground state of $La_{2-x}Sr_xCuO_{4+y}$.

studies, we conclude that the superconducting phase is always accompanied by the incommensurate dynamical spin correlation. To elucidate the role of such spin fluctuations for the superconductivity, the characteristic temperature as well as energy of the commensurate-incommensurate crossover or transition should be studied as a function of doping and be compared with the so-called spin-gap temperature and energy.

In addition to the spin fluctuations, we newly observed a static incommensurate magnetic order around the 1/8-doping for both Sr-doped and oxygen-doped superconducting samples. The detail and posibly the role of the magnetic order in the superconducting state, coexistence/phase separation and competition/cooperation with the superconductivity, will be clarified in near future. In order to elucidate these issues the study of T_C-suppression in the oxygen-doped sample is very important.

In the superconducting states, an energy-gap opens for the optimally and slightly overdoped samples. The reason for the missing of energy gap in the other doping region is not elucidated at present. The search for the energy-gap around the 1/8-doping as well as lightly Zn-doped superconductor is highly required.

At x=0.05, just below the lower critical concentration x_{LC}, we observed a mixture of a broad commensurate elastic peak and an incommensurate peak with the spin modulation along [1,1] in contrast to along [1,0] in the superconducting phase. This observation should trigger a reinvestigation of the doping dependence of q-spectrum of the commensurate spin correlation for x<0.05.

Finally, we note no one have observed any superlattice peak caused by the hole-ordering in LTO structure in contrast to $La_{1.6-x}Nd_{0.4}Sr_xCuO_4$ in LTT structure.

Acknowledgment

The authors wish to thank H. Fukuyama, O. Sakai, V. J. Emery and J.M. Tranquada for valuable discussions. We also acknowledge M. Onodera and K. Nemoto for their technical assistance. The US-Japan collaboration program on neutron scattering also provided support for the neutron scattering experiment at NIST. The present work in part was supported by Grant-in-Aid for Scientific Research from the Japanese Ministry of Education, Science, Sports and Culture, and by a Grant for the Promotion of Science from the Science and Technology Agency and by CREST. Work at Brookhaven National Laboratory was carried out under Contract No. DE-AC0298-CH10886, Division of Material Science, U. S. Deperment of Energy. The research at MIT was supported by the NSF under Grant No. DMR97-04532 and by the MRSEC Program of the National Science Foundation under Award No. DMR94-00334.

References

[1] See e.g., M. Imai, A. Fujimori and Y. Tokura: Rev. Mod. Phys., in press.
[2] H. Yoshizawa, S. Mitsuda, H. Kitazawa and K. Katsumata: J. Phys. Soc. Jpn. **57**, 3686; T.R.Thurston, R.J.Birgeneau, M.A.Kastner, N.W.Preyer, G.Shirane, Y.Fujii, K.Yamada, Y.Endoh, K.Kakurai, M.Matsuda,Y.Hidaka and T.Murakami: Phys. Rev. B**40** (1989) 4585.
[3] K. Yamada, C.H. Lee, K. Kurahashi, J.Wada, Y. Kimura, S.Ueki, Y. Endoh, S. Hosoya, G. Shirane, R.J. Birgeneau, M.Greven, M.A. Kastner and Y. Kim:Phys. Rev. B**57** (1998) 6165.
[4] T. Suzuki, T. Goto, T. Shinoda, T. Fukase, H. Kimura, K. Yamada, M. Ohashi, Y. Yamaguchi: Phys. Rev. B**57** (1998) R3229.
[5] H. Kimura, K. Hirota, K. Yamada, Y. Endoh, S.H.Lee, C. H. Majkrzak, R. Erwin, G. Shirane, M. Greven, Y. S. Lee, M. A. Kastner and R.J. Birgeneau: submitted to Phys. Rew. B.
[6] I. Tanaka and H. Kojima: Nature (London) **337** (1989) 21.
[7] B. O. Wells, R. J. Birgeneau, F. C. Chou, Y. Endoh, D.C. Johnston, M. A. Kastner, Y.S. Lee, G. Shirane, J.M. Tranquada and K. Yamada: Z. Phys. B**100** (1996) 535; B. O. Wells, Y. S. Lee, M. A. Kastner, R. H. Christianson, R. J. Birgeneau, K. Yamada, Y. Endoh and G. Shirane: Science **277** (1997) 1067.
[8] K. Kurahashi, S. Wakimoto, C. H. Lee, K. Yamada and S. Hosoya: J. Phys. Soc. Jpn. **65** (1996) 3994.
[9] S.Wakimoto, K.Yamada, Y.Endoh and S.Hosoya: J. Phys. Soc. Jpn. **65** (1996) 581.
[10] R. J. Birgeneau, R. J. Christianson, Y. Endoh, M.A. Kastner, Y.S. Lee, G. Shirane, B.O. Wells and K.Yamada : Physica B **237-238** (1997) 84.
[11] J.M. Tranquada, J.D. Axe, N. Ichikawa, A.R. Moodenbaugh, Y.Nakamura and S. Uchida: Phys. Rev. Lett. **78** (1997) 338.
[12] H.A. Mook, P. Dai, S.M.Hayden, G. Aeppli, T.G. Perring and F. Dogan: submitted.
[13] H. Yoshizawa, T. Kakeshita, R. Kajimoto, T. Tanabe, T. Katsufuji, and Y. Tokura,Physica B **241-243**, 880 (1998).
[14] K.Yamada: Advances in Superconductivity X (1998) 37.
[15] For a review, see N.P.Ong(1990) In: Ginsberg DM (ed) Physical Properties of High- Temperature Superconductors II. World Scientific, Singapore, pp 459
[16] A. Fujimori, A. Ino, T. Mizokawa, C. Kim, Z-X. Shen, T. Sasagawa, T. Kimura, K. Kishio, M. Takaba, K. Tamasaku, H. Eisaki, S.Uchida (1997) International Conference on Spectroscopies in Novel Superconductors.
[17] M. Matsuda, R.J. Birgeneau, H. Chou, Y. Endoh, M.A. Kastner, H. Kojima, K. Kuroda, G. Shirane, I. Tanaka, K. Yamada : J. Phys. Soc. Jpn. **62** (1993) 443.
[18] C.-H. Lee, K.Yamada, Y. Endoh, G. Shirane, R. J. Birgeneau, M. Greven, Y.M. Kim and M..A. Kastner, in preparation.
[19] K. Kumagai, K. Kawano, I. Watanabe, K. Nishiyama and K. Nagamine : J. Supercond. **7** (1994) 63.

[20] G. M. Luke, L.P. Le, B.J. Sternlieb, W.D. Wu, Y.J. Uemura, J.H. Brewer, T.N. Riseman, S. Ishibashi and S. Uchida; Physica C **185-189** (1991) 1175.
[21] T. Goto, S. Kazama, K. Miyagawa and T. Fukase: J. Phys. Soc. Jpn. **63** (1994) 3494.
[22] K. Hirota, K. Yamada, I. Tanaka and H. Kojima, Physica B **241-243**, 819 (1998).
[23] Y.S. Lee (unpublished data).
[24] S. Wakimoto, R.J. Birgeneau, Y. Endoh, P.M. Gehring, K. Hirota, M.A. Kastner, S.H. Lee, Y.S. Lee, G. Shirane, S. Ueki and K. Yamada, in preparation.
[25] Ch. Niedermayer, C. Bernhard, T. Blasius, A. Golnik, A. Moodenbaugh and J.I. Budnick: Phys. Rev. Lett. **80** (1998) 3843.

Experimental Studies on Singlet Formation in High-T_c Oxides and Quantum Spin Systems

M. Sato

Department of Physics, Division of Material Science, Nagoya University, Furo-cho, Chikusa-ku, Nagoya 464-8602 and CREST, Japan Science and Technology Corporation (JST)

Abstract. Detailed studies on the singlet formation in high-T_c oxides and quantum spin systems have been carried out by various experimental methods including neutron scattering and NMR. The results can clarify how the anomalous metallic phase of high-T_c oxides are formed with decreasing temperature and present important information on the mechanism of the superconductivity.

1. Introduction

In the study of high temperature superconductors, it is very important to understand the nature of their anomalous normal state, which all kinds of pysical arguments are based on. Figure 1 shows the phase diagram originally introduced to schematically indicate the region of the "anomalous metallic phase (AMP)" [1-3], where T is the temperature and δ is the concentration of carriers doped into the "half-filled" Mott insulating state. Because the superconductivity appears in the low temperature region of the AMP surrounded by the broken line which indicates one of the characteristic temperatures, $T_0(\delta)$, it is believed to be primarily important for the understanding of the origin of the high-T_c superconductivity, to clarify how the AMP is formed as T is lowered. The present paper reports results of several experimental studies which present information on the microscopic origin(s) of the AMP formation which is (are) closely related to the occurrence of the high-T_c superconductivity.

2. Description of the Anomalous Metallic Phase

In Fig. 1, two characteristic temperatures T_{SG} and T_0, where various physical properties exhibit crossover-like change with varying T are shown. The temperature T_{SG} is well-known as the spin-gap temperature, where the NMR relaxation rate of Cu nuclei in high-T_c oxides divided by T, $1/T_1T$ has maximum value as a function of T [4-5]. Because $1/T_1T$ is propertional to the low energy spectral weight of the magnetic excitation, T_{SG} can be considered as the characteristic temperature where the gap-like structure begins to appear

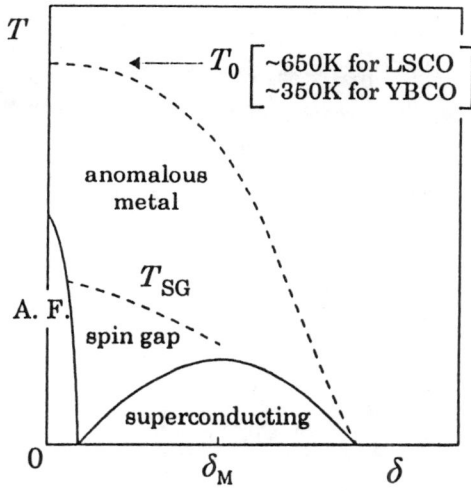

Figure 1: Schematic phase diagram which shows the region of the anomalous metallic phase below T_0.

with decreasing T in the spectral function $\chi''(\omega)$, $\hbar\omega$ being the energy of the excitation. Much attention has been paid to this temperature T_{SG}, because it seems to correspond to T_{RVB}, the temperature of the transition to the singlet RVB state in the phase diagram of the mean-field $t-J$ model [6]. Direct observation of the gap-like structure of $\chi''(q,\omega)$ at q around (π,π) in the reciprocal space was first reported by Rossat-Mignod et al. [7] and then by the group of BNL and Nagoya Univ. [8] for YBa$_2$Cu$_3$O$_{6+y}$ (YBCO or YBCO$_{6+y}$).

The δ-dependence of the electronic specific heat coefficient γ of high-T_c oxides shown in Fig. 2 [3] has been argued by Imada [9] as an experimental fact which might be related to the spin-gap fromation: He has pointed out a possibility that the metal-insulator (M-I) transition which occurs as the systems approach the "half-filled" electronic state, is not due to the increase of the electronic effective mass m^* but due to the decrease of the carrier number caused by the singlet (or spin-gap) formation. On this point, it is interesting to note that the Hall coefficient R_H of La$_{2-x}$Sr$_x$CuO$_4$ (LSCO) exhibits rather drastic increase with decreasing T as shown in Fig. 3(a) [10,11], suggesting the existence of the crossover-like change of R_H from the small-R_H (or "large carrier number") high-temperature region to the large-R_H (or "small carrier number") low-temperature one. This crossover temperature is defined as T_0 [1-3]. At around T_0, the resistivirity ρ exhibits an anomalous decrease with decreasing T and the spin susceptibility becomes maximum [1-3,12,13]. For YBCO$_{6+y}$, the T_0 value near the M-I phase boundary is smaller than that of LSCO system [11,14] (It is indicated in Fig. 1). It is worth noting that the thermoelectric powers of high-T_c systems also exhibit quite characteristic x- and T-dependences [15]. In order to identify the origin(s) of the anomalous behavior observed in various quantities below T_0, the T-dependence of the square of the

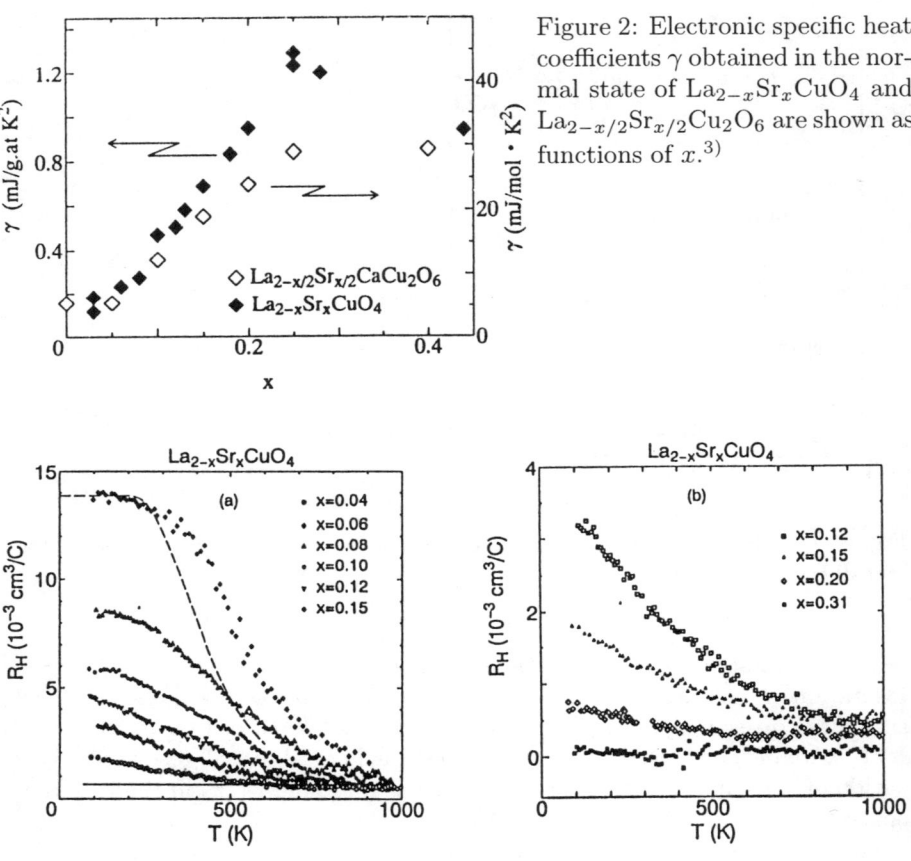

Figure 2: Electronic specific heat coefficients γ obtained in the normal state of $La_{2-x}Sr_xCuO_4$ and $La_{2-x/2}Sr_{x/2}Cu_2O_6$ are shown as functions of x.[3)]

Figure 3: Hall coefficients R_H of $La_{2-x}Sr_xCuO_4$ are plotted against T for various x values [11]. In (a), the $\xi^2 - T$ curve is shown by the broken line. The horizontal line in (a) roughly shows the R_H value corresponding to a value estimated from the band picture.

magnetic corelation length ξ of $La_{2-x}Sr_xCuO_4$ ($x = 0.04$) [16] shown by the broken line in Fig. 3(a) is compared with that of R_H for $x = 0.04$, where the vertical scale of ξ^2 is properly chosen for the comparsion. The resemblance of the $\xi^2 - T$ and $R_H - T$ curves indicates the important role of the growth of the antiferromagnetic (AF) correlation in realizing the AMP which exists below T_0. The theoretical calulation of $R_H - T$ carried out by Miyake and Narikiyo [17] by considering the AF correlation, seems to support the idea. Observed effects of doping of magnetic and nonmagnetic atoms into the Cu atom sites [18] and the presence of the strong low energy magnetic excitations [19] support the idea, too.

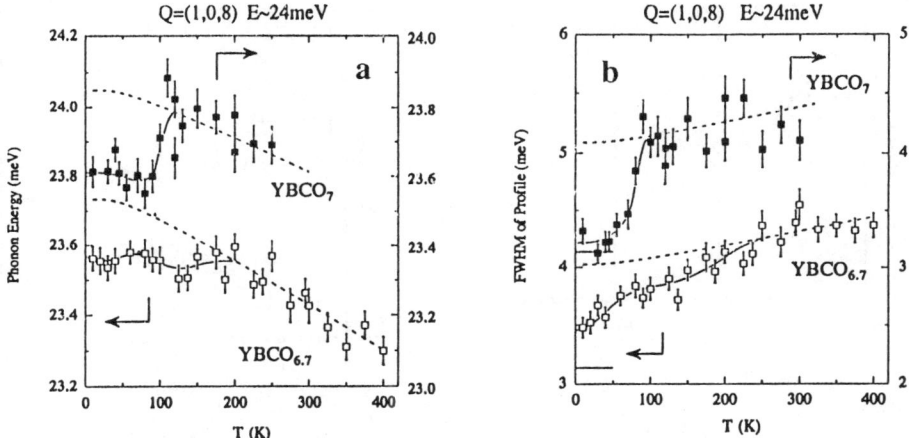

Figure 4: Energies (a) and profile widths (b) of the B_{2u}-phonon mode of YBCO$_7$ and YBCO$_{6.7}$ are plotted as functions of T.

3. Observation of the Singlet Formation Process and Discussion

Now, it is not easy for us to distinguish by using only the information stated above what mechanism(s), the singlet formation or growth of the AF correlation or both of them is (are) relevant to the anomalous behavior observed in the AMP of high-T_c oxides. Then, in order to obtain further information on this point, several experimental studies have been carried out on YBCO$_{6+y}$ and low dimensional quantum spin systems.

Here, we first present results of the studies on one of two B-symmetry phonon modes with c-axis polarized motions of oxygen atoms in the CuO$_2$ planes: The mode is in this case used as a probe of the singlet formation. Normand et al. [20] have pointed out that B-symmetry modes of YBCO system strongly couple to the superconducting order parameter with (x^2-y^2)-like symmetry. Actually, the strong effect of the superconductivity on the B_{1g} mode with the energy $\hbar\omega \sim 41$ meV was experimentally reported by Pyka et al. [21] for YBCO$_7$ with $T_c \sim 90$ K before the suggestion of ref. 20. We have studied the B_{2u}-mode of YBa$_2$Cu$_{3-x}$Zn$_x$O$_{6+y}$ by means of neutron inelastic scattering. The mode is optically silent and for neutrons, it is much easier to study than the B_{1g}-mode, because the energy (~ 24 meV) is much lower than that of the latter. Results are shown in Fig. 4 [22-24] for YBCO$_7$ and YBCO$_{6.7}$ ($T_c \sim 62$ K) where the broken lines indicate the T-dependence expected from ordinary unharmonic effects. A sharp anomaly can be observed in the T-dependence of the phonon energy of YBCO$_7$ at around $T \sim T_c$. For YBCO$_{6.7}$, the anomalous downward deviation of the phonon energy from the broken line appears at a temperature higher than 200 K and the deviation gradually grows with decreasing T, indicating that the singlet formation itself gradually grows from the temperature far above T_c. Similar behavior has also been observed in the T-dependence of the profile widths. Because the observed phonon anomalies

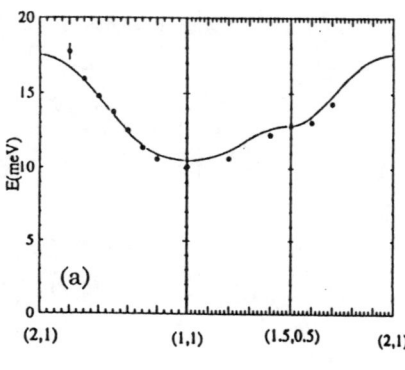
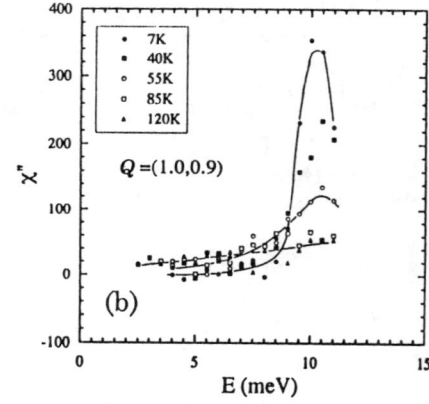

Figure 5: (a) Dispersion curves of the triplet excitation of CaV$_4$O$_9$ along the high-symmetry cuts at 7 K and (b) examples of the observed spectral functions of the triplet excitation $\chi"$ of CaV$_4$O$_9$. Solid lines are guides for the eyes.

are expected only in the temperature region where the maximum gap energy $2\Delta_0$ of the singlet pairs with the anisotropy of $2\Delta(\boldsymbol{k}) = \Delta_0(\cos k_x - \cos k_y)$ is larger than the phonon energy (~ 24 meV), the singlet formation should begin, with decreasing T, at a temperature far above 200 K (of course, much higher than $T_{SG} \sim 150$ K), and it is likely that the singlet correlation persists up to the characteristic temperature T_0, which indicates two kinds of correlations, the singlet and AF ones coexist below the temperature $T \sim T_0$.

This kind of coexistence has been clearly observed in low dimensional spin-gap systems, CaV$_4$O$_9$ [25,26] and CuNb$_2$O$_6$ [27-30]. For CaV$_4$O$_9$, the spin-gap behavior has been found by measuring the magnetic susceptibility χ, NMR $1/T_1T$ and magnetic excitation spectra $\chi"(\boldsymbol{q}, \omega)$. In Fig. 5(a), the dispersion curves of the triplet excitaion obtained by neutron inelastic scattering are shown, detailed analyses of which have shown that the spin-gap ground state can be considered to be a two-dimensional linkage of the interacting plaquette-singlet units [26]. For this system, the spectral function of the triplet excitation $\chi"(\omega)$ measured at $\boldsymbol{Q}$=(1.0,0.9) at several temperatures are shown in Fig. 5(b), where the concave behavior of the $\chi" - \omega$ curve in the region of small ω remains appreciable up to the temperature of the spin susceptibility maximum (We call this temperature T_0, too. $T_0 \sim 100$ K). On the other hand, the value of $\chi"$ at a fixed energy ω much smaller than the gap energy Δ (~ 10 meV) becomes maximum at a temperature $T_{SG} \sim 50$ K significantly lower than T_0. Because $\chi"(\omega)$ is expected in the ω-region, $\omega << \Delta$ to be proportional to the $1/T_1T$, it suggests that the temperature of the $1/T_1T$-maximum or T_{SG} is much lower than T_0. The relationship between T_0 and T_{SG} is very smilar to that of high-T_c systems, and here we know that the concave behavior or the effect of the singlet formation can be observed up to the temperature $\sim T_0$.

The monoclinic (M-) CuNb$_2$O$_6$ has quasi-one-dimensitional zig-zag chains of quantum spins (S=1/2) of Cu^{2+} ions. For this system, the spin susceptibility,

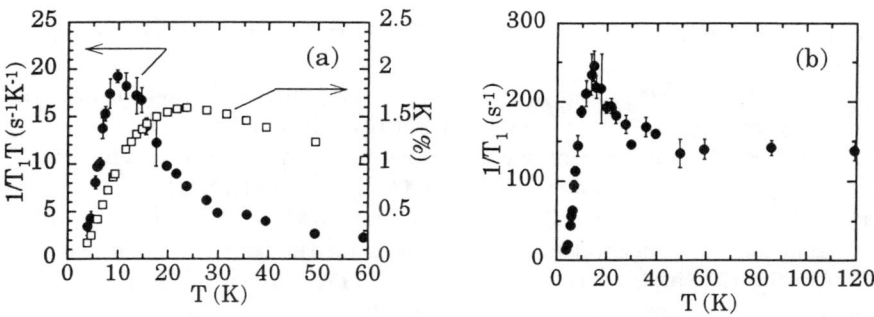

Figure 6: (a) NMR Knight shift and $1/T_1T$ of M-CuNb$_2$O$_6$ are shown against T. (b) NMR relaxation rate $1/T_1$ of M-CuNb$_2$O$_6$ is plotted as a function of T.

the spin contribution of the electronic specific heat C_{spin}, the NMR Knight shifts K and $1/T_1T$ of ^{93}Nb and the magnetic excitation spectra $\chi''(\omega)$ have been measured, where the spin-gap value of $\Delta \sim 20$ K has been consistently obtained. The minimum of the dispersion curve appears at $Q = (0, 0, 1/2)$, indicating that this system has the ferromagnetic (F)-AF alternationg bond [28,30] (note that the distance between nearest neighbor Cu site within a chain is c/2) and therefore the mechanism of the gap formation is considered to be similar to that of Haldane systems. For this system, the temperature of the $1/T_1T$-maximum or T_{SG} (~ 10 K) is found, as shown in Fig. 6(a), to be much lower than that of the susceptibility (or Knight shift)-maximum or T_0 (~ 24 K) similarly to the case of high-T_c oxides. Now, it is emphasized that the gap-like structure of $\chi''(\omega)$ or the concave behavior of the $\chi'' - \omega$ curve persists up to the temperature $T \sim T_0$ (see Fig. 7). It should be also emphasized that as shown in Fig. 6(b), $1/T_1$ begins to deviate upwards from a constant value at

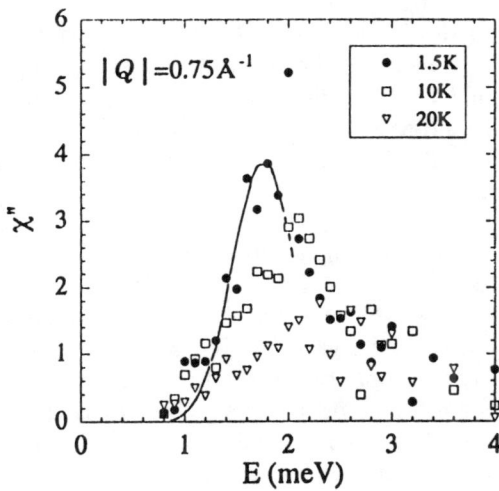

Figure 7: The spectral functions χ'' measured at several temperatures for polycrystalline sample of M-CuNb$_2$O$_6$ are shown against the energy $\omega(= E)$. The line indicates the low energy edge of the profile expected for the spin-gap value of about 20 K.

around T_0 with decreasing T, indicating the growth of the AF correlation of Cu spins.

Now, all the above findings indicate that the singlet formation and the AF correlation appears almost simultaneously at around T_0 with decreasing T in the present spin systems with spin-gapped ground state. The results of the B_{2u}-phonon measurements for YBCO$_{6+y}$ also indicate that the singlet formation appears at around T_0 almost simultaneously with the growth of the AF correlation. It is interesting here to note that the coexistence of the long range AF and the singlet correlations have been established in impurity doped CuGeO$_3$ [31,32]. As we have pointed out in our previous papers [1-3,24,27-30], T_0 is not just a temperature where the AF correlation begins to develop but also a temperature where the fluctuating singlet pairs appear with decreasing T. As a result of the appearance of the singlet and AF correlations, the AMP is, we believe, formed below T_0 with decreasing T. It is an answer to our initial question what mechanism(s) is (are) important for realizing the AMP (The answer does not contradict the idea of ref. 9 and the model of ref. 17, either).

4. Further Consideration and Concluding Remarks

We have shown that both the singlet and AF correlations begin to grow almost simultaneously at around T_0 with decreasing T, resulting in the formation of the AMP. Recently, the existence of the pseudo gap has been directly observed by photoemission [33,34] and by electron tunneling studies [35,36] on Bi$_2$Sr$_2$CaCu$_2$O$_{8+\delta}$. The results of the tunneling studies clearly indicate that the gap-structure of the quasi particle tunneling spectra persists up to the temperature $T \sim T_0$. The photoemission results seem to be slightly different: The gap-like structure persists just up to the temperatures $T \sim T_{SG}$. However, the very large broadenig of the quasi particles above T_c [37.38] as well as the possible broad distribution of the excitation gap Δ of singlet pairs above T_c may make it rather difficult to identify the "leading edge" of the spectra which is important to estimate the pseudo-gap value. We believe that the nonzero value of $|\Delta|$ is found even in the temperature region higher than T_{SG} possibly up to T_0. The exchange coupling J_s among the electrons induces both the singlet and AF correlations with decreasing T and as the result of the appearance of these correlations, the AMP is formed. With further decreasing T, the singlet correlation dominates at $T \sim T_{SG}$ and the superconductivity appears when T is lowered further.

Now, the exchange interaction J_s which commonly exists in high-T_c systems is considered to be playing an important role in realizing the AMP and the superconductivity. Then, a question arises why the $2\Delta_0$ value observed at low temperatures do not have a clear correlation with T_0. There are several possible origins. First, randomness effects are well-known to drastically charge the magnetic excitation spectra [39,40] and to suppress the superconductivity. We think that the possible difference of degree of randomness among various high-T_c systems is one of the origins of the different T_0 and $2\Delta_0$ values. The existence of the tendency of the charge ordering, which is, for example, known as the 1/8 problem in $(La,Nd)_{2-x}Sr_xCuO_4$ [41] also considered to be one of the origins of the difference. Very large decrease in the quasi-particle broadening with decreasing T [37,38] is also an important candidate mechanism of the difference, because the electronic state may change drastically with temperature and therefore the value of the low temperature energy gap may not necessarily have an accurate correlation with the high temperature pseudo gap.

The present work has been carried out in collaboration with the members of Nagoya University group and Prof. K. Kakurai's group of ISSP of The Univ. of Tokyo. The present work was supported by a Grant-in-Aid for Scientific Research on Priority Area from the Ministry of Education, Science, Sports and Culture.

References

[1] The similar phase diagram was first presented in Koon Chodendo no Kagaku, Report on the Research supported by Grant-in Aid for Scientific Research on Priority Areas from the Ministry of Education, Science, Sports and Culture (in Japanese), ed. by M. Tachiki, March 1994, p. 133
[2] M. Sato: Physica C **263** (1996) 271
[3] M. Sato: *Studies of High Temperature Superconductors* vol.19, ed. A. Narlikar, Nova Science Publishers INC, April 1996, p. 213
[4] H. Yasuoka, T. Imai and T. Shimizu: *Strong Correlation and Superconductivity*, ed. H. Fukuyama, S. Maekawa and A.P. Malozemoff (Springer Series 1989) p. 254
[5] M. Takigawa, A.P. Hammel, J.D. Thompson, R.H. Heffner, A. Fisk and K.C. Ott: Phys. Rev. B **43** (1991) 247
[6] Y. Suzumura, Y. Hasegawa and H. Fukuyama: J. Phys. Soc. Jpn. **57** (1988) 2768
[7] J. Rossat-Mignod, L.T. Regnault, C. Vettier, P. Bourges, J. Bossy, J.Y. Henry and G. Lapertot: Physica B **169** (1991) 58
[8] B.J. Sternlieb, M. Sato, S. Shamoto, G. Shirane and J.M. Tranquada: Phys. Rev. B **47** (1993) 5320
[9] M. Imada: J. Phys. Soc. Jpn. **62** (1993) 1105
[10] T. Nishikawa, J. Takeda and M. Sato: J. Phys. Soc. Jpn. **62** (1993) 2568
[11] T. Nishikawa, J. Takeda and M. Sato: J. Phys. Soc. Jpn. **63** (1994) 1441
[12] H.Y. Hwang, B. Batlogg, H. Takagi, H.L. Kao, J. Kwo, R. Cava, J.J. Krajewski and W.F. Peak. Jr: Phys. Rev. Lett. **72** (1994) 2636
[13] T. Nakano, M. Oda, C. Manabe, N. Momono, Y. Miura and M. Ido: Phys. Rev. B **49** (1994) 16000
[14] T. Ito, K. Takano and S. Uchida: Phys. Rev. Lett. **70** (199) 3995
[15] J. Takeda, T. Nishikawa and M. Sato: Physica C **231** (1994) 293
[16] G. Shirane, R.J. Birgeneau, Y. Endoh and M.A. Kastner: Phys. Rev. B **197** (1994) 158-174
[17] K. Miyake and O. Narikiyo: J. Phys. Soc. Jpn. **63** (1994) 3821

[18] J. Takeda, K. Kodama, M. Sato, T. Nishioka and M. Kontani: J. Phys. Soc. Jpn. **63** (1994) 3564
[19] For example, see K. Kodama, S. Shamoto, H. Harashina, M. Sato, M. Nishi and K. Kakurai: Physica C **263** (1996) 333
[20] B. Normand, H. Kohno and H. Fukuyama: Phys. Rev. B **53** (1996) 856
[21] N. Pyka, W. Reichardt, L. Dintschovious, G. Engel, J. Rossat-Mignod and J.Y. Henry: Phys. Rev. Lett. **70** (1993) 1457
[22] H. Harashina, K. Kodama, S. Shamoto, M. Sato, K. Kakurai and M. Nishi: J. Phys. Soc. Jpn. **64** (1995) 1462
[23] H. Harashina, K. Kodama, S. Shamoto, M. Sato, K. Kakurai and M. Nishi: Physica C **263** (1996) 257
[24] H. Harashina, H. Sasaki, K. Kodama, M. Sato, S. Shamoto, K. Kakurai and M. Nishi: J. Phys. Soc. Jpn. **67** (1998) No. 9
[25] S. Taniguchi, T. Nishikawa, Y. Yasui, Y. Kobayashi, M. Sato, T. Nishioka, M. Kontani and K. Sano: J. Phys. Soc. Jpn. **64** (1995) 2758
[26] K. Kodama, H. Harashina, H. Sasaki, Y. Kobayashi, M. Kasai, S. Taniguchi, Y. Yasui, M. Sato, K. Kakurai, T. Mori and M. Nishi: J. Phys. Soc. Jpn. **66** (1997) 793
[27] K. Kodama, T. Fukamachi, M. Kanada, H. Harashina and M. Sato: J. Phys. Soc. Jpn. **67** (1998) 57
[28] T. Nishikawa, M. Kato, T. Fukamachi, M. Kanada, H. Harashina and M. Sato: J. Phys. Soc. Jpn. **67** (1998) 1988
[29] K. Kodama, H. Harashina, H. Sasaki, M. Kato, M. Sato, K. Kakurai and M. Nishi: J. Phys. Soc. Jpn. submitted
[30] T. Fukamachi, Y. Kobayashi, M. Kanada, M. Kasai, Y. Yasui and M. Sato: J. Phys. Soc. Jpn. **67** (1998) 2107
[31] L.P. Regnault, J.P. Renard, G. Dhalenne and A. Revcoluscchi: Europhys. Lett. **32** (1995) 579
[32] H. Fukuyama, T. Tanimoto and M. Sato: J. Phys. Soc. Jpn. **65** (1996) 1182
[33] A.G. Loser, D.S. Dessau and Z.-X. Shen: Physica C **263** (1996) 208 and D.S. Marshall, D.S. Dessau, A.G. Loser, C.-H. Park, A.Y. Matsiira, J.K. Eckstein, I. Bozovic, P. Founier, A. Kapitulnik, W.E. Spicer and Z.-X. Shen: Phys. Rev. Lett. **76** (1996) 4841
[34] H. Ding, T. Yokoya, J.C. Campuzano, T. Takahashi, M. Randeria, M.R. Norman, T. Mochiku, K. Kadowaki and J. Giapintzakis: Nature **382** (1996) 51
[35] C. Renner, B. Revaz, J.-Y. Genoud, K. Kadowaki and ϕ. Fischer: Phys. Rev. Lett. **80** (1998) 149
[36] A. Matsuda: private communications
[37] D.A. Bonn, R. Liang, T.M. Riseman, D.J. Baar, D.C.C. Moris, J.H. Brewer W.N. Hardy, C. Kallin and Berlinsky: Phys. Rev. B **47** (1993) 11314
[38] R.C. Yu, M.B. Salamon, J.P. Lu and W.C. Lee: Phys. Rev. Lett. **69** (1992) 1431.

[39] K. Kakurai, S. Shamoto, T. Kiyokura, M. Sato, J.M. Tranquada and G. Shirane: Phys. Rev. B **48** (1993) 3485
[40] H. Harashina, S. Shamoto, T. Kiyokura, M. Sato, K. Kakurai and G. Shirane: J. Phys. Soc. Jpn. **62** (1993) 4009
[41] J.M. Tranquada, B.J. Sternlieb, J.D. Axe, Y. Nakamura and S. Uchida: Nature **375** (1995) 561

Thermal Conductivity as a Probe of Quasi-Particles in the Cuprates

N. P. Ong, K. Krishana, Y. Zhang, and Z. A. Xu[†]

Joseph Henry Laboratories of Physics, Princeton University, Princeton, New Jersey 08544, U. S. A.

Abstract. In underdoped $YBa_2Cu_3O_x$ ($x = 6.63$), the low-T thermal conductivity κ_{xx} varies steeply with field B at small B, and saturates to a nearly field-independent value at high fields. The simple expression $[1 + p(T)|B|]^{-1}$ provides an excellent fit to $\kappa_{xx}(B)$ over a wide range of fields. From the fit, we extract the zero-field mean-free-path, and the low temperature behavior of the QP current. The procedure also allows the QP Hall angle θ_Q to be obtained. We find that θ_Q falls on the $1/T^2$ curve extrapolated from the electrical Hall angle above T_c. Moreover, it shares the same T dependence as the field scale $p(T)$ extracted from κ_{xx}. We discuss implications of these results.

1 Introduction

Thermal conductivity is potentially a very useful probe of the quasi-particle excitations in the superconducting state of the cuprates because it is capable of detecting the quasi-particle (QP) current in the bulk [1, 2, 3, 4]. In addition, measurements of its field dependence may yield quantitative information on the QP mean free path. At present, this seems to be the best way to investigate the low-lying excitations of the condensate. However, the task of disentangling the QP current from the larger phonon current in the cuprates poses a difficult problem for experiment.

We report recent experiments in which the direct separation of the QP current is achieved by detailed analysis of the field dependence of the longitudinal conductivity κ_{xx}. This line of approach was motivated by the observation of plateau features in high-purity $Bi_2Sr_2CaCu_2O_8$ (Bi 2212) [5]. The existence of the plateaus at low T (where $\partial \kappa_{xx}/\partial H = 0$) implies that, in the cuprates, vortices are essentially transparent to the phonons. Extensions of these measurements to underdoped $YBa_2Cu_3O_x$ (YBCO) reveal that this result may be a rather general feature of the cuprates in the clean-limit. This seems to us a significant finding since it allows a direct separation of the QP current by the application of an intense field. In addition, we find that the zero-field mean-free-path ℓ_0 of the QP may be estimated to within a factor equal to the vortex scattering cross-section σ_{tr}.

The isolation of the QP current allows more specific information to be extracted from the thermal Hall conductivity κ_{xy} (Righi-Leduc effect) [4]. With the electronic current independently determined, we may now obtain the Hall angle $\tan \theta_Q$. The QP Hall angle uncovers a number of interesting features which we discuss below.

2 Experiment

Measurements of κ_{xx} in the mixed state of the cuprates have been reported by several groups [3, 4, 5, 6, 7, 8]. Experiments in intense magnetic fields **H** are complicated by problems such as cleaving of the crystal by the large torques generated. The most serious problem, however, seems to stem from the field sensitivity of the thermometers. At the resolution needed, the field dependence of the sensors is serious (in thermocouples, moreover, the field sensitivity is also history dependent). Previously, we employed a bridge-balance method to get around the field-sensitivity problem [5].

In our present approach, we adopt a single-heater, two-sensor method in which the temperature difference δT between the ends of the sample is detected by two closely matched resistive sensors (cernox). The thermal gradient $-\nabla T$ is applied in the ab plane, and **H** is parallel to **c**. The temperature is regulated (with a third base-cernox), and measurements are taken after waiting about 10 min. for the field to stabilize. The readings of the two sensors are recorded at three values of the heater current $I = 0, 0.4, 0.6$ mA (typically). The $I = 0$ readings are used to calibrate the field dependence of the two cernox sensors, while the values of δT determined with I at the two values provide a check on the linearity of the sample response. By testing with a standard material that has no intrinsic field dependence in κ_{xx} (nylon), we have found that at temperatures above 8 K this method provides an accurate and highly reproducible determination of the intrinsic conductance to a resolution of 1 in 10^3. Although the bridge-balance method is capable of higher resolution, the present technique lends itself to full automation. A higher density of points may be obtained, and checks (e.g. for linearity) can be made *in situ*. A pair of thermocouple junctions are used to detect the H-antisymmetric (Hall) gradient to obtain κ_{xy}.

With the high sensitivity, we have found that, in untwinned, optimally-doped $Ba_2Cu_3O_x$ ($T_c = 93$ K, $x = 6.95$), κ_{xx} in a field **H** $\parallel$ **c** becomes increasingly hysteretic below 35 K. Although the hysteresis is small (about 5% of the total κ_{xx} at 8 K), it greatly complicates the extraction of the QP current (we discuss this later). In underdoped crystals, however, the hysteresis is unobservable up to 14 tesla (less than 10^{-3}), and the observed κ_{xx} vs. H is a faithful representation of its intrinsic dependence on B.

In this report, we discuss data from a twinned, underdoped crystal in which $T_c = 63$ K, and $x = 6.63$. The zero-field temperature profile of κ_{xx} is shown in Fig. 1. The relative magnitude of the anomaly in κ_{xx} is only about a quarter of that in the 93-K YBCO, but larger than in optimum Bi 2212 [5] and $La_{2-x}Sr_xCuO_4$ (LSCO) [7]. Also shown (solid symbols) is our new estimate of the phonon conductivity (κ_B). One of our main results is that the entire field dependence of κ_{xx} derives from the QP current, while the phonon current is unaffected by H.

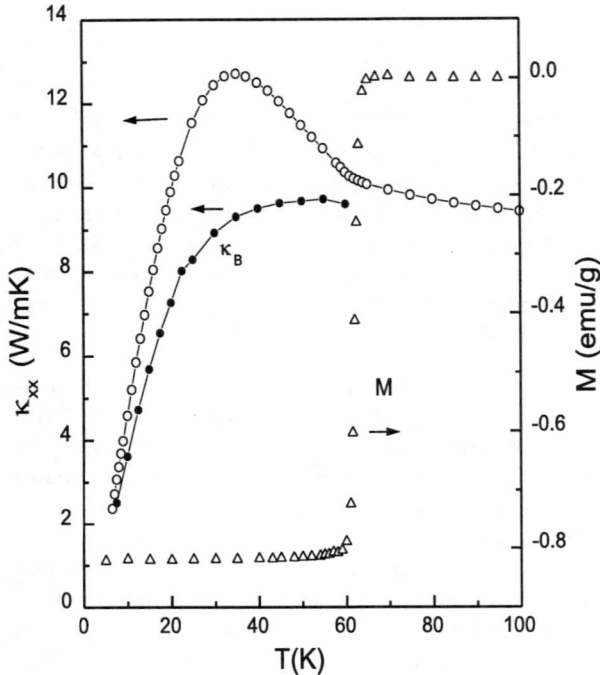

Fig. 1. The T dependence of the in-plane thermal conductivity $\kappa_{xx}(0)$ in zero field (open symbols) in underdoped YBCO with $x = 6.63$ and $T_c = 63$ K. The solid symbols are the phonon conductivity κ_B estimated by the fit of the thermal magnetoconductance to Eq. 1. The dc magnetization of the crystal is shown as open triangles ($H = 5$ Oe).

3 Results and Analysis

Figure 2 displays the field dependence of κ_{xx} at selected T. With decreasing T, the initial slope of κ_{xx} increases rapidly. Below 15 K, the rapid decrease crosses over to an almost flat dependence, which recalls the plateau features observed in single-domain Bi 2212 [5]. Within our resolution, there is no resolvable hystereses at the temperatures investigated. The higher precision and larger range of the new data enable us to compare the observed field dependence with various expressions. We find that the field dependence is accurately fitted (Fig. 2) to the expression

$$\kappa_{xx}(B,T) = \frac{\kappa_e(T)}{(1 + p(T)|B|)} + \kappa_B(T), \tag{1}$$

where the entire B dependence resides in the denominator of the first term, and the term κ_B is a field- independent background. At each temperature, the fit yields the two parameters $\kappa_e(T)$ and $p(T)$ associated with the QP current, and the term κ_B which we identify with the phonon term, viz. $\kappa_B = \kappa_{ph}$ (plus any small residual electronic term that is H *independent*).

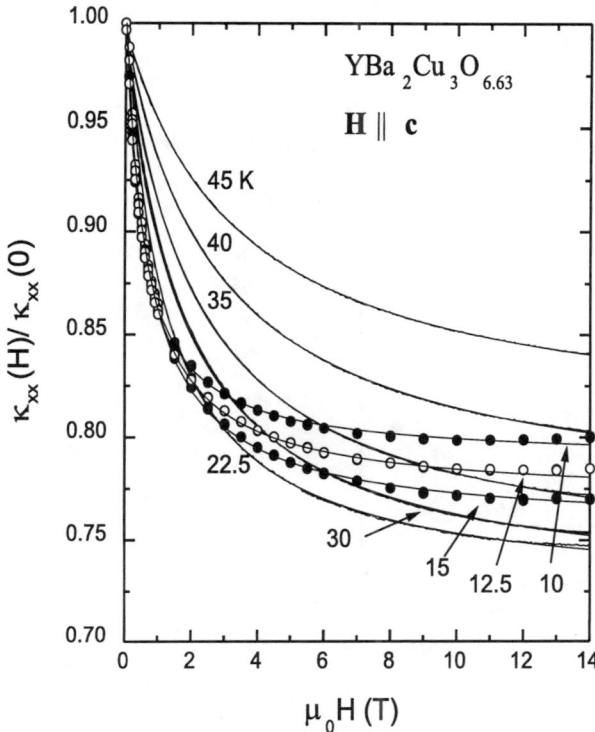

Fig. 2. Variation of κ_{xx} with field **H** ∥ **c** in underdoped YBa$_2$Cu$_3$O$_{6.63}$ at indicated temperatures. All curves are non-hysteretic to 1 part in 10^3 (the curves above 15 K are superpositions of sweep-up and sweep-down traces). A fit to the data at 22.5 K is also shown superposed. Curves at 15 K and below (discrete symbols) show the approach of κ_{xx} to the H-independent value (κ_B).

In our previous experiments on optimally-doped YBCO [4] and LSCO [7], fits to Eq. 1 were ambiguous because of strong hystereses (the distortions introduced caused the extracted $p(T)$ to be non-monotonic in T). These extraneous effects led us to consider alternate expressions (see below), as well as a possibly field dependent κ_{ph}. However, the present results have clarified the problem. In addition to the absence of observable hysteresis, the curves for κ_{xx} at low T display pronounced curvature in moderate fields, corresponding to a strong attenuation of the QP current. The attenuation uncovers a nearly field-independent background that we identify with the phonon thermal conductivity. In the main panel of Fig. 3, we display the T dependence of the parameters κ_e and $p(T)$.

The motivation for Eq. 1 is that, near the vortex core, the steep variation of the pair potential and the circulating superfluid together present a strong scattering potential for an incident QP [10]. Expressing the scattering rate as a transport cross-section σ_{tr}, we assume additivity of the rates, and write the QP mean-free-path in a field as $\ell(B) = \ell_0/[1 + \ell_0 \sigma_{tr}|B|/\phi_0]$, where ℓ_0 is

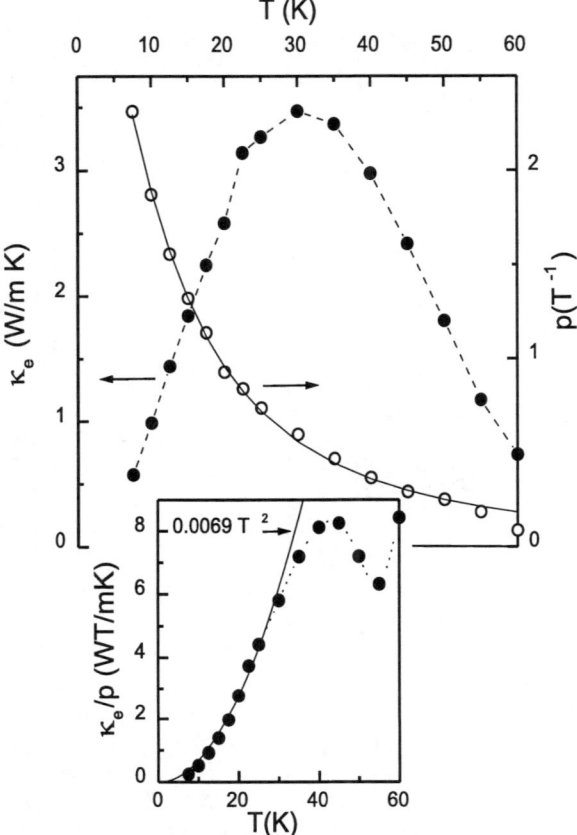

Fig. 3. The T dependence of the zero-field QP conductivity κ_e and the field scale $1/p(T)$ extracted from fits to κ_{xx} vs. H in underdoped YBCO. The T^2 dependence of $1/p(T)$ suggests that it is related to a scattering rate. The solid line is $c_0 + c_1T + c_2T^2$, with $c_0 = 0.25$, $c_1=0.016$, and $c_2= 0.0012$ (SI units). The inset plots the T dependence of the ratio $L_e(T) \equiv \kappa_e(T)/p(T)$ (in which ℓ_0 cancels). In the vortex scattering model, L_e should vary as T^2 in a d-wave superconductor at low T (Eq. 3)

the *zero-field* value of the mean-free-path, and ϕ_0 the flux quantum. Thus, in this model, the QP conductivity is given by Eq. 1 with [10, 11, 4]

$$p(T) = \ell_0 \sigma_{tr}/\phi_0. \tag{2}$$

In the Boltzmann equation approach, we may write the QP thermal conductivity (in zero field) as [12]

$$\kappa_e(T) = \frac{1}{T}\sum_{\mathbf{k}}(-\frac{\partial f}{\partial E})E(\mathbf{k})^2 v_x(\mathbf{k})^2 \tau(\mathbf{k}), \tag{3}$$

where $E(\mathbf{k})$ and $\tau(\mathbf{k})$ are, respectively, the energy and lifetime of a QP in the state $\mathbf{k}$, and $\mathbf{v}(\mathbf{k}) = \hbar^{-1}\nabla E(\mathbf{k})$ its group velocity. In a d-wave superconductor at low T, the excitations are confined to Dirac cones at the nodes where the energy may be parametrized as $E(k_1, k_2) = \hbar\sqrt{(k_1 v_f)^2 + (k_2 v_2)^2}$ (k_1 and k_2 are the components of $\mathbf{k}$ normal and parallel to the Fermi Surface), κ_e reduces to

$$\kappa_e(T) = \frac{\eta}{\pi} \frac{k_B^3 T^2}{\hbar}^2 \frac{\ell_0}{v_2}, \tag{4}$$

where $\ell_0 = v_f \tau_0$ is the mean free path at the nodes, and $\eta \equiv \int_0^\infty dx\, x^3(-df/dx) \sim 5.41$.

The T^2 dependence of κ_e in Eq. 4 is masked by the strong T dependence of ℓ_0. However, in our experiment, the latter is obtained independently from the field dependence (with $p(T)$ given by Eq. 2). We may divide out the T dependence of ℓ_0, to isolate the quantity $L_e(T) = \kappa_e(T)/p(T)$. Comparing Eqs. 2 and 3, we are left with an expression that contains only two material-specific parameters v_2 and σ_{tr}, viz.

$$L_e(T) = \frac{\eta}{\pi} \frac{k_B^3 T^2}{\hbar^2} \frac{\phi_0}{v_2 \sigma_{tr}}, \tag{5}$$

Figure 3 (inset) reveals that the measured $L_e(T)$ displays a nearly T^2 dependence at low T, that may be fitted to give $L_e(T) = 6.9 \times 10^{-3} T^2$ WT/mK. Comparing this expression with Eq. 5, we determine from our experiment

$$v_2 \sigma_{tr} = 2.11 \times 10^{-4} \text{m}^2/\text{s}. \tag{6}$$

4 The Hall angle

The field parameter $p(T)$ has been extracted from κ_{xx} alone. While its nominally $1/T^2$ variation (Fig. 3) is consistent with the identification $p(T) \sim \ell_0$ (Eq. 2), it is important to see if this is consistent with a separate experiment. We turn next to the Hall conductivity κ_{xy}. In the previous Hall study on optimal YBCO [4], κ_{xy} was analyzed without the benefit of information on the diagonal electronic current. From the analysis above, we may now extract the thermal Hall angle $\tan\theta_Q(H) = \kappa_{xy}(H)/\kappa_e(H)$ as a continuous function of H at each temperature. In general, $\tan\theta_Q(H)$ displays strong negative curvature vs. H [13]. Here, we restrict our discussion to the weak-field value $\theta_Q(0)$.

In underdoped YBCO, the small QP population generates a weak thermal Hall current, and the uncertainties in determining θ_Q are quite large (compared to optimum YBCO) [13]. Nevertheless, we find two interesting features of the Hall angle (see Fig. 4). First, $\theta_Q(0)$ (solid triangles) and p (open circles) share the same T-dependence from T_c to about 20 K. Secondly, we recall that the normal-state electrical Hall angle θ_N^e (open triangles) follows a $1/T^2$ dependence. The new values for $\tan\theta_Q(0)$ lie on the curve for θ_N^e extrapolated below T_c. As shown in Fig. 4, the three quantities p, $\tan\theta_Q(0)$ and $\tan\theta_N^e$

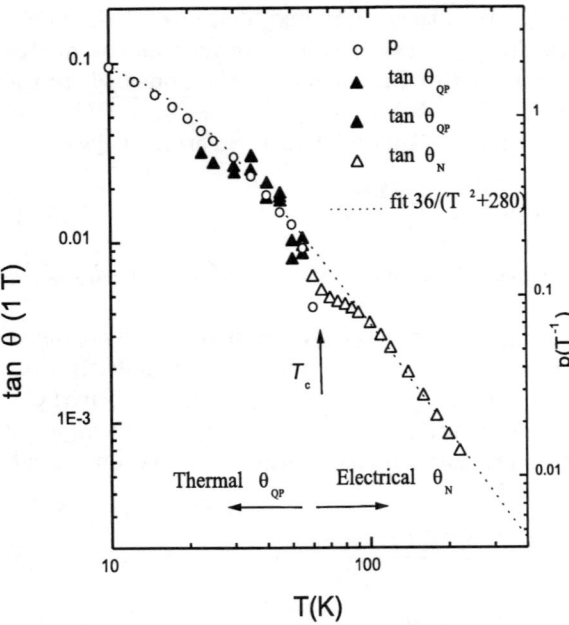

Fig. 4. The temperature dependences of the field scale $p(T)$ (open circles) and the weak-field QP Hall angle below T_c (closed triangles) and the electrical Hall angle $\tan\theta_N^e$ above (open triangles). The close similarity of T-dependences of p and $\tan\theta_Q$ strongly suggests that p is proportional to the QP lifetime. The broken line shows that the QP weak-field Hall angle is numerically equal to the extrapolation of $\tan\theta_N^e$.

fall on the same curve over about 2.5 decades (in the plot p and $\tan\theta_Q(0)$ are related by a constant scale factor). Just above T_c, $\tan\theta_N^e$ displays a slight dip associated with fluctuation effects. The similarity between θ_Q and θ_N^e has also been pointed out by Zeini et al. [14].

We briefly discuss our interpretation. These results suggest that the $1/T^2$ dependence is, in fact, intrinsic to the QP's below T_c. The ubiquitous T^2 dependence of $\cot\theta_N^e$ in the normal state appears to be an extension of the low-temperature behavior into the normal state. The similarity between the T-dependences of p and θ_Q imply that the diagonal and the Hall channels relax with the same T-dependence, $\sim 1/T^2$, consistent with simple Drude behavior. Just as in conventional metals, the thermal Hall resistivity in the mixed state, $W_{xy} \equiv \kappa_{xy}/\kappa_e^2 = \tan\theta_Q(0)/\kappa_e$, should provide a measure of the heat capacity of the QP population, as it does (see inset of Fig. 3).

This conventional behavior is abruptly altered when we cross T_c into the normal state. The Hall angle continues to relax at the same numerical rate. By contrast, the scattering rate in the diagonal conductivity undergoes a dramatic change [13]. The transport mean-free-path ℓ_0 decreases abruptly by a factor of 4-6 (10 in optimal crystals) across T_c. This sharp decrease, followed by a nominally T-linear scattering rate, is responsible for the anomalously strong T- dependence of the Hall coefficient in the normal state. Thus, the anomalous channel responsible for most of the strange-metal properties is the diagonal conductivity. The Hall channel appears to be quite conventional. A detailed discussion of these results including optimal crystals will appear elsewhere [13].

5 Discussion

We have discussed how the field dependence of κ_{xx} in underdoped YBCO may be analyzed to extract electronic parameters, such as ℓ_0 and σ_{tr}. The analysis derives from two striking features of κ_{xx} intrinsic to high-purity 60-K YBCO crystals at low T, namely the steep decrease of κ_{xx} in weak field followed by a saturation at high field, and the absence of resolvable hysteresis.

The first feature allows us to fit a much broader range of field scales (as expressed by the dimensionless parameter $p(T)B$). The higher density of measurements also helps. We illustrate this point as follows. In a previous attempt [7] to analyze similar measurements in LSCO, it was found that the κ_{xx} vs. H curves were equally well-fitted by Eq. 1 and the expression $G(B) = \psi(1/2 + B_0/B) + \ln(B_0/B)$, with $\psi(x)$ the digamma function. With the data scatter and the smaller range of reduced fields B/B_0 in LSCO, the two fits could not be distinguished, and Ong et al. [7] argued the case for adopting $G(B)$ to describe κ_{xx} in LSCO. However, with the larger field scale pB here, we find that Eq. 1 provides a much better fit (this is evident if the two fits are compared in a plot versus $\ln H$). The physics underlying the two fit expressions is of course quite different. In light of the present work, we now favor adopting Eq. 1, instead of the digamma function fit, for analyzing κ_{xx} vs. H curves in cuprates.

The second feature (no observable hysteresis) relates to the issue of remanence and vortex pinning in cuprates. It is known that the relaxation of non-equilibrium flux distributions may produce a slow drift in κ_{xx} if it is observed a few seconds after a change in H is made [3, 6]. Further, vortex pinning effects at low T can lead to step-like jumps in κ_{xx} when the field sweep direction is changed. In optimum YBCO and in strongly overdoped Bi 2212, we find that κ_{xx} increases step-wise when the sweep direction of H is reversed from up to down. Recently, however, a hysteretic loop of the opposite sign has been reported by Aubin et al. [8] in Bi 2212 (κ_{xx} decreases step-wise when H is swept down). At present, the origin of the hystereses in κ_{xx} is not understood (especially the existence of hystereses with different signs). We note that the magnitude of the hystereses in is much smaller than that observed in the magnetization M vs. H. In our measurements on underdoped YBCO, no hysteresis in κ_{xx} is observed for fields as large as 14 T at temperatures down to 6 K, even though hystereses are sizeable in the M vs. H curves. In particular, the magnitude of κ_{xx} at the plateau-like region is not hysteretic. The absence of hysteresis implies that the magnetization is too small to influence the measured κ_{xx}, so we may assume that $B = \mu_0 H$, as tacitly assumed in the fits.

By contrast, hysteretic effects cannot be neglected in optimally doped YBCO (as discussed above). Stronger vortex pinning is clearly responsible for the larger hysteresis in the 93-K crystal. Below 35 K in this crystal, the hysteresis steadily increases to about 5% at 8 K. Although the hysteresis is small, the remanence produces in the trace of κ_{xx} vs. H both a broadening at small H and an asymmetry about $H = 0$ that strongly distort the fit to Eq. 1. The distortions preclude a meaningful extraction of below about 35 K. Thus, of the various phases of the cuprates we investigated (YBCO, Bi 2212 and LSCO), the underdoped phase of YBCO appears to be the most suitable for our purpose of isolating the QP current from the total thermal current.

In high-purity single-domain crystals of Bi 2212, the field dependence of κ_{xx} displays a distinct break in slope in κ_{xx} at a characteristic field H_k, followed by a plateau region in which κ_{xx} is nearly independent of H [5]. The field H_k, which varies approximately as T^2, was interpreted as a field-induced phase transition, possibly involving a new order parameter. We compare the present results with the two findings in Bi 2212, i.e. the kink feature at H_k and the existence of the plateau. As shown in Fig. 2, κ_{xx} in 60-K YBCO smoothly crosses over into the field-independent region at low T, instead of displaying a sharp kink. While the plateau regime is similar in the two systems, the kink feature signalling a phase transition is absent in the YBCO crystal. A difference between the two systems is the electronic anisotropy. From the resistivity anisotropy ρ_c/ρ_{ab} ($\sim 10^5$ compared with 10^3), Bi 2212 is much closer to the $2D$ limit than underdoped YBCO. Whether this is a significant factor is a subject for future investigation.

In their experiment Aubin et al. [8] observed the value of κ_{xx} in Bi 2212 at the plateau (at 8 K) to be 1% higher in the field sweep-up direction than in sweep-down. They suggest that the plateau may be associated with, or reflect a specific state of the vortex system. The hysteresis in Aubin's sample is about 5 times larger (at 8 K) than in the two crystals used by Krishana et al. [5, 9]. At higher temperatures, 15-20 K, where the plateau is just as prominent, the hysteresis is almost unresolved. The hysteresis is an extrinsic effect possibly associated with stronger flux pinning in overdoped crystals. The present results in YBCO show that, whenever the QP current is very strongly suppressed by the available field, the plateau that remains is non-hysteretic. Hence, it seems unlikely to be related to a specific state of the vortices. However, to access the plateau, it seems necessary to work in the underdoped regime and to use high-purity crystals with weak pinning (as discussed above, we are unable to access the plateau region in 90-K YBCO). In an interesting calculation, Franz [17] has shown recently that the plateau in our YBCO experiment could result from exact cancellation between scattering from a disordered vortex array and the increase in quasiparticle density produced by the field (from the Doppler-shift term $\mathbf{v_s} \cdot \mathbf{v_k}$). At the plateau, κ_e attains the universal value [18] ($\kappa_{00}^e \sim 0.1$ W/mK at 10 K). In principle, this final electronic value should be added to κ_{ph} in our empirical background term κ_B. Further experiments to test this idea are underway.

Simultaneous measurement of both κ_{xy} and κ_{xx} also provides consistency checks on the interpretation of the field-dependence of the thermal conductivity. The weak-field Hall angle may be expressed as a 'Hall' mean-free-path $\ell_H \equiv \theta_Q \hbar k_F/e$. The value of θ_Q at 10 K gives $\ell_H \sim 4,200 \text{Å}$. The similarity of the T dependences in θ_Q and $p(T)$ provides strong evidence that the entire field dependence of $\kappa_x x(H,T)$ derives from the scattering of QP by the vortex array. The phonon current, although sizeable, seems to play no role in the field dependence. We acknowledge support from the U.S. Office of Naval Research and the U.S. National Science Foundation. Useful conversations with Philip Anderson, Marcel Franz, Duncan Haldane, Patrick Lee, and Shin-Ichi Uchida are gratefully acknowledged.

†*Permanent address: Department of Physics, Zhejiang University, Hangzhou, China.*

References

1. S.J. Hagen, Z.Z. Wang, and N.P. Ong, Phys. Rev. B **40**, 9389 (1989).
2. R. C. Yu, M. B. Salamon, Jian Ping Lu, and W. C. Lee, Phys. Rev. Lett. **69**, 1431 (1992).
3. R. A. Richardson, S. D. Peacor, C. Uher, and Franco Nori, J. Appl. Phys. **72**, 4788 (1992); S. D. Peacor, J. L. Cohn and C. Uher, Phys. Rev. B **43**, 8721 (1991).
4. K. Krishana, J. M. Harris, and N. P. Ong, Phys. Rev. Lett. **75**, 3529 (1995).
5. K. Krishana, N. P. Ong, Q. Li, G. D. Gu, and N. Koshizuka, Science **277**, 83 (1997).
6. R. A. Richardson, S. D. Peacor, Franco Nori, and C. Uher, Phys. Rev. Lett. **67**, 3856 (1991); Phys. Rev. B **44**, 9508 (1991).
7. N. P. Ong, K. Krishana, and T. Kimura, Physica C **282-287**, 244 (1997).
8. H. Aubin, K. Behnia, S. Ooi, and T. Tamegai, Science **280**, 11 (1998).
9. K. Krishana et al., Science **280**, 11 (1998).
10. Robert M. Cleary, Phys. Rev. **175**, 587 (1968).
11. W.F. Vinen, E. M. Forgan, C.E. Gough and M. J. Hood, Physica **55**, 94 (1971).
12. J. Bardeen, G. Rickayzen, and L. Tewordt, Phys. Rev. **113**, 982 (1959).
13. K. Krishana, N. P. Ong, Z.A. Xu, Y. Zhang, R. Gagnon and L. Taillefer, to be published.
14. B. Zeini et al., preprint 1998. Zeini et al. find a factor of 2 discrepancy between the two Hall angles. This may relate to their assumption that $\tan\theta_Q$ is strictly linear in H, whereas it has significant curvature in our data.
15. W. N. Hardy, D. A. Bonn, D. C. Morgan, R. Liang, and K. Zhang, Phys. Rev. Lett. **70**, 3999 (1993).
16. Xiao-Gang Wen and Patrick A. Lee, Phys. Rev. Lett. **80**, 2193 (1998); P. A. Lee and X. G. Wen, *ibid.***78**, 4111 (1997).
17. Marcel Franz, preprint 1998.
18. P. A. Lee, Phys. Rev. Lett. **71**, 1887 (1993).

A New-Type Superconducting State Locally Induced Around a Vortex in the Two-Dimensional t-J Model

Masao OGATA*

Department of Basic Science, Graduate School of Arts and Sciences, University of Tokyo, Komaba, Meguro-ku, Tokyo 153

We study ground states in the two-dimensional t-J model and develop a microscopic theory for vortices in $d_{x^2-y^2}$-wave superconducting state, using a variational theory within a Gutzwiller approximation. A new superconducting state is found near half-filling. This state has Cooper pairs (with finite total momentum) between the residual quasiparticles near the nodes of $d_{x^2-y^2}$-wave state. Consequently there is no node in the gap function of the new state, which can be observed experimentally. For $\delta < \delta_c$ this state is a bulk state, and for $\delta > \delta_c$ it is also induced around vortex cores when a magnetic field is applied. This means that a vortex plays a role as a nucleation center of the new state with full gap. The induced state causes the splitting of the zero-energy peak, which gives a possible explanation for the experimental data of scanning tunneling spectroscopy.

I. INTRODUCTION

Symmetry of Cooper pairs is considered as a key to elucidate the mechanism of high-T_c superconductivity. Many extensive experiments suggest that the unconventional $d_{x^2-y^2}$ anisotropic pairing is realized in high-T_c oxides [1]. Interesting phenomena due to this anisotropic pairing have been studied phenomenologically. Actually the two-dimensional (2D) t-J model, which is considered as an effective Hamiltonian for the high-T_c oxides [2], will probably have a ground-state with the $d_{x^2-y^2}$-wave pairing state for the reasonable parameters.

Recently, however, there are some experiments which cannot be explained in a simple $d_{x^2-y^2}$-wave pairing [3–6]. These are (1) scanning tunneling spectroscopy (STS) near the vortex core [3,4], (2) STS at the surface of a sample [5] and (3) thermal conductivity $\kappa(H)$ under an external magnetic field [6]. In (1), a splitting or an absence of the zero-energy conductance peak is found near the vortex core, which suggests a kind of gap opening. Because both (1) and (3) are related to the quasiparticle behaviors in the vortex core or around the vortices, it is urgent to develop a theory which gives reliable results at least about the vortex states in the superconducting ground state to understand these anomalous behaviors observed in high-T_c materials.

In this paper we develop a microscopic theory of vortices in the $d_{x^2-y^2}$-wave superconducting state in the 2D t-J model. So far the extended Hubbard model or Ginzburg-Landau theory have been used to study the vortex states. In these models, it has been shown that the electronic state around a vortex core looks

---

*Electronic address: ogata@sola.c.u-tokyo.ac.jp

like a conventional one [7]. However, as we will show shortly, there appear special situations in the 2D t-J model near half-filling. First we will show that a new kind of superconducting state with spatially-oscillating order parameters is stabilized close to the half-filling within the Gutzwiller approximation [8]. This new state breaks the translational symmetry. Since all the previous calculations assumed spatially uniform order parameters, stabilization of the present new state was not found before. Moreover it turns out that the new state has no node of the gap function in contrast to the $d_{x^2-y^2}$-wave symmetry case. We find that the existence of the $d_{x^2-y^2}$-wave nodes close to $(\pm\frac{\pi}{2}, \pm\frac{\pi}{2})$ is essential for the instability toward the new state. Then we will show that this kind of new state is also induced around a vortex core. Since the new state has a finite gap everywhere on the Fermi surface, the zero-energy conductance peak near the vortex core splits into two levels [9]. Our results give a possible explanation for the STS observation.

In §2 we first show that a new mean-field-type solution (bulk) is possible in the low-doping region, i.e., for $\delta < \delta_c$. And then in §3 we discuss that this new state is induced around a vortex even for $\delta > \delta_c$. Apparently the doping dependence is important, which can be investigated only when we consider realistic models for high-T_c materials, for example, 2D t-J model.

II. FORMULATION AND A NEW MEAN-FIELD SUPERCONDUCTING STATE NEAR HALF-FILLING

The t-J model is derived by regarding a singlet of a Cu spin and a doped hole as a mobile vacancy in the Heisenberg spin system [2]

$$\mathcal{H}_{tJ} = -\sum_{\langle ij \rangle \sigma} P_G(tc_{i\sigma}^\dagger c_{j\sigma} + \text{h.c.}) P_G + \sum_{\langle ij \rangle} J \boldsymbol{S}_i \cdot \boldsymbol{S}_j, \tag{2.1}$$

where $\boldsymbol{S}_i = c_{i\alpha}^\dagger (\frac{1}{2}\boldsymbol{\sigma})_{\alpha\beta} c_{i\beta}$, and the Gutzwiller projection operator P_G is defined as $P_G = \Pi_i(1 - \hat{n}_{i\uparrow}\hat{n}_{i\downarrow})$, which prohibits double occupied sites.

For this Hamiltonian, we assume a variational state defined as

$$\Psi = P_G \Phi, \tag{2.2}$$

where the function Φ is a one-body mean-field-type wave function, in particular a BCS-type variational state with various pairing. In order to study all the possible states, we assume a variational state $P_G|\text{BCS}(\Delta_{ij}, \chi_{ij\sigma})\rangle$ with site-dependent order parameters Δ_{ij} $(= \langle c_{i\uparrow}^\dagger c_{j\downarrow}^\dagger \rangle)$ and $\chi_{ij\sigma}$ $(= \langle c_{i\sigma}^\dagger c_{j\sigma} \rangle)$, which are to be determined so as to minimize the variational energy

$$E_{\text{var}} = \langle \text{BCS}(\Delta_{ij}, \chi_{ij\sigma})|P_G \mathcal{H}_{tJ} P_G|\text{BCS}(\Delta_{ij}, \chi_{ij\sigma})\rangle. \tag{2.3}$$

It is usually difficult to estimate E_{var} due to the Gutzwiller projection. Here, we use a Gutzwiller approximation [10] in which the effect of projection is taken into account as statistical weights.

In this approximation the variational energy is rewritten as

$$E_{\text{var}} = \langle \text{BCS}(\Delta_{ij}, \chi_{ij\sigma})|\mathcal{H}_{\text{eff}}|\text{BCS}(\Delta_{ij}, \chi_{ij\sigma})\rangle. \tag{2.4}$$

In the effective Hamiltonian $\mathcal{H}_{\text{eff}}$, the parameters t and J are replaced with $t_{\text{eff}} = g_t t$ and $J_{\text{eff}} = g_s J$, respectively with [10]

$$g_t = \frac{2\delta}{1+\delta}, \quad g_s = \frac{4}{(1+\delta)^2}. \tag{2.5}$$

Here δ is the hole concentration, $\delta = 1 - n$. One of the important effects of correlation is the replacement of t by t_{eff}; and approaching the half-filling the effective hopping is reduced. For the case of uniform $|\text{BCS}(\Delta, \chi)\rangle$ state, the variational energies were compared with those obtained in the variational Monte Carlo method [11]. It was found that the two methods give very similar variational energies and thus we think the Gutzwiller approximation gives a fairly reliable estimate of E_{var} even quantitatively.

Minimizing E_{var}, we obtain a Bogoliubov-de Gennes equation and a set of self-consistent equations in a similar way to the BCS mean-field theory:

$$\begin{pmatrix} H_{ij} & F_{ij} \\ F_{ji}^* & -H_{ji} \end{pmatrix} \begin{pmatrix} u_j^\alpha \\ v_j^\alpha \end{pmatrix} = E^\alpha \begin{pmatrix} u_i^\alpha \\ v_i^\alpha \end{pmatrix}, \tag{2.6}$$

with

$$H_{ij} = -\sum_\delta \left\{ t_{\text{eff}} + \frac{3}{4} J_{\text{eff}} \chi_{ji} \right\} \delta_{j=i+\delta} - \mu \delta_{ij},$$

$$F_{ij}^* = -\sum_\delta \frac{3}{4} J_{\text{eff}} \Delta_{ij} \delta_{j=i+\delta}, \tag{2.7}$$

and self-consistent equations

$$\Delta_{ij} = \langle c_{i\uparrow}^\dagger c_{j\downarrow}^\dagger \rangle = -\frac{1}{4} \sum_\alpha (u_i^{\alpha*} v_j^\alpha + u_j^{\alpha*} v_i^\alpha) \tanh \frac{\beta E^\alpha}{2},$$

$$\chi_{ij} = \langle c_{i\sigma}^\dagger c_{j\sigma} \rangle = -\frac{1}{4} \sum_\alpha (u_i^{\alpha*} u_j^\alpha - v_j^{\alpha*} v_i^\alpha) \tanh \frac{\beta E^\alpha}{2}. \tag{2.8}$$

Here, we assumed $\chi_{ij\uparrow} = \chi_{ij\downarrow} = \chi_{ij}$ and that Δ_{ij} is a singlet pair, i.e., $\Delta_{ij} = \Delta_{ji}$.

We have tried various spatial patterns for the order parameters Δ_{ij} and χ_{ij} to determine the self-consistent solution without an external magnetic field. As a result we find a new type of superconducting state as shown in Fig. 1 near half-filling [8]. In Fig. 1, solid lines represent the bonds where Δ_{ij} and χ_{ij} have larger absolute values Δ_1, χ_1 than on the bonds represented by dashed lines ($\Delta_1 > \Delta_2, \chi_1 > \chi_2$).

Δ_{ij} on the horizontal bonds ($\Delta_{i,i\pm\hat{x}}$) have alternating values in the x-direction, but they are always real and positive, while those on the vertical bonds ($\Delta_{i,i\pm\hat{y}}$) are always real and negative. Here, the lattice constant is set to be unity, and $\hat{x}$ ($\hat{y}$) represents a unit vector in the x (y) direction, respectively. $\chi_{ij} = \chi_1, \chi_2$ are real and positive for all the bonds. Since all the order parameters are real, this state does not break the time-reversal symmetry, but instead it breaks the

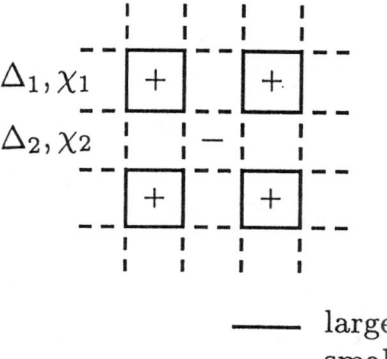

FIG. 1. Spatial configuration of new superconducting state.

translational symmetry. The pattern of order parameters shown in Fig. 1 is interpreted as follows: a spatially-oscillating component is induced in addition to the original $d_{x^2-y^2}$-wave order parameter. As we can see from Fig. 1, all the plaquettes are classified into four kinds. In the plaquette which is depicted as "+", the $d_{x^2-y^2}$-wave component is enhanced, while in the plaquette shown by "−", it is suppressed.

For $J/t = 0.3$ where high-T_c superconductivity occurs, this state is realized as a uniform bulk ground-state in the region $\delta < \delta_c$ ($\delta_c = 7.0\%$). In the following section we show that this state is induced around a vortex even for $\delta > \delta_c$.

Calculating the density of states in the new state, we find that the original V-shape density of states of the $d_{x^2-y^2}$-wave superconductivity is destroyed near the Fermi energy and it has a finite gap. This means that the nodes of the $d_{x^2-y^2}$-wave disappear in this new state and there is a finite gap everywhere on the Fermi surface. We check all the eigenvalues in the Bogoliubov-de Gennes equation inside the whole Brillouin zone to find that the lowest excited state is located at $\bm{k} = (\frac{\pi}{2}, \frac{\pi}{2})$, where $d_{x^2-y^2}$-wave has the node.

In order to see the mechanism of the gap opening, we define induced order parameters as $\Delta_{ij}^{d} + \Delta_{ij}^{\text{induced}}$ where Δ_{ij}^{d} is the original uniform $d_{x^2-y^2}$-wave order parameters. Fourier transformation of $\Delta_{ij}^{\text{induced}}$ shows that the induced Cooper pairs have total momenta $\bm{Q}_x = (\pi, 0)$ and $\bm{Q}_y = (0, \pi)$, i.e.,

$$\langle c^\dagger_{\bm{k}\uparrow} c^\dagger_{-\bm{k}+\bm{Q}_x\downarrow}\rangle^{\text{induced}} = 2i\Delta^{\text{induced}} \sin k_x,$$
$$\langle c^\dagger_{\bm{k}\uparrow} c^\dagger_{-\bm{k}+\bm{Q}_y\downarrow}\rangle^{\text{induced}} = -2i\Delta^{\text{induced}} \sin k_y. \qquad (2.9)$$

Although $\sin k_x$ and $\sin k_y$ factors look like a p-wave pairing, the Cooper pairs are singlet because of the finite total momentum. The spatial oscillation of the order parameter in Fig. 1 results in the Cooper pairs with finite total momenta $\bm{Q}_x$ and $\bm{Q}_y$.

These Cooper pairs are formed between the *residual* quasiparticles which exist in the $d_{x^2-y^2}$-wave superconducting state around its nodes at $\bm{k} = (\pm\frac{\pi}{2}, \pm\frac{\pi}{2})$. Figure 2 shows the the Fermi surface of the cosine band near half-filling and the induced Cooper pair with finite total momentum $\bm{Q}_x = (\pi, 0)$. The solid circles

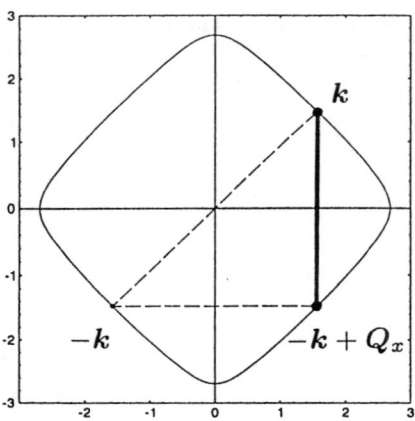

FIG. 2. Fermi surface and the Cooper pairing with finite total momentum appearing in the new state.

indicate the momenta k and $-k + Q_x$ which are located close to the positions of the $d_{x^2-y^2}$-wave nodes. Actually $\sin k_x$ in eq. (2.9) has maximum at $k_x = \frac{\pi}{2}$ and the pair shown in Fig. 2 has the maximum amplitude. Due to this pairing between the residual quasiparticles, the original nodes of $d_{x^2-y^2}$-wave symmetry disappear in the new state and the superconducting gap opens everywhere on the Fermi surface.

In this sense, the new state is special near half-filling. Also the spatial oscillation of χ_{ij} (shown in Fig. 1) helps to stabilize the new state, since the phase space for the residual quasiparticles around the nodes is small [8]. The presence of the order parameter χ_{ij} is also characteristic in the t-J model.

III. ELECTRONIC STATES AROUND A VORTEX IN THE 2D T-J MODEL

Next we study the vortex in the t-J model when $\delta > \delta_c$. For the study of the vortex, we use the Bogoliubov-de Gennes equation (2.6) and the self-consistency equations (2.8) under a uniform magnetic field [9]. The uniform magnetic field is introduced in terms of Peierls phase of the hopping term as

$$t_{ij} = t_{\text{eff}} \exp(i \frac{e}{\hbar c} \int_i^j \bm{A} \cdot d\bm{r}). \tag{3.1}$$

We assume a square vortex lattice and each $N \times N$ site has one vortex; typically we use $N = 18$. In order to make the Peierls phase compatible with vortex lattice symmetry, we need a magnetic unit cell with $2N \times N$ sites including 2 vortices. In this case, by the appropriate choice of gauge [12], the Peierls phase and the order parameter Δ_{ij} have a translational symmetry with respect to the magnetic unit cell. We solve the Bogoliubov-de Gennes equation numerically

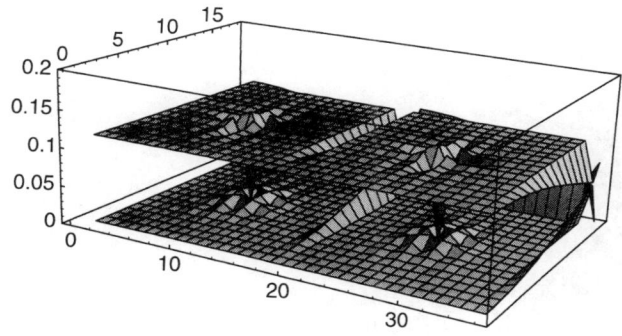

FIG. 3. Spatial dependence of the amplitudes of d-wave (upper) and s-wave (lower) components of the order parameter around vortex cores in the t-J model. The doping rate is $\delta = 0.125$ and $J/t = 0.3$. The order parameter in each plaquette is defined by averaging $\Delta_{ij}(= \langle c_{i\uparrow}^\dagger c_{j\downarrow}^\dagger \rangle)$'s on the two parallel sides of the plaquette.

and carry out an iteration until the self-consistent equations for Δ_{ij} and χ_{ij} are satisfied.

Figure 3 shows the obtained superconducting order-parameters near the vortex core for $\delta = 0.125$ and $J/t = 0.3$.

The center of the vortex is located in the middle of a plaquette so that we define $\Delta_x(\boldsymbol{r})$ and $\Delta_y(\boldsymbol{r})$ for each plaquette by averaging Δ_{ij}'s on the two parallel sides of the plaquette. The obtained $\Delta_x(\boldsymbol{r})$ and $\Delta_y(\boldsymbol{r})$ are decomposed into extended s-wave and d-wave components as

$$\Delta_x(\boldsymbol{r}) = \left(\Delta_d(\boldsymbol{r}) + e^{i\alpha}\Delta_s(\boldsymbol{r})\right) e^{i\theta},$$
$$\Delta_y(\boldsymbol{r}) = \left(-\Delta_d(\boldsymbol{r}) + e^{i\alpha}\Delta_s(\boldsymbol{r})\right) e^{i\theta}, \tag{3.2}$$

where θ is the angle from the nearest vortex center.

We can see that the similar states as in Fig. 1 is locally induced around the core. Since the vortex core is located in the middle of a plaquette, $d_{x^2-y^2}$-wave component Δ_d is suppressed at that plaquette. This situation is very close to the situation in Fig. 1 and actually we can expect that the vortex center corresponds to the plaquette with "$-$" symbol in Fig. 1. The oscillating behavior of the $d_{x^2-y^2}$-wave component around the vortex core in Fig. 3 is close to the configuration in Fig. 1, which is characteristic to the new state. This indicates that a vortex plays a role as a *nucleation center* of the new state under a magnetic field.

Next we discuss the relationship to the STS experiment. In the same approximation we calculate the local density of states near the vortex core. This quantity is directly related to the tunnel conductance of the STS experiment.

For the conventional $d_{x^2-y^2}$-wave case, the local density of states in the vicinity of the core has a zero-energy peak [7]. This zero-energy peak is interpreted as follows: In the quasi-classical picture [13] the quasiparticles inside the vortex core can propagate in the direction of 45 degrees, since they do not feel the

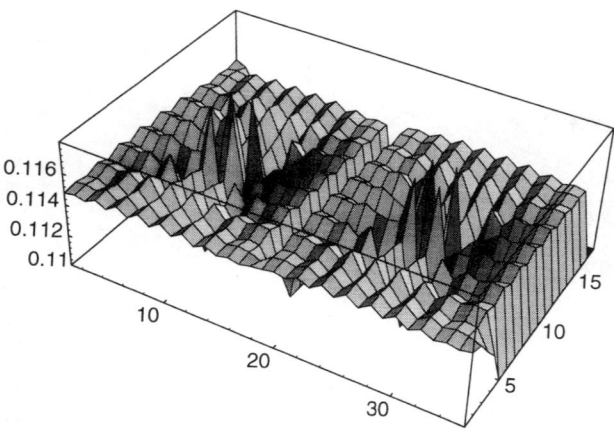

FIG. 4. Close-up of the spatial dependence of the amplitudes of d-wave component of the order parameter around vortex cores in the t-J model.

pair potential because of the gap node (i.e. $\Delta(\boldsymbol{k}) = 0$ in this direction). This extended state has the zero energy and forms the zero-energy conductance peak. This actually happens when the doping is increased in the t-J model.

On the contrary, in the low-doping case of the t-J model, the quasiparticle inside the core is approximately localized [9]. This localization is understood in terms of the new state discussed in the previous section. The quasiparticles which try to propagate in the 45-degree directions are blocked by the new state with full gap. As a result, the situation becomes similar to the s-wave superconducting state and thus the zero-energy peak splits into two levels. We expect that this explains the splitting of the zero-energy peak observed in the STS of $YBa_2Cu_3O_{7-\delta}$ [3].

As the doping rate increases, the induced spatial oscillation becomes small and as a result the vortex core approaches the conventional one in the pure $d_{x^2-y^2}$-wave superconductivity. The enhancement of the new state is characteristic in the low-doping region.

In order to see to what extent the vortex induces the new state around the core, we show a close-up picture of the $d_{x^2-y^2}$ component in Fig. 4.

The spatial oscillation is apparent in particular in the x and y direction from the core. Although the amplitude of the oscillation is small, we think that each vortex induces this kind of gap state within a range of about ten lattice constants around its core. This induced new phase will be related to the recent discovery of field-induced transition into the new superconducting state [6].

Recently another possibility of $d_{x^2-y^2} + id_{xy}$ state is proposed [14,15], which can be induced by a magnetic field or magnetic impurities. In our approximation, we check the d_{xy} component around the vortex in the 2D t-J model. For each plaquette, d_{xy} component is defined as

$$\Delta_{d_{xy}} = \langle c^\dagger_{i\uparrow} c^\dagger_{i+\hat{x}+\hat{y}\downarrow} \rangle - \langle c^\dagger_{i+\hat{x}\uparrow} c^\dagger_{i+\hat{y}\downarrow} \rangle. \tag{3.3}$$

We find that the expectation value $\Delta_{d_{xy}}$ is actually induced around a vortex core, but its amplitude is very small ($|\Delta_{d_{xy}}| < 0.001$). Thus we conclude that $d_{x^2-y^2}+ id_{xy}$ state is not plausible in 2D t-J model. Moreover, since there is no interaction term between next-nearest-neighbor sites in the t-J model (such as $J' \boldsymbol{S}_i \cdot \boldsymbol{S}_j$ term), the finite expectation value $\Delta_{d_{xy}}$ does not lead to an excitation gap. Even if we include J' term, the amplitude of the energy gap is proportional to $J'\Delta_{d_{xy}}$, which becomes very small.

IV. SUMMARY AND DISCUSSION

In summary we studied the variational states in the 2D t-J model at $T = 0$. In the above, the Gutzwiller approximation is used but the slave-boson mean-field theory also gives the similar result.

We find that the new state with spatially oscillating order parameters can be realized near half-filling. This state has Cooper pairs with finite total momenta, which are formed between the residual quasiparticles near the nodes of the $d_{x^2-y^2}$-wave superconductivity. As a result the new state has a finite gap everywhere on the Fermi surface. For $\delta < \delta_c$ this state is a bulk state in the Gutzwiller approximation, and for $\delta > \delta_c$ this state is induced around vortex cores. In this sense, the vortex plays a role as a nucleation center of the new state with full gap. We think that this mechanism explains the splitting of the zero-energy peak in STS observed in YBCO and the field-induced gap state observed in $\kappa(H)$.

However there are only a few experiments so far, so that a more extensive comparison between the theory and experiment is necessary. In particular, a systematic study on the doping-dependence is important. The anomalous behavior in the low-doping region is characteristic in the t-J model. We predict that the splitting of zero-energy peak does not show up in a overdoped case. This doping dependence can be used to check the validity of the t-J model as well as the importance of strong correlation for understanding the microscopic properties of superconductivity in high-T_c materials. For the vortex cores in BSCCO materials, the STS data [4] are still mysterious. In any case, the material- and doping-dependences of the zero energy peak are to be studied in further detail experimentally as well as theoretically.

The authors wish to thank H. Shiba, H. Fukuyama, H. Takagi, N. P. Ong, R. B. Laughlin, A. V. Balatsky, H. Yasuoka, Ø. Fisher, Ch. Renner, H. Ebisawa, K. Machida, M. Sigrist, K. Yonemitsu, T. Nishio, and M. Ichioka for their useful discussions.

REFERENCES

[1] For a review, see D. J. Scalapino: Phys. Rep. **250**, 329 (1995).
[2] F. C. Zhang and T. M. Rice: Phys. Rev. B **37**, 3759 (1988).
[3] I. Maggio-Aprile, Ch. Renner, A. Erb, E. Walker and Ø. Fischer: Phys. Rev. Lett. **75**, 2754 (1995).
[4] Ch. Renner, B. Revaz, K. Kadowaki, I. Maggio-Aprile, and Ø. Fischer: Phys. Rev. Lett. **80**, 3606 (1998).

[5] M. Covington, M. Aprili, E. Paraoanu, L. H. Greene, F. Xu, J. Zhu, C. A. Mirkin, Phys. Rev. Lett. **79**, 277 (1997).
[6] K. Krishana, N. P. Ong, Q. Li, G. D. Gu and N. Koshizuka: Science **277**, 83 (1997).
[7] Y. Wang and A. H. MacDonald: Phys. Rev. B **52**, R3876 (1995).
[8] M. Ogata: J. Phys. Soc. Jpn. **66**, 3375 (1997).
[9] A. Himeda, M. Ogata, Y. Tanaka, and S. Kashiwaya: J. Phys. Soc. Jpn. **66**, 3367 (1997).
[10] F. C Zhang, C. Gros, T. M. Rice, and H. Shiba: Supercond. Sci. Technol. **1**, 36 (1988).
[11] H. Yokoyama and M. Ogata: J. Phys. Soc. Jpn. **65**, 3615 (1996).
[12] T. Nishio, K. Yonemitsu and H. Ebisawa: J. Phys. Soc. Jpn. **66**, 953 (1997).
[13] M. Ichioka, N. Hayashi, N. Enomoto and K. Machida: Phys. Rev. B **53**, 15316 (1996).
[14] R. B. Laughlin: preprint.
[15] A. V. Balatsky: Phys. Rev. Lett. **80**, 1972 (1998).

Instability of a 2-Dimensional Landau-Fermi Liquid Due to Umklapp Scattering

N. Furukawa[1] and T.M. Rice

Theoretische Physik, ETH-Hönggerberg, CH–8093 Zürich, Switzerland

Abstract. A model for the breakdown of Landau theory for a 2-dimensional Fermi liquid is proposed drawing on the behavior of ladder systems, where a partial or complete truncation of the Fermi surface can be driven by opening spin and charge gaps in the absence of symmetry breaking. The latter is driven by a divergence of Umklapp scattering. The possibility of similar behavior in two dimensions is examined. For the cuprates, the spin and charge gaps would spread out from the saddle points and reduce the Fermi surface to four disconnected segments, a form which agrees with several recent phenomenological models and recent experiments.

1. Introduction

The generic phase diagram of the high-T_c cuprate superconductors as a function of the variables temperature, T, and hole doping, δ, is by now generally accepted. But a comprehensive theory that covers the evolution with doping from the stoichiometric insulating antiferromagnet through the underdoped spin-gap phase and the anomalous metallic phase at optimal doping, to the more conventional overdoped regime, is still lacking. Here we will concentrate on one aspect, namely the breakdown of the Landau-Fermi liquid behavior in the overdoped regime as the hole density is lowered, or equivalently the electron density is increased. This evolution occurs rapidly with decreasing δ, and it is especially noteworthy that altho' the overdoped Fermi liquid has the highest normal state conductivity, it does not become superconducting at all. All these points toward an instability of the conventional Landau-Fermi liquid that is density driven, e.g., triggered by a critical hole or equivalently electron density.

A number of proposals have appeared which ascribe this breakdown to the proximity to some critical point and a transition to a long range ordered state. But these have the difficulty that the only symmetry breaking observed is to the d-wave superconducting state or to long range antiferromagnetic (AF) order

---

[1] on leave from I.S.S.P., Univ. of Tokyo, 7-22-1 Roppongi, Minato-ku, Tokyo 106, Japan

which only appears at a much smaller hole density. This suggests that the instability of the Landau-Fermi liquid is not associated with a symmetry breaking and that we should look elsewhere for its origin. In particular the instability could be a precursor to the Mott insulating state that occurs at half-filling. The charge gap of a Mott insulator signals the complete breakdown of the Fermi surface at the stoichiometric electron density. In the scenario where the magnetic interactions are dominant, the Fermi surface will be already partially truncated at finite hole density through the appearance of incommensurate AF order. The partial truncation becomes more complete until finally at half-filling there is a charge gap over the whole Fermi surface. But as we have said in the cuprates, experiments show a different behavior and point towards a partial truncation of the Fermi surface without any obvious symmetry breaking when the hole density is reduced below optimal doping.

Recently a lot of progress has been made on the understanding of ladder systems. These are intermediate between one and two dimensions and can be analyzed reliably in detail [1]. Their properties at low energies, or temperatures, are crucially dependent on their width. The single chain forms the well-known Tomonaga-Luttinger state which at half-filling develops a charge gap but not a spin gap. By contrast the 2-leg ladder forms a Luther-Emery liquid with holes bound in relative 'd-wave' pairs. At half-filling there is a charge gap but also a spin gap. In this case the spin-spin correlations are purely short range and one speaks of a spin liquid although one should be clear that this is not a disordered state but a unique quantum coherent groundstate. This behavior has been found in strong coupling, e.g. in numerical investigations of the t-J model, which is equivalent to the large U Hubbard model, but also in weak coupling qualitatively similar behavior is found [2]. This insulating spin liquid (ISL) can be regarded as a form of short range resonant valence bond (RVB) state. It is specially interesting to examine the weak coupling behavior. Away from half-filling the key criterion to obtain the Luther-Emery behavior is the presence of two bands (bonding and anti-bonding) at the Fermi energy. At half-filling additional Umklapp scattering processes enter the one-loop renormalization group (RG) equations and these scale to a strong coupling fixed point with a charge and spin gap [2]. Note that because of the purely short range order in the spin system, one cannot associate these gaps with symmetry breaking or long range order. Instead one has a Fermi system with a truncated Fermi surface which is not associated with a broken translational symmetry, — just the sort of behavior discussed earlier.

The 3-leg ladder is also very interesting. At half-filling the strong coupling limit is equivalent to the 3-leg Heisenberg $S = 1/2$ AF ladder. This has been well studied and reduces in the low energy sector to an effective single chain Heisenberg model with longer range but unfrustrated AF coupling [3]. The weak coupling limit shows a similar behavior (C0S1 in the Balents-Fisher notation (zero charge and one spin gapless modes)). When holes are introduced one finds in strong coupling calculations that these enter the channel with odd parity w.r.t.

reflection about the central leg and form a single channel Tomonaga-Luttinger liquid [4, 5]. The even parity channels remain at the stoichiometric filling and continue to form an ISL. Only after a critical hole density is reached, do the holes enter the even parity channels. The result is a finite region of hole doping where the original Fermi surface with 3 bands (or 6 Fermi points) is partially truncated to 2 Fermi points, but again without a broken translational symmetry. The key is again Umklapp scattering which has scaled to a strong coupling fixed point and which has introduced a charge gap in the even parity channels. This is then a clear example of a partially truncated Fermi surface through the formation of an ISL over part of the Fermi surface.

Recently we have examined the possibility of obtaining similar behavior in a two-dimensional system [6, 7]. The key to forming a charge gap lies in Umklapp scattering which clearly is responsible for the formation of a Mott insulator in one dimension. In two dimensions, the shape of the Fermi surface enters and there are two distinct possibilities that we will consider in turn. Either the Fermi surface bulges out towards the diagonals $(\pm 1, \pm 1)$ in the Brillouin zone or towards the saddle points. The former case is more straightforward to analyze but the latter case is the one which occurs in the cuprates and we will discuss them in turn.

2. 4–Patch Model

We start with a general 2-dim. dispersion relation, e.g. a form, $\varepsilon(\mathbf{k}) = -2t(\cos k_x + \cos k_y) - 4t' \cos k_x \cos k_y$ with $t(t')$ as (next) nearest neighbor hopping matrix elements. Taking $t > 0$ and $t' > 0$ and increasing the electron density, $n = 1 - \delta$, leads to a Fermi surface which touches the 4 points $(\pm \pi/2, \pm \pi/2)$. Following Haldane [8], we divide the Fermi surface into patches and examine the patches near the 4 points $(\pm \pi/2, \pm \pi/2)$. These 4 patches on the Fermi surface are connected through Umklapp processes which leads us to examine the renormalization group (RG) equations for the coupling constants [6]. The RG equations have similarities to those for a 2-leg ladder at half-filling which also has 4 Fermi surface points and which are known to scale to a strong coupling solution.

The 4 patches around $(\pm \pi/2, \pm \pi/2)$ are sketched in Fig. 1. The size of a patch is defined by a wave vector cutoff, k_c, and within each patch α ($\alpha = 1, \ldots, 4$) the electron energy relative to the chemical potential, μ is expanded as

$$\varepsilon_\alpha(\mathbf{q}) - \mu = vq_\alpha + uq_{\perp,\alpha}^2 \qquad (1)$$

where $\mathbf{q}$ is the wavenumber measured from the center of α-th patch, and $q_\alpha(q_{\perp,\alpha})$ is the component of $\mathbf{q}$ normal (tangent) to the Fermi surface at the center of the patch. The Fermi velocity is given by v, and the energy cutoff is $E_0 = vk_c$.

The linear dispersion relation leads to logarithmic anomalies in the particle-hole (Peierls) and particle-particle (Cooper) channels as in one dimension but

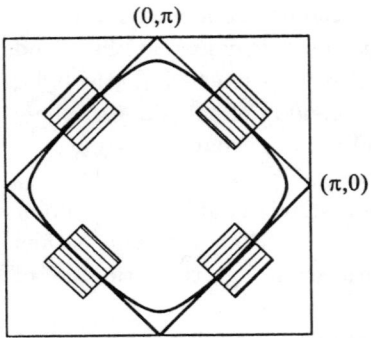

Fig. 1: Definitions of 4 patches, shown as hatched rectangular areas. The bold curve represents the 2-dimensional Fermi surface which touches the points $(\pm\pi/2, \pm\pi/2)$.

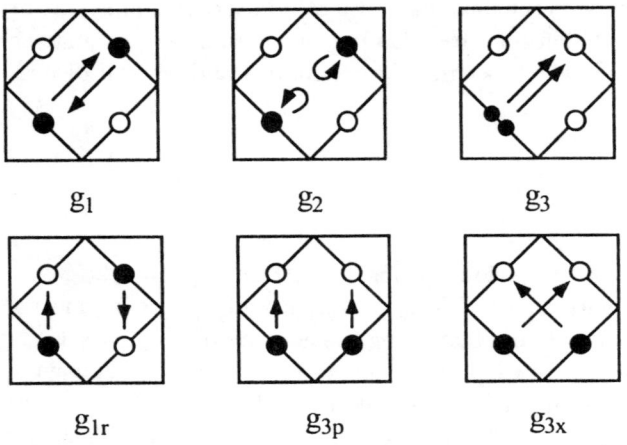

Fig. 2: The definitions of vertices for the 4-patch model.

the transverse dispersion introduces an infrared cutoff, $E_T \approx u k_c^2$. The non-interacting susceptibility in the Peierls channel takes the form $\chi_p(\omega) = 1/2 \ln (\max(\omega, E_T)/E_0)$. In the parameter region $\omega > E_T$, the infrared cutoff from the transverse dispersion can be ignored and a set of RG equations can be derived as in one dimension.

In Fig. 2 we define the normal vertices g_1, g_2 and g_{1r} as well as Umklapp vertices g_3, g_{3p} and g_{3x}. Other interactions are not treated here since they are irrelevant within the framework of a one-loop approximation. Summing up all one-loop diagrams, we obtain the RG equations

$$\dot{g}_1 = g_1^2 + g_{1r}^2 + 2g_{3x}^2 - 2g_{3x}g_{3p}, \qquad (2)$$

$$\dot{g}_2 = \frac{1}{2}\left(g_1^2 + 2g_{1r}^2 - g_3^2 - 2g_{3p}^2\right), \qquad (3)$$

$$\dot{g}_{1r} = (g_1 + g_2)\, g_{1r}, \tag{4}$$
$$\dot{g}_3 = (g_1 - 2g_2)\, g_3 + 2g_{3x}^2 - 2g_{3x}g_{3p} - g_{3p}^2, \tag{5}$$
$$\dot{g}_{3x} = 2g_1 g_{3x} - g_1 g_{3p} - g_2 g_{3x} + g_3 g_{3x} - g_3 g_{3p}, \tag{6}$$
$$\dot{g}_{3p} = -(g_2 + g_3)\, g_{3p}. \tag{7}$$

Here $\dot{g}_i \equiv x(dg_i)/(dx)$ and $x = \omega/E_0$.

These RG equations coincide with those obtained by Houghton and Marston in their study of a lightly doped flux state [9]. They differ from those studied by Zheleznyak et al. [10], who examined a 4-patch model with patches oriented along $(1,0)$ and $(0,1)$ directions. Umklapp scattering does not connect perpendicular patches in that case, so a close resemblance to single chain behavior follows.

We take repulsive Umklapp interactions, as $g_3 = g_{3x} = g_{3p} = U$, and treat g_1, g_2 and g_{1r} as parameters. The fixed points are obtained by numerically integrating the RG equations. In a wide region around $g_1 \sim g_2 \sim g_{1r} \sim U$, we find a strong coupling fixed point where both normal and Umklapp vertices diverge and a singularity appears at $\omega \sim \omega_c = E_0 \exp(-1/\Lambda)$ where $\Lambda \propto U$. In two dimensions, such an anomaly at finite ω is an artifact of the one-loop calculation and higher order terms will shift it to $\omega = 0$. Nevertheless, ω_c represents the energy scale where the system crosses over from weak coupling to strong coupling. We will explicitly assume that the interactions $\sim U$ are strong enough so that $\omega_c > E_T$ in which case the existence of a finite curvature becomes irrelevant at the strong coupling fixed point. In contrast, if the system had scaled to weak coupling, then E_T would always remain relevant. There is a limit with weak interactions and a dispersion relation with $t' \ll t$ where both conditions, $E_T \left(= 2t' k_c^2\right) \ll \omega_c$ and $U \ll E_0$, are satisfied and our approach based on one-loop RG equations is justified. We speculate that the qualitative nature of the anomaly obtained in this weak coupling region is also present in the strong coupling region $U \gtrsim t, t'$.

We examine the nature of the fixed point through the anomalies which these 4 patches contribute to the susceptibilities. At the fixed point with strong Umklapp coupling described above, the leading divergence is observed in the spin susceptibility $\chi_s(\mathbf{q})$ at $\mathbf{q} = (\pi, \pi)$ with the exponent $\alpha_s = -1.782$, while the exponents for charge and superconducting susceptibilities are positive so that these susceptibilities do not have a divergent contribution. The uniform spin ($\chi_s(0)$) and charge (κ) susceptibilities are also of interest. In the case of a 1d chain system, we have spin gap behavior when there is a divergence in g_1 and charge gap behavior from g_3. In the present case, both g_1 and g_3 flow to strong coupling which indicates a tendency to open up both spin and charge gaps.

We now compare the present results to those of a two-leg ladder at half-filling. In this case, as Balents and Fisher have shown [2], there are 9 vertices which are relevant within a one-loop calculation. Again the flow is too strong coupling in backward and Umklapp scattering channels. In this case the properties of the strong coupling fixed point are well established. The system is an insulating spin liquid (ISL) with both spin and charge gaps (C0S0 in the Balents-Fisher notation) and is an example of a short range RVB state, first proposed by Anderson for a $S = 1/2$ Heisenberg model [11]. The spin susceptibility $\chi_s(\pi, \pi)$ is strongly enhanced but remains finite.

In the present case we cannot be sure of the spin properties from the one-loop calculations especially since the spin susceptibility at (π, π) and $(0, 0)$ behave in a contradictory way. What is certain is the scaling to strong coupling with diverging Umklapp scattering. This gives us confidence in the result that the compressibility $\kappa = dk_{F,\alpha}/d\mu \to 0$ at the fixed point as it does in the two-leg ladder at half-filling. This has several profound consequences. First the condensate that forms is pinned and insulating. Secondly when additional electrons are added to the system, the Fermi surface does not simply expand along the $(\pm 1, \pm 1)$ directions beyond the $(\pm \pi/2, \pm \pi/2)$ points as would happen for non-interacting electrons. Instead the charge gap and vanishing $dk_{F\alpha}/d\mu$ force the additional electrons to be accommodated in the rest of Fermi surface.

3. 2–Patch (Saddle Point) Model

In the cuprates $t' < 0$ so that critical behavior can now be expected when the FS touches the saddle points at $(\pi, 0)$ and $(0, \pi)$. The leading singularity arises from electron states near the saddle points which leads one to consider a 2-patch model as illustrated in Fig. 3a [7]. The Cooper channel now has a log-square divergence and the Peierls channel at $\mathbf{q} = (\pi, \pi)$ has a similar form but crosses over to a single log form at a scale $\approx t'$. These singularities may be treated within a Wilson RG scheme [12]. There are four relevant interaction vertices $g_i (i = 1, 4)$ illustrated in Fig. 4. Note that normal and Umklapp processes are indistinguishable here since the patches are at the zone edge. The one-loop RG eqns. in terms of the scaling variable $y = \ln^2(\omega/E_0)$ were obtained first by Lederer, Montambaux and Poilblanc [13] and take the form,

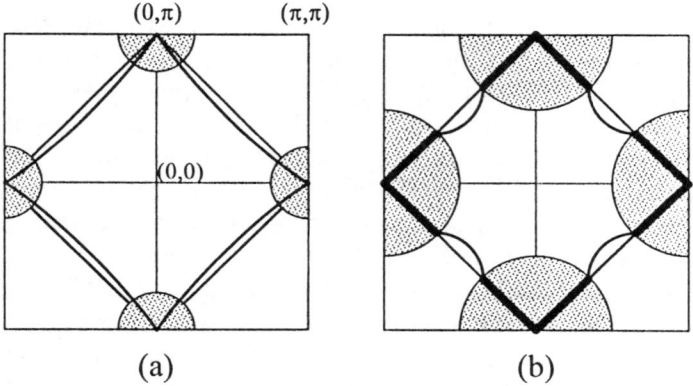

Fig. 3: Fermi surface (FS). (a) Two patches of the FS at the saddle points. (b) Truncated FS as electron density is increased.

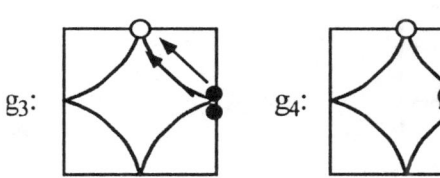

Fig. 4: The definitions of vertices for the 2-patch model.

$$\dot{g}_1 = 2d_1(y)g_1(g_2 - g_1), \tag{8}$$
$$\dot{g}_2 = d_1(y)(g_2^2 + g_3^2), \tag{9}$$
$$\dot{g}_3 = -2g_3g_4 + 2d_1(y)g_3(2g_2 - g_1), \tag{10}$$
$$\dot{g}_4 = (g_3^2 + g_4^2). \tag{11}$$

The function $d_1(y)$ describes the relative weight of the Cooper and Peierls channels with asymptotic forms, $d_1(y) \to 1$ at $y \approx 1$ and $d_1(y) \sim \ln(t/t')/\sqrt{y}$ as $y \to \infty$.

The case $d_1 \equiv 1$ arises in the limit $t' = 0$ and was studied by Schulz [14] and Dzyaloshinskii [15]. It was shown that $\chi_s(\pi, \pi)$ has the same exponent as d-wave pairing but is dominant to the next leading divergent terms. The fixed point is understood as a Mott insulator with long range AF order. The limit $d_1 = 0$ was treated by Dzyaloshinskii [16]. In this case (10) and (11) combine to give the simple flow equation

$$\dot{g}_- = -g_-^2 \tag{12}$$

with $g_- = g_4 - g_3$. Dzyaloshinskii considered the weak-coupling fixed point $g_- \to 0$ which arises from a starting value $g_- \geq 0$, and discussed the Tomonaga-Luttinger liquid behavior that results.

We concentrate on the RG equations with $0 < d_1(y) < 1$ which enables us to consider finite values of the ratios t'/t and U/t. We assume $t'/t \ll 1$ so that we are close to half-filling. The one-loop RG equations are solved numerically. Starting from a Hubbard-model initial value $g_i = U(i = 1 \sim 4)$, the vertices flow to a strong coupling fixed point with $g_2 \to +\infty$, $g_3 \to +\infty$ and $g_4 \to -\infty$, at a value $y_c \sim t/U_c$.

Although one cannot solve for the strong coupling fixed point using only one-loop RG equation, a qualitative description of the behavior comes from the examination of the susceptibilities. In Fig. 5 we show the exponents for d-wave pairing, $\chi_s(\pi, \pi)$, $\chi_s(0)$ and charge compressibility. Comparison of the values of the exponents shows us that the most divergent susceptibility is that of the

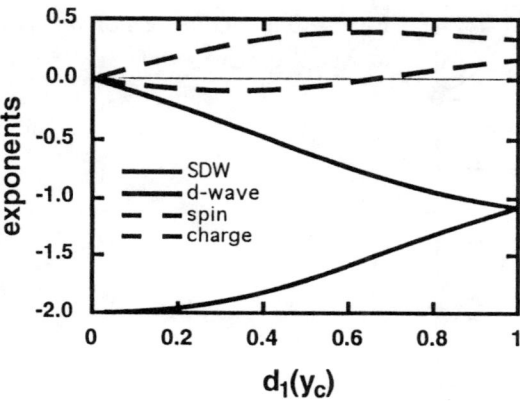

Fig. 5: Exponents for various susceptibilities. For uniform spin and charge susceptibilities, exponents are scaled by d_2/d_1

d-wave pairing throughout the parameter region of $d_1(y_c)$, as found previously by Lederer et al. [13]. Note $\chi_s(0)$ has a positive exponent indicating spin gap behavior.

The exponent for the charge compressibility changes its sign at $d_1(y_c) \sim 0.6$. Namely there exists a critical interaction strength U_c such that for $U > U_c$ the charge compressibility is suppressed to zero. The critical point U_c is determined by t' in the form $U_c/t \propto \ln^{-2}(t/t')$. This implies a transition from a superconducting phase at $U < U_c$ with its origin in enhanced Cooper channel due to the van Hove singularity, to a phase with a charge gap at $U > U_c$ which can be regarded as a precursor of the Mott transition. The fixed point at $U > U_c$ resembles closely that of the half-filling two-leg ladder which has spin and charge gaps while the most divergent susceptibility is the d-wave pairing. The characteristics of this fixed point (C0S0 in the Balents-Fisher [2] notation) are well understood as an ISL of short range RVB form. The close similarity between the fixed points leads us to assign them to the same universality class.

Next we consider increasing the electron density. One possibility would be to follow the non-interacting FS which expands beyond the saddle points. But the flow to strong coupling and the opening of a charge and spin gap leads us to consider a second possibility, namely that the FS is pinned at the saddle point and does not expand beyond it. This proposal was put forward in ref. [7] after an examination of 8 FS-patches on the Umklapp surface (US) defined by lines joining the saddle points. Further support for this proposal comes from the lightly doped 3-leg ladder [4, 5] where in strong coupling a C1S1 phase occurs with an ISL with exactly half-filling in the even parity channels and an open FS only in the odd parity channel. This contrasts with the one-loop RG results which gives a C2S1 phase with holes immediately entering both odd and even parity channels [2]. Our proposal is sketched in Fig. 3(b) and it is based on a lateral spread of the spin and charge gap along the US leading to a set of 4 open

FS segments consisting of arcs centered at the points ($\pm\pi/2, \pm\pi/2$). Note the area enclosed by the surface defined by the US and these 4 arcs contain the full electron density, consistent with a generalized form of Luttinger's Theorem.

4. Discussion

Since the ISL is not characterized by any simple broken symmetry or order parameter, the resulting state cannot be described by a simple mean field or Hartree-Fock factorization. The proposal of a FS consisting of 4 disconnected arcs has strong parallels to the results of recent gauge theory calculation for the lightly doped strong coupling t-J model by Lee and Wen [17]. Signs of such behavior are also evident in a recent analysis of the momentum distribution using a high temperature series by Putikka et al. [18]. Note models which include only nearest-neighbor terms in the kinetic energy (i.e. $t' = 0$) are a special limit from the present point of view.

The proposal that an ISL truncates the FS along the US in the vicinity of the saddle points has some interesting consequences. There will be a coupling to the open segments in the Cooper channel through the scattering of electron pairs out of the ISL to the open FS segments. This process is reminiscent of the coupling of fermions to bosonic preformed pairs in the Geshkenbein-Ioffe-Larkin model [19]. They argued for an infinite mass for such pairs to suppress their contribution to transport properties. Such scattering process will be an efficient mechanism for d-wave pairing on the open FS segments.

In the normal state there is a close similarity to a phenomenological model proposed by Ioffe and Millis [20], to explain the anomalous transport properties. Here also the FS segments have usual quasi-particle properties (i.e. there is no spin-charge separation) but the scattering rate is assumed to vary strongly along the FS arcs. In our case we can expect the scattering rate will vary strongly since Umklapp processes will lead to the strong scattering at the end of the FS arcs where they meet the US. Ioffe and Millis justified their model by a comparison to the tunneling and ARPES experiments [21] which show a single particle gap opening in the vicinity of the saddle points similar to the form in Fig. 3b. Especially in underdoped cuprates there are clear signs in the ARPES experiments of single particle energy gaps at the saddle points at low temperature [21]. Lastly we refer the reader to the very recent preprint by Balents, Fisher and Nayak [22] which introduces the concept of a Nodal liquid with properties similar to the ISL discussed above.

Acknowledgment The work on the 2-path model was performed in collaboration with M. Salmhofer and we acknowledge useful conversations with S. Haas, R. Hlubina, D. Khveshchenko, M. Sigrist, E. Trubowitz and F.C. Zhang. N. F. is supported by a Monbusho Grant for overseas research.

References

1. For a review see, E. Dagotto and T.M. Rice, Science **271**, 618 (1996).
2. L. Balents and M.P.A. Fisher, Phys. Rev. B **53**, 12133 (1996); H.-H. Lin, L. Balents and M.P.A. Fisher, Phys. Rev. B **56**, 6569 (1997).
3. B. Frischmuth, S. Haas, G. Sierra and T.M. Rice, Phys. Rev. B **55**, R3340 (1997).
4. T.M. Rice, S. Haas, M. Sigrist and F.C. Zhang, Phys. Rev. B **56**, 14655 (1997).
5. S.R. White and D.J. Scalapino, Phys. Rev. B **57**, 3031 (1998).
6. N. Furukawa and T.M. Rice, J. Phys. Cond. Mat. **10**, L381 (1998).
7. N. Furukawa, T.M. Rice and M. Salmhofer, cond-mat/9806159.
8. F.D.M. Haldane, Proc. Int. School Phys. 'Enrico Fermi' Course 121 (1991) ed. J.R. Schrieffer and R.A. Broglia (New York, North Holland); also J. Fröhlich and R. Götschmann, Phys. Rev. B **55**, 6788 (1997).
9. A. Houghton and J.B. Marston, Phys. Rev. B **48**, 7790 (1993).
10. A.T. Zheleznyak, V.M. Yakorenko and I.E. Dzyaloshinskii, Phys. Rev. B **55**, 1200 (1997).
11. P.W. Anderson, Science **235**, 1196 (1987).
12. J. Feldman, M. Salmhofer and E. Trubowitz, J. Stat. Phys. **84**, 1209 (1996).
13. P. Lederer, G. Montambaux and D. Poilblanc, J. Physique **48**, 1613 (1987).
14. H.J. Schulz, Europhys. Lett. **4**, 609 (1987).
15. I.E. Dzyaloshinskii, Sov. Phys. JETP **66**, 848 (1987).
16. I.E. Dzyaloshinskii, J. Phys. I France **6**, 119 (1996).
17. see P.A. Lee this volume.
18. W.O. Putikka, M.U. Luchini and R.R.P. Singh, preprint, cond-mat/9803141.
19. V.G. Geshkenbein, L.B. Ioffe and A.I. Larkin, Phys. Rev. B **55**, 3173 (1997).
20. L.B. Ioffe and A.J. Millis, preprint, cond-mat/9801092.
21. see J.C. Campuzano and Z.X. Shen this volume.
22. L. Balents, M.P.A. Fisher and C. Nayak, preprint, cond-mat/9803086.

Antiferromagnetism, Singlet and Disorder

Hidetoshi Fukuyama and Hiroshi Kohno

Department of Physics, University of Tokyo, Bunkyo-ku, Tokyo 113-0033, Japan

Abstract. Roles of disorder on the competition between antiferromagnetism and singlet d-wave superconductivity in high-T_c cuprates are analyzed theoretically by considering the case of $La_{2-x}Sr_xCuO_4$ as a typical example.

1 Introduction

It is widely accepted that the high T_c superconductivity in cuprates is realized next to Mott (to be precise charge transfer type) insulators, where the Heisenberg localized spin model with spin $S = \frac{1}{2}$ on the square lattice is a sound basis for theoretical studies [1]. This implies that the small number of holes doped into charge transfer type insulators first of all destroy antiferromagnetism and result in superconductivity with high critical temperature, T_c, and the $d_{x^2-y^2}$ symmetry. Hence the understanding of this process of the destruction of antiferromagnetism (AF) and the emergence of the $d_{x^2-y^2}$-wave superconductivity (d-SC) due to carrier doping into the Mott insulator is the core of the problem of high T_c superconductivity. Such a view had been stressed by Anderson [2] even soon after the discovery of high T_c cuprates, and has been adopted by various researchers. Many of these theoretical studies have considered clean systems. In actual cuprates, however, there exist more or less effects of disorder, which influence the physical properties in an important way. By studying physical and chemical processes of the effects of disorder in these regions, one may be able to extract the essence of carrier doping which is the purpose of this article. We take unit of $\hbar = k_B = 1$.

2 Experimental Facts

Various experiments have revealed that there exist several characteristic temperatures in the underdoped region, of e.g. LSCO, as shown schematically in Fig.1, where T_N = the Néel ordering temperature, T_g = the spin-glass transition temperature, T_c = the critical temperature of the superconductivity, T_R = the characteristic temperature where the in-plane resistivity takes a minimum, and T^* = the temperature where the spin susceptibility has a maximum. In Fig.1 the spin-gap temperature, T_{SG}, defined as the temperature where the NMR rate has a maximum [3] is also indicated, though this has not been clearly identified so far in LSCO in the temperature region indicated in this Fig.1. The possible reason for this apparent absence of T_{SG} in

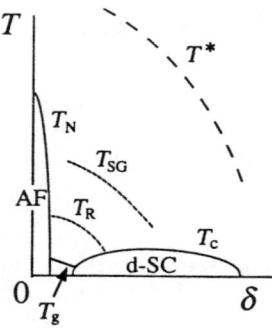

Fig. 1. Experimental phase diagram of high-T_c cuprates in the plane of temperature, T, and the hole doping rate, δ.

LSCO will be discussed later. It was found that the Hall coefficient has strong temperature dependences at around T^* [4]. The general trend of the doping dependences of these characteristic temperatures are now believed to be universal among the high T_c cuprates, except for the existence of clear anomalies at around $\delta \sim \frac{1}{8}$ in LSCO, which will not be addressed here explicitly.

3 Spin-Gap and Pseudo-Gap

The indication of the formation of the gap in the spin excitation spectra has been noted by Yasuoka et al.[3] in the NMR rate at Cu sites in the underdoped YBCO, which is called the spin-gap phenomenon and sets in at $T_{\rm SG}$. It is to be noted that this $T_{\rm SG}$ reflects the gap formation of the spin excitations at very low energy region, e.g. $\omega \sim$ 10-100 MHz in the case of NMR, and with the wave vector, q, at around (π, π). On the other hand, the uniform susceptibility χ and the NMR shift, K, which reflect the spin excitations at $q = 0$, have different temperature dependences, i.e., the shift has maxima at temperature generally higher than $T_{\rm SG}$ and its suppression as the temperature is lowered is more gradual than those of the NMR rate below $T_{\rm SG}$, the latter of which is almost of the activation type [5, 6]. The pseudo-gap with $d_{x^2-y^2}$ symmetry in the electron spectra disclosed by ARPES appears to set in at $T_{\rm SG}$ [7, 8].

The existence of the spin-gap and the pseudo-gap is considered to be the most conspicuous feature of high T_c cuprates, whose theoretical understanding will sort out the true physics behind the high T_c in cuprates.

4 Slave Boson Mean Field Theory of the t-J Model

It has been proposed [5, 6, 9] based on the slave boson mean field theory for the t-J model [10] that the spin-gap phenomenon is due to the formation of spinon-singlet, which has the $d_{x^2-y^2}$ symmetry. In this slave-boson theory the

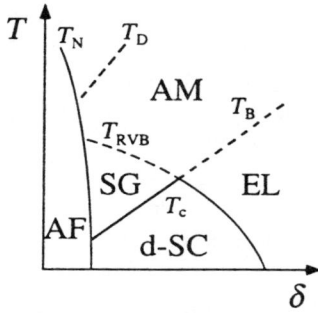

Fig. 2. Theoretical phase diagram (schematic) of the t-J model based on the slave-boson scheme. Antiferromagnetic (AF), $d_{x^2-y^2}$-wave superconducting (d-SC) and metallic phases are separated by phase transition lines. In the metallic phase, there are three characteristic regions, *i.e.*, spin-gap (SG), anomalous metal (AM) and electron-liquid (EL), which are continuously connected to each other without phase boundaries separating them.

spin and charge degrees of freedom of electrons are assumed to be separated at the mean field level [11] and is represented as spinons and holons with a local constraint [12]. However, depending on the roles played by the fluctuations around this mean field theory, which are represented as $U(1)$ gauge field [13, 14, 15], spinons and holons are bound to form back to electrons, which is termed as the gauge confinement [16].

The results of this mean field approximation together with the gauge fluctuations are shown schematically in Fig.2 in the absence of disorder, where T_N and T_c are the Néel ordering temperature and the superconducting critical temperature, while T_{RVB} and T_B are the crossover temperatures of spinon singlet RVB formation and the holon Bose condensation, respectively. (The doping dependences of T_{RVB} and T_B have been noted by several authors [17, 18] soon after the paper by Anderson [2].) It is expected that below T_N and T_B holons and spinons are bound, *i.e.* spin and charge are not separated, while they are separated otherwise. Hence spin and charge will not be separated in the ground states, $T \equiv 0$. As has been first suggested by Rice [5], the theoretical T_{RVB} has naturally been identified with the experimantal spin-gap temperature, T_{SG}, which is only a crossover temperature and is not a critical temperature of the phase transition [19].

There are theoretical studies [20, 21] that the gauge fluctuations could suppress T_{RVB} down to T_c and some theoretical efforts [22] have been devoted to elaborate the theory, which respects $SU(2)$ symmetry. We take a view, however, that T_{RVB} in the $U(1)$ theory is stabilized by taking the damping effects on spinons due to the gauge fluctuations into account, *i.e.*, the original conclusion by Nagaosa and Lee [19] is essentially valid that the short range orders of singlet RVB but not true ordering is formed at around T_{RVB}. This damping effects have not been taken into account in refs.[20, 21], and are now under investigations. This view is corroborated by the experimental fact that not only the NMR rate but also the oxygen phonon modes with the B-symmetry [23] show the onset of the anomalous frequency shift around the same temperature, T_{RVB}, as the theory predicts [24]. Moreover the assumption of the persistence of T_{RVB} is in accordance with the results of ARPES, *i.e.*, pseudo-gaps, as we will see in the following.

5 Pseudo-Gap and Electron Spectral Weight

The electron spectral weight, $A(\mathbf{k}, \varepsilon)$, in the present mean field theory is given by the convolution of those of spinons $A_{\mathbf{k}}^{\text{spinon}}(\varepsilon)$ and holons $A_{\mathbf{q}}^{\text{holon}}(\omega)$;

$$A(\mathbf{k}, \varepsilon) \equiv -\frac{1}{\pi} \text{Im} G^R(\mathbf{k}, \varepsilon)$$

$$= \frac{1}{2} \sum_{\mathbf{q}} \int_{-\infty}^{\infty} d\omega \left[\text{cth} \frac{\omega}{2T} + \text{th} \frac{\varepsilon + \omega}{2T} \right] A_{\mathbf{q}}^{\text{holon}}(\omega) A_{\mathbf{k}+\mathbf{q}}^{\text{spinon}}(\varepsilon + \omega).$$

In Fig.3 are shown the results of numerical calculation [25] for the ε-dependences of $A(\mathbf{k}, \varepsilon)$ for several choices of $\mathbf{k}$ along the two different symmetry axes as

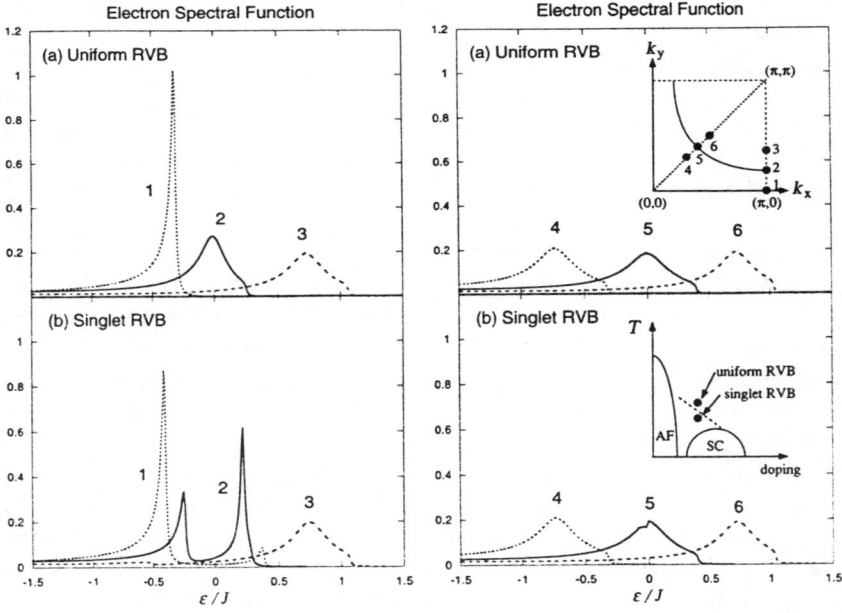

Fig. 3. Spectral function of electron, $A(\mathbf{k}, \varepsilon)$, as a function of energy ε for several k points on the two symmetry axes in the Brillouin zone as indicated in the inset. The upper panels are for uniform RVB state (a), and the lower for the singlet RVB state (b). The calculation is done with YBCO-type Fermi surface.

indicated in the inset (upper panel). The calculation was done, based on the above equation, with mean-field parameters corresponding to two points in the phase diagram indicated in the inset (lower panel), but at much higher temperature $T = 0.5J$ to simulate the inelastic broadening due to gauge fluctuations. In Fig.3, (a) and (b) correspond to $T > T_{\text{RVB}}$ and $T < T_{\text{RVB}}$, respectively, the latter being applicable to the pseudo-gap region. As expected, the electron spectra for $T < T_{\text{RVB}}$ show pseudo-gaps, which have $d_{x^2-y^2}$ symmetry reflecting that of singlet RVB of spinons. Note that once in the superconducting state, $T < T_c$, the gauge confinement is realized resulting in the essential suppression of the gauge fluctuations, which leads to the sharp energy dependence of spectral weight of electrons.

6 Ground States in the Underdoped Region

The ground state without any doping is the Néel state. With a small doping ($\delta \leq 0.02$ in LSCO), the ground state maintains the Néel ordering but with the reduced ordering temperature as in the diluted antiferromagnets such as $\text{Rb}_2(\text{Co,Mg})\text{F}_4$ [26] but the suppression of T_N by doping is stronger here probably because the wave functions of doped carriers are extended. With the higher doping but before the onset of superconductivity, the ground state is insulating from the viewpoint of transport properties. Regarding the magnetic properties, the true long range order of antiferromagnetism had been expected as in the disordered spin-Peierls systems [27], since disorder in the singlet ground state can lead to the true staggered spin orderings with the spatially modulated amplitude due to the intrinsic quantum coherence of the singlet ground state [28]. The calculations of the excitation spectra for such disordered spin-Peierls systems [29], whose typical example is CuGeO_3 with small amount of Zn replacing Cu, or Si or Mg replacing Ge, are schematically shown in Fig.4. Here in Fig.4 (a) is shown the spectral weight in the clean systems that has a well-defined mode with a gap around the antiferromagnetic wave vector, Q. With a small amount of disorder, however, the gapped modes

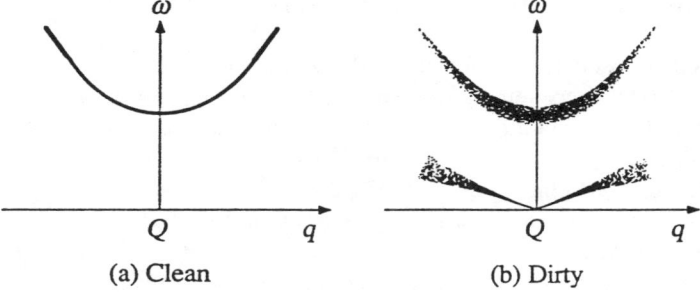

Fig. 4. Magnetic excitation spectra in the clean (a) and disordered (b) spin-Peierls systems.

shift downwards with appreciable broadening together with the introduction of new spectral weight at low energy, $\omega \to 0$, at Q, as shown in Fig.4(b). This disorder-induced low-energy spectral weight becomes a well-defined Goldstone mode of antiferromagnetic spin wave as experimentally detected by ESR [30]; a surprizing fact. In the disorder-induced Néel ordered state the spin waves exist in a very limited region of $q \sim Q$ and $\omega \sim 0$ with very small total spectral weight, in proportion to the concentration of the impurities. At some critical temperature the long-range Néel ordering will disappear, but the spectral weight still remains at around $q \sim Q$, $\omega \simeq 0$. A similar situation is expected to be realized in the square lattice in high T_c cuprates [31, 32], since the system does not have intrinsic magnetic frustrations. For example, the fact [33] that the replacement of Cu by Zn in $YBa_2Cu_4O_8$ resulted in the complete destruction of the spin gap phenomenon in the NMR rate keeping the NMR shift unchanged can be understood naturally based on this notion. It is possible that similar processes are present in LSCO.

Experimentally, however, for $0.02 < \delta \leq 0.05$ in LSCO, the ground state has turned out to be a spin-glass state instead of the true long range order of antiferromagnetism [34, 35]. A possible reason for the preference of the spin-glass state to the Néel state is the extended wave functions of doped holes resulting in the longer-ranged effective exchange interactions between Cu spins; i.e., the localization length of holes, ξ_{loc}, is expected to diverge at $\delta = \delta_c$ where the superconductivity sets in. This indicates the essential interplay between magnetic and transport properties near and at the critical doping rate at which both properties undergo dramatic changes, i.e., magnetic to singlet, and insulating to superconducting [36].

Recent proposal of $SO(5)$ theory [37], though so far considered only for clean cases, offers an illuminating pictute of the phase transition between AF and d-SC in terms of 'superspin.' Since this $SO(5)$ theory does not assume spin-charge separation and works with physical electrons [38], the conclusions can be different from the present studies in general. At $T = 0$, however, where the spin and charge are no longer separated in the present scheme, these two theories have common features.

7 Spin-Charge Separation at Finite Temperature

In the present slave-boson theory for the t-J model, we have seen that in the ground state there is no spin-charge separation. As the temperature is raised in the underdoped region, however, the spin-charge separation is expected once either $T > T_N$ or T_c. In this region the quasi-particles, which are composites of spinons and holons, have strong damping because of the spin-charge separation and the resultant strong gauge fluctuations. However once for $T < T_c$, where spin and charge are no longer separated and the gauge fluctuations get massive and then greatly reduced, the damping of quasi-particles will be reduced dramatically. Hence a drastic change of the intrinsic damping of quasi-particles is expected through T_c. Actually such reduction

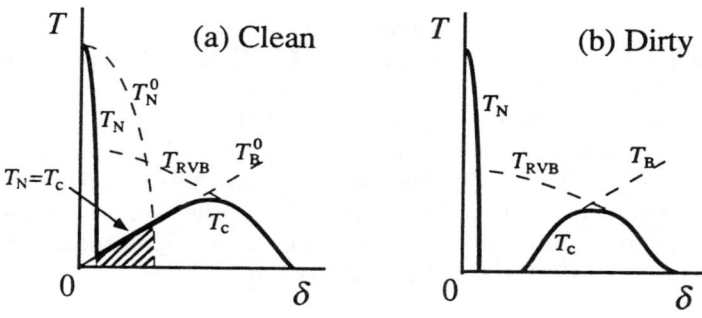

Fig. 5. Proposed phase diagram for high-T_c cuprates. (a) clean case (b) the case with strong disorder.

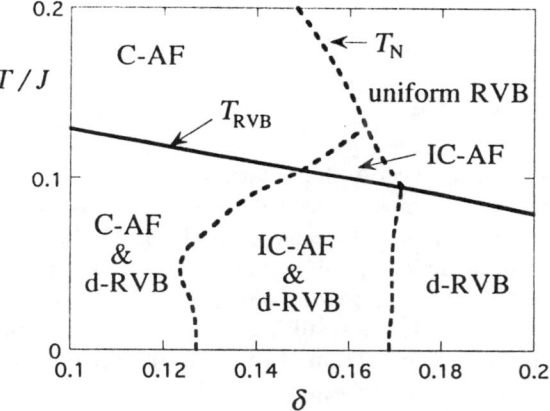

Fig. 6. The region of coexistence of incommensurate antiferromagnetism (IC-AF) and $d_{x^2-y^2}$-wave singlet RVB (d-RVB) together with that of commensurate antiferromagnetism (C-AF) and d-RVB in the mean field phase diagram of the t-J model with LSCO-type Fermi surface used in refs.[9,40]. The solid line represents $T_{\rm RVB}$, while $T_{\rm B}$, which determines T_c in the underdoped region, is not drawn explicitly. The magnetic phase boundaries (dashed lines) are determined by the condition that inverse static spin susceptibility, $\chi_{\rm RPA}(\mathbf{q})^{-1} \equiv \chi_0(\mathbf{q})^{-1} + J(\cos q_x + \cos q_y)$, takes its minimum value of zero at some wave vector $\mathbf{q} = \mathbf{q}_0$, where $\mathbf{q}_0 = (\pi, \pi)$ in the C-region and $\mathbf{q}_0 = (\pi, \pi \pm \eta), (\pi \pm \eta, \pi)$ with $\eta \neq 0$ in the IC region. The $\chi_0(\mathbf{q})$ is evaluated in the uniform RVB or d-RVB state without any magnetic ordering.

of the damping of the quasi-particles has been inferred experimentally [39]. Based on this we expect the mean field phase diagram to be modified by the gauge fluctuations as in Fig.5(a), where incommensurate antiferromagnetism (IC-AF) and d-SC coexist in the shaded region. Actually our former RPA calculation [40] based on the mean field theory for the t-J model extended to represent the particular features of LSCO by including the transfer integral between the next nearest neighbors leads to instability of $d_{x^2-y^2}$-wave singlet RVB (d-RVB) state towards such coexistence. We show in Fig.6 the mean field phase diagram showing such coexistence of IC-AF and d-RVB together with that of commensurate antiferromagnetism (C-AF) and d-RVB. (The coexistence of C-AF and d-SC (or d-RVB) in the t-J model has been demonstrated by Inaba et al.[41].) In Fig.5(a), we assumed that once $T < T_c$ the results of mean field calculations get valid because of the reduced damping while T_N is greatly reduced for $T > T_c$. Fig.5(a) affords a possible explanation of the recent remarkable experimental findings [35] that the Bragg peak intensity for the incommensurate spin structure sets in at the superconducting critical temperature in La_2CuO_{4+y} and $La_{1.88}Sr_{0.12}CuO_4$.

In the presence of strong disorder the phase diagram of Fig.5(a) will be modified to that shown in Fig.5(b) because of the disorder-induced localization of holes (in the ground state this is the same as the Anderson localization in the presence of strong correlations). This will be the case for LSCO in the underdoped region.

8 Summary and Discussions

In this paper the possible origins of several characteristic temperatures in the underdoped region, which are the most remarkable features, have been discussed based on the theoretical results of the slave boson mean field theory together with the possible effects of the $U(1)$ gauge fluctuations. This theory is based on the assumption of spin-charge (spinon-holon) separation to start with. Once this assumption is accepted, the resulting theory is very simple and transparent, and gives a possible explanation of the spin-gap phenomena seen in NMR rate and phonon anomalies together with the pseudo-gap observed in ARPES. Though the effects of gauge fluctuations are not yet fully explored, we assumed T_{RVB} survives the fluctuations effects possibly due to the strong damping effects on the spinons, which obviously need further detailed studies.

Acknowledgment

We would like to thank Professors Y. Endoh and N. Nagaosa for fruitful discussions. This work is supported by a Grant-in-Aid for Scientific Research from the Ministry of Education, Science, Sports and Culture of Japan.

References

[1] S. Chakravarty, B.I. Halperin and D. Nelson: *Phys. Rev.* **B39** (1988) 7443.
[2] P.W. Anderson; *Science* **235** (1987) 1196.
[3] H. Yasuoka, T. Imai and T. Shimizu: in *Strong Correlation and Superconductivity*, eds. H. Fukuyama, S. Maekawa and A. P. Malozemoff (Springer-Verlag, Berlin, 1989), p.254.

[4] T. Nishikawa, J. Takeda and M. Sato: *J. Phys. Soc. Jpn.* **62** (1993) 2568; See also, M. Sato: This proceedings; H.Y. Hwang et al.: *Phys. Rev. Lett.* **12** (1994) 2636.
[5] T.M. Rice: in *The Physics and Chemistry of Oxide Superconductors*, eds.Y. Iye and H. Yasuoka (Springer-Verlag, Berlin, 1992), p.313.
[6] H. Fukuyama: *Prog. Theor. Phys.* Suppl.**108** (1992) 287.
[7] A.G. Loeser, Z.-X. Shen, D.S. Dessau, D.S. Marshall, C.H. Park, P. Fournier and A. Kapitulnik: *Science* **273** (1996) 325.
[8] H. Ding, T. Yokoya, J.C. Campuzano, T. Takahashi, M. Randeria, M.R. Norman, T. Mochiku, K. Kadowaki and J. Giapintzakis: *Nature* **382** (1996) 51.
[9] T. Tanamoto, H. Kohno and H. Fukuyama: *J. Phys. Soc. Jpn.* **63** (1994) 2739.
[10] F.C. Zhang and T.M. Rice: *Phys. Rev.* **B37** (1988) 3759.
[11] G. Baskaran, Z. Zou and P.W. Anderson: *Solid State Commun.* **63** (1987) 973.
[12] For a theoretical review on high-T_c cuprates as doped Mott insulator based on the present framework, H. Fukuyama: *J. Phys. Chem. Solids* **59** (1998) 447.
[13] G. Baskaran and P.W. Anderson: *Phys. Rev.* **B37** (1988) 580.
[14] L.B. Ioffe and A.I. Larkin: *Phys. Rev.* **B39** (1989) 8988.
[15] N. Nagaosa and P.A. Lee: *Phys. Rev. Lett.* **64** (1990) 2450; P.A. Lee and N. Nagaosa: *Phys. Rev.* **B46** (1992) 5621.
[16] N. Nagaosa: *Phys. Rev. Lett.* **71** (1993) 4210, and this proceedings.
[17] G. Kotliar and J. Liu: *Phys. Rev.* **B38** (1988) 5142.
[18] Y. Suzumura, Y. Hasegawa and H. Fukuyama: *J. Phys. Soc. Jpn.* **57** (1988) 401; *ibid* 2768.
[19] N. Nagaosa and P.A. Lee: *Phys. Rev.* **B45** (1992) 966.
[20] M. Ubbens and P.A. Lee: *Phys. Rev.* **B49** (1994) 6853.
[21] P. Lederer and E. Abrahams: *Phys. Rev.* **B53** (1996) 10680.
[22] P.A. Lee, N. Nagaosa, T.K. Ng and X.G. Wen: *Phys. Rev.* **B57** (1998) 6003.
[23] M. Käll, A.P. Litvinchuk, P. Berastegui, L.-G. Johansson and L. Börjesson: *Physica* **C 225** (1994) 317; H. Harashina, K. Kodama, S. Shamoto, M. Sato, K. Kakurai and M. Nishi: *Physica* **C 263** (1996) 257.
[24] B. Normand, H. Kohno and H. Fukuyama: *Phys. Rev.* **B53** (1996) 856.
[25] H. Kohno, M. Mitra and H. Fukuyama: in preparation.
[26] H. Ikeda, K. Iwasa and K.H. Andersen: *J. Phys. Soc. Jpn.* **62** (1993) 3832.
[27] L.P. Regnault, J.P. Renard, G. Dhalenne and A. Revcolevschi: *Europhys. Lett.* **32** (1995) 579; M. Hase, K. Uchinokura, R.J. Birgeneau, K. Hirota and G. Shirane: *J. Phys. Soc. Jpn.* **65** (1996) 1392.
[28] H. Fukuyama, T. Tanimoto and M. Saito: *J. Phys. Soc. Jpn.* **65** (1996) 1182.
[29] M. Saito and H. Fukuyama: *J. Phys. Soc. Jpn.* **66** (1997) 3259.
[30] M. Hase, M. Hagiwara and K. Katsumata, *Phys. Rev.* **B54** (1996) R3772; H. Nojiri, T. Hamamoto, Z.J. Wang, S. Mitsudo, M. Motokawa, S. Kimura, H. Ohta, A. Ogiwara O. Fujita and J. Akimitsu: *J. Phys.: Condensed Matter* **9** (1997) 1331.
[31] H. Fukuyama and H. Kohno: *Czech. J. Phys.* **46** (1996) Suppl. S6, 3146.
[32] G. Khaliullin, R. Kilian, S. Krivenko and P. Fulde: *Phys. Rev.* **B56** (1997) 11882.
[33] G.-q. Zheng, T. Odaguchi, T. Mito, Y. Kitaoka, K. Asayama and Y. Kodama: *J. Phys. Soc. Jpn.* **62** (1993) 2591.
[34] B. Keimer, N. Belk, R.J. Birgeneau, A. Cassanho, C.Y. Chen, M. Greven, M.A. Kastner, A. Aharony, Y. Endoh, R.W. Erwin and G. Shirane: *Phys. Rev.* **B46** (1992) 14034.

[35] H. Kimura, K. Hirota, H. Matsushita, K. Yamada, Y. Endoh, S.-H. Lee, C. F. Majkrzak, R. Erwin, G. Shirane, M. Greven, Y. S. Lee, M. A. Kastner, and R. J. Birgeneau: preprint; Y. Endoh: private communications.
[36] H. Fukuyama: *Rev. High Pressure Sci. Technol.*, **7** (1998) 465.
[37] S.-C. Zhang: *Science* **275** (1997) 1089.
[38] S. Rabello, H. Kohno. E. Demler and S.-C. Zhang: *Phys. Rev. Lett.* **80** (1998) 3586; C. Henley: *ibid* 3590.
[39] M.C. Nuss, P.M. Mankiewich, M.L. O'Malley, E.H. Westerwick and P.B. Littlewood: *Phys. Rev. Lett.* **66** (1991) 3305; D.A. Bonn, P. Dosanjh, R. Liang and W.N. Hardy: *Phys. Rev. Lett.* **68** (1992) 2390; R.C. Yu, M.B. Salamon, J.P. Lu and W.C. Lee: *Phys. Rev. Lett.* **69** (1992) 1431.
[40] T. Tanamoto, H. Kohno and H. Fukuyama: *J. Phys. Soc. Jpn.* **62** (1993) 717.
[41] M. Inaba, H. Matsukawa, M. Saito and H. Fukuyama: *Physica* **C 257** (1996) 299.

The Spin Gap and Superconducting States of Underdoped Cuprates

Patrick A. Lee

Department of Physics, Massachusetts Institute of Technology, Cambridge, MA 02139

Abstract. We review our $SU(2)$ formulation of the t-J model as a description of the underdoped cuprates. The model incorporates spin-charge separation and successfully explains many unusual normal state properties, including the existence of a spin gap and the appearance of a Fermi surface segment. Very recently, we extended our theory to the superconducting state and show how BCS-like quasiparticles evolve out of the Fermi surface segment.

It has become clear in the past seveal years that the cuprates show many highly unusual properties both in the normal and superconducting (SC) states. These unusual features are related to the fact that the cuprates are doped Mott insulators. It is then not surprising that the unusual behaviors are most striking in the underdoped region, when the concentration of doped holes, x is small. In the normal state a pseudogap is obseved in a temperature range considerably above the SC transition temperature T_c. The gap is seen in NMR relaxation rate $1/T_1$, Knight shift and specific heat. It is also seen in c-axis conductivity and photoemission experiments which reveal that the pseudogap is roughly of the same size and **k** depdendence as the d-wave SC gap. Furthermore, the gap size is essentially independent of x even when T_c is reduced with decreasing x. On the other hand, the in-plane transport properties are only slightly affected by the pseudogap. The resistivity shows a small decrease and, more importantly, the spectral weight of the Drude part of $\sigma(\omega)$ is not affected at all by the presence of the pseudogap and is proportional to x. We believe this is strong experimental evidence supporting the notion of spin-charge separation [1] in these materials. The spins form RVB singlets so that it costs energy (spin gap) to make triplet excitations. However the in-plane conductivity is carried by x holes, which remain gapless. In c-axis conductivity and photoemission, a physical electron is removed from the plane, which carries both spin and charge. It then follows that the spin gap should appear in these experiments.

The physics of spin charge separation appears naturally in a class of theory which starts with the t-J model and enforces the constraint of no double occupation by decomposing the electron into a fermion and a boson. The fermion carries spin index and the boson keeps track of the charge degrees of freedom. At the mean field (MF) level, the phase diagram includes a spin gap phase where the fermions form d-wave pairs.[2, 3] As the temperature is lowered, the boson condenses, resulting in a d-wave SC state. Fluctuations about the MF solution leads to a $U(1)$ gauge theory and the fermion and boson are strongly coupled to the gauge field.[4]

Recently, we developed an improved version of the $U(1)$ theory, called the $SU(2)$ slave boson theory.[5, 6] We introduce an $SU(2)$ doublet of boson fields $b^T = (b_1, b_2)$, in addition to the fermion doublet $\psi^\dagger = (\psi_\uparrow, \psi_\downarrow^\dagger)$. The physical electron is represented by the $SU(2)$ singlet formed out of these two doublets, $c_\uparrow = \frac{1}{\sqrt{2}} b^T \psi$, $c_\downarrow = \frac{1}{\sqrt{2}} b^\dagger \bar{\psi}$ where $\bar{\psi} = i\tau^2 \psi^*$. We are motivated by the observation made by Affleck et al.[7] that at half-filling ($x = 0$) the fermion representation of the t-J model has the $SU(2)$ symmetry in that a spin-up electron can be represented by a spin-up fermion or the absence of a spin-down fermion. In the $U(1)$ formulation this symmetry is broken as soon as $x \neq 0$, and out of a infinte degeneracy of states, the d-wave fermion pairing state is picked out as the MF solution. In contrast, even at the mean field level, the low lying states which are missing in the $U(1)$ mean field theory are included in the new $SU(2)$ formulation. For example, the spin gap state can be described equally well as the d-wave pairing state, or a staggered flux phase, where the fermions see gauge fluxes which alternate from plaquette to plaquette. The $SU(2)$ gauge transformation relates these states and guarantees that there is no breaking of the translational symmetry. The fermion spectrum exhibits a d-wave type gap, with maximum gap at $(\pi, 0)$ and nodes at $(\pi/2, \pi/2)$. We compute the physical electron spectral function, which at the mean field level, is a convolution between the fermion and boson spectra. We further introduced a residual interaction between the fermions and bosons. The resulting spectra can be compared with photoemission experiments on exact diagonalization studies and have the following features.

1. The spectra consist of a coherent part with spectral weight x and dispersion of order J and a broad incoherent part. The coherent part closely resembles the fermion dispersion. The residual interaction broadens and shifts the nodes at $(\pi/2, \pi/2)$ so that we obtain a "Fermi surface segment" near $(\pi/2, \pi/2)$. Away from this segment a gap appears in the excitation spectrum which grows to its maximal magnitude near $(0, \pi)$. This behavior is in qualitative agreement with the angle-resolved photoemission experiment.[8, 9]

2. In our theory the gap appears in the fermion spectrum while the bosons, which represent charge excitations, remain gapless. Thus a gap appears

occupation by decomposing the electron into a fermion and a boson. The fermion carries spin index and the boson keeps track of the charge degrees of freedom. At the mean field (MF) level, the phase diagram includes a spin gap phase where the fermions form d-wave pairs.[2, 3] As the temperature is lowered, the boson condenses, resulting in a d-wave SC state. Fluctuations about the MF solution leads to a $U(1)$ gauge theory and the fermion and boson are strongly coupled to the gauge field.[4]

Recently, we developed an improved version of the $U(1)$ theory, called the $SU(2)$ slave boson theory.[5, 6] We introduce an $SU(2)$ doublet of boson fields $b^T = (b_1, b_2)$, in addition to the fermion doublet $\psi^\dagger = (\psi_\uparrow, \psi_\downarrow^\dagger)$. The physical electron is represented by the $SU(2)$ singlet formed out of these two doublets, $c_\uparrow = \frac{1}{\sqrt{2}} b^T \psi$, $c_\downarrow = \frac{1}{\sqrt{2}} b^\dagger \bar{\psi}$ where $\bar{\psi} = i\tau^2 \psi^*$. We are motivated by the observation made by Affleck et al.[7] that at half-filling ($x = 0$) the fermion representation of the t-J model has the $SU(2)$ symmetry in that a spin-up electron can be represented by a spin-up fermion or the absence of a spin-down fermion. In the $U(1)$ formulation this symmetry is broken as soon as $x \neq 0$, and out of a infinte degeneracy of states, the d-wave fermion pairing state is picked out as the MF solution. In contrast, even at the mean field level, the low lying states which are missing in the $U(1)$ mean field theory are included in the new $SU(2)$ formulation. For example, the spin gap state can be described equally well as the d-wave pairing state, or a staggered flux phase, where the fermions see gauge fluxes which alternate from plaquette to plaquette. The $SU(2)$ gauge transformation relates these states and guarantees that there is no breaking of the translational symmetry. The fermion spectrum exhibits a d-wave type gap, with maximum gap at $(\pi, 0)$ and nodes at $(\pi/2, \pi/2)$. We compute the physical electron spectral function, which at the mean field level, is a convolution between the fermion and boson spectra. We further introduced a residual interaction between the fermions and bosons. The resulting spectra can be compared with photoemission experiments on exact diagonalization studies and have the following features.

1. The spectra consist of a coherent part with spectral weight x and dispersion of order J and a broad incoherent part. The coherent part closely resembles the fermion dispersion. The residual interaction broadens and shifts the nodes at $(\pi/2, \pi/2)$ so that we obtain a "Fermi surface segment" near $(\pi/2, \pi/2)$. Away from this segment a gap appears in the excitation spectrum which grows to its maximal magnitude near $(0, \pi)$. This behavior is in qualitative agreement with the angle-resolved photoemission experiment.[8, 9]

2. In our theory the gap appears in the fermion spectrum while the bosons, which represent charge excitations, remain gapless. Thus a gap appears

The fact that $d\rho_s/dT$ is independent of x and that both ρ_s and T_c are proportional to x means that a scaled plot of $\rho_s(T)/\rho_s(0)$ vs T/T_c should be independent of x for small T/T_c. In fact, such a scaled plot for YBCO$_{6.95}$ and YBCO$_{6.60}$ shows a remarkable universality over the entire temperature range.[13] We can use the data to extract the ratio v_f/v_2 using Eq. (2). Using the YBCO$_{6.95}$ data, we obtain a velocity anisotropy $v_f/v_2 = 6.8$.

It was recently pointed out [14] that, in conventional BCS superconuctors developed out of a Fermi liquid, the Fermi liquid correction to the qp current appears, so that, in general,

$$\mathbf{j}(\mathbf{k}) = -e\alpha \mathbf{v}_F . \tag{3}$$

For example, if only a single Fourier component of the Landau parameter F_{1s} is important, $\alpha = 1 + F_{1s}/3$, but more complicated anisotropic Landau parameters are generally possible. With the more general assumption Eq. (3), the phenomenological model now predicts that

$$\frac{\rho_s(T)}{m} = \frac{x}{ma^2} - \frac{2\ln 2}{\pi}\alpha^2 \left(\frac{v_F}{v_2}\right) T . \tag{4}$$

We have seen that, in order to agree with experiments, α near the nodes is either exactly unity or close to it, and must be independent of x. On the other hand, if one attempts to describe the normal state of underdoped cuprates by Fermi liquid theory, one faces the dilemma that the area of the Fermi surface is $1 - x$ while the spectral weight of the Drude peak (which develops into the superfluid density in the SC state) is proportional to x. In Fermi liquid theory this can be accommodated by assuming $1 + F_{1s}/3 = x$. From Eq. (4) we see that, within this scenario, the T dependence of ρ_s is too small by a factor of $\alpha^2 = x^2$. Thus a proper microscopic theory must explain in a natural way why the spectral weight is x while $\alpha \approx 1$. We believe this requirement is a central issue in the high-T_c problem, and lies at the heart of the debate of spin-charge separation vs Fermi liquid theory in the normal state.

To expand on this point further, we note that in the original U(1) gauge field formulation of the $t-J$ model, the prediction for $\rho_s(T)$ takes the form of Eq. (4) with $\alpha = x$ and therefore is in strong disagreement with experiment. This follows simply from the Ioffe-Larkin rule which states that the inverse of the response function of the fermion and boson should add to give the physical inverse response. In the superconducting state, the fermion and boson acquire superfluid densities ρ_F and ρ_s so that

$$\rho_s^{-1}(T) = \rho_F^{-1}(T) + \rho_B^{-1}(T) \tag{5}$$

where $\rho_F \approx (1-x)$ and $\rho_B \approx x$. However, only the temperature dependence of ρ_F depends on the qp gap structure and is expected to be of the form

$\rho_F(T) \approx (1-x)(1-T/\Delta_0)$, whereas the temperature dependence of ρ_B arises only through the excitation of sound mode and should be higher power in T, which can be ignored. Inserting these into Eq. (5) we see that $\rho_s(T)$ is predicted to be $x - x^2 T/\Delta_0$. Basically in the U(1) gauge theory the mismatch of the Fermi surface area and the Drude spectral weight (or ρ_s in the superconducting state) is solved by a Landau parameter, so that $\alpha = x$. Thus we may conclude that it is not sufficient to treat the gauge fluctuation only to quadratic order as in the Ioffe-Larkin theory.

We believe this difficulty is tied to the notion of Bose condensation as a way of achieving superconductivity. The reason is the following. The electron operator c_k is a convolution of the fermion and boson operator in momentum space. Let us suppose that the external $\mathbf{A}$ field couples only to the boson (this is true in the SU(2) formulation and is approximately true in some gauge choice in the U(1) formulation). In the presence of $\mathbf{A}$, $b_\mathbf{q} \to b_{\mathbf{q}+\mathbf{A}}$ so that after the convolution $c_\mathbf{k} \to c_{\mathbf{k}+\mathbf{A}}$ and $\epsilon_\mathbf{k} \to \epsilon_{\mathbf{k}+\mathbf{A}}$ as expected. Thus $j_\mathbf{k} = -e\partial\epsilon/\partial\mathbf{A} = -e\partial\epsilon/\partial\mathbf{k}$. Let us see what happens in the superconducting state. If we assume that the fermions are already paired, superconductivity can be driven by the condensation of bosons $< b_{\mathbf{k}=0} > = 0$. However, in the presence of $\mathbf{A}$, the Bose condensate remains rigid and stays in the $k = 0$ state. This is clearly seen in the Ginsburg Landau theory for free energy $|(\nabla - 2e\mathbf{A}/c)b|^2$ where $< b_{\mathbf{k}=0} > \neq 0$ in the presence of $\mathbf{A}$ is responsible for the Higgs mechanism and the London penetration depth. Upon convolution, we see that for the electron operator, $\mathbf{k}$ is not shifted by $\mathbf{A}$ so that $\epsilon(\mathbf{k})$ is independent of $\mathbf{A}$. The qp now carries no current! In the U(1) formulation, the gauge field a causes a small shift in the Fermion spectrum and leads to Eq. (3) with $\alpha = x$. This is clearly an unacceptable situation and can be seen most acutely for the qp at the Fermi surface along the (π,π) direction. Here the energy gap vanishes so that the qp in the superconducting state is basically the same state above T_c. Yet, according to the Bose condensation scenario, the current carried by this qp drops abruptly below T_c.

Now that we have identified the problem, we can see that there are two possible ways to avoid it. The first is to argue that due to fluctuations, only a small fraction of the bosons are in the condensate and we can reduce the problem, but not eliminate it. We call this the single boson condensation (SBC) scenario. The result is that α can lie anywhere between x and 1, and most likely somewhere in between. A second possibility is allowed in the SU(2) formulatin but not in the U(1) formulation. In SU(2) theory there are two species of bosons b_1 and b_2 and we can pair them to form a gauge singlet pair $< b_1(\mathbf{i}) b_2(\mathbf{j}) > \neq 0$. We shall call this the boson pair condensation (BPC) scenario. Since $< b_1 > = < b_2 > = 0$, the problem is avoided and we find that $\alpha = 1$. This is really a consequence of continuity because in this scenario the superconducting qp along (π,π) is smoothly connected to the electron state above T_c. This result comes out of an explicit calculation which we outline below.[15]

In SU(2) theory we go beyond MF theory by calculating the electron propagator through a ladder diagram [5, 6] to include effects of pairing between the boson and the fermion. Here we will consider only the simplest on-site interaction $V\left(c_\uparrow^\dagger c_\uparrow + c_\downarrow^\dagger c_\downarrow\right)$, which, when written in terms of bosons and fermions, generates an attraction between bosons and the fermions if $V > 0$. There are also other pairing interactions, but they will not modify our results qualitatively. The resulting electron propagator is given by

$$\mathbf{G_A}(\omega, \mathbf{k}) \equiv \begin{pmatrix} -i\langle c_\uparrow c_\uparrow^\dagger \rangle & -i\langle c_\uparrow c_\downarrow \rangle \\ -i\langle c_\downarrow^\dagger c_\uparrow^\dagger \rangle & -i\langle c_\downarrow^\dagger c_\downarrow \rangle \end{pmatrix} \qquad (6)$$

$$= \left[\begin{pmatrix} G_{0,\mathbf{A}}(\omega, \mathbf{k}) & F_{0,\mathbf{A}}(\omega, \mathbf{k}) \\ F_{0,\mathbf{A}}(\omega, \mathbf{k}) & -G_{0,\mathbf{A}}(-\omega, -\mathbf{k}) \end{pmatrix}^{-1} - V\tau^3\right]^{-1} \qquad (7)$$

We first consider the second scenario where there are no SBC, but there is a nonzero $F_{0,\mathbf{A}}$ proportional to the boson pair parameter x_{pc}. For $\mathbf{A} = 0$, the poles of $G_{11}(\omega, \mathbf{k})$ comes in pairs of opposite signs, just as in BCS theory. However the total residue is $\frac{x}{2(1-VG_{in})^2}$, significantly reduced from the BCS value. There are two positive branches which determine the qp excitations

$$E_\pm^{(sc)}(\mathbf{k}) = \sqrt{\tilde{E}_\pm^2 + \left(\frac{x_{pc}}{x}\Delta\right)^2} \qquad (8)$$

where

$$\tilde{E}_\pm = \pm\sqrt{(\varepsilon - \tilde{\mu})^2 + \Delta^2 - \left(\frac{x_{pc}}{x}\Delta\right)^2} - \tilde{\mu} \qquad (9)$$

and $\tilde{\mu} = -\frac{xV}{4(1-VG_{in})}$. In order to interpret those results, let us first consider the normal state which is recovered by setting $x_{pc} = 0$ in Eq. (8) and Eq. (9), yielding the normal state dispersion $E_\pm^N \equiv \tilde{E}_\pm(x_{pc} = 0)$. This corresponds to a massless Dirac cone initially centered at $(\pm\pi/2, \pm\pi/2)$ when $V = 0$ which is the MF fermion spectrum of the staggered-Flux (s-Flux) phase. The effect of V (the boson-fermion pairing) is two-fold. The $\tilde{\mu}$ inside the square-root shift the location of the node towards $(0,0)$ by a distance $\Delta k = -\tilde{\mu}/v_F$ while the last term shift the spectrum upwards. The cone intersects the Fermi energy to form a small Fermi pocket with linear dimension of order x. As shown in Fig. 1(a), the spectral weight is concentrated on one side of the cone, so that only a segment of FS on the side close to the origin carries substantial weight. This is the origin of the notion of "FS segment" introduced in Ref. [5, 6].

Now let us see what happens in the SC state when $x_{pc} \neq 0$. Equation (8) takes the standard BCS form if $\tilde{E}_\pm$ is interpreted as the normal state dispersion. However, $\tilde{E}_\pm$ differs from the normal state spectrum $E_\pm^N$ by the

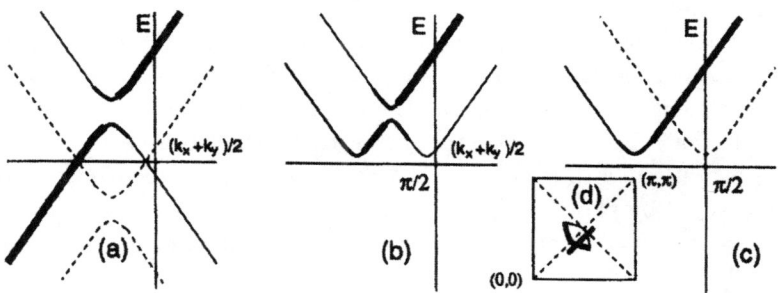

Figure 1: Schematic illustration of the qp dispersion (the pole locations of **G**) for (a) normal state, and SC state with (b) $0 < x_{pc} < x$ and (c) $x_{pc} = x$. The line thickness indicates the size of residue of G_{11}, and the dashed line indicates vanishing residue. The momentum scan is along the straight line in (d), where the curved segment is the FS segment in the normal state.

appearance of the term $-(x_{pc}\Delta/x)^2$ in Eq. (9). Close to the node this term is small so that qualitatively the spectrum develops from the normal state in a BCS fashion, as shown in Fig. 1(b). This is particularly true if the higher energy gap between the two branches is smeared by lifetime effects. Thus we see that the "FS segment" is gapped in a BCS-like fashion. However, the velocity v_2 in the $(1,-1)$ direction, being proportional to x_{pc}/x, does not extrapolate to the gap at $(0,\pi)$ (which is essentially independent of x_{pc}), but crosses over to it at the edge of the FS segment. It is worth remarking that in the special case $x_{pc} = x$, $E_{\pm}^{(sc)}$ reduces to the standard BCS form with the normal state dispersion $\varepsilon(\mathbf{k})$, a chemical potential $2\tilde{\mu}$ and a SC gap $\Delta(\mathbf{k})$. The high energy gap closes and spectral weight on one branch vanishes, yielding a BCS spectrum as shown in Fig. 1(c).

We have also calculated the effect of constant **A** on the qp dispersion, to linear order of **A**. This adds a term $\frac{1}{c}\mathbf{j}_{\pm} \cdot \mathbf{A}$ to Eq. (9) where $\mathbf{j}_{\pm}$ is interpreted as the current carried by the qp. We recall that in standard BCS theory, the current is given in term of the normal state spectrum by $c\partial_{\mathbf{A}}\varepsilon_{\mathbf{A}} = e\partial_{\mathbf{k}}\varepsilon$ because $\varepsilon_{\mathbf{A}}(\mathbf{k}) = \varepsilon(\mathbf{k} + \frac{e}{c}\mathbf{A})$. Remarkably this is almost true in our case in the sense that $\mathbf{j}_{\pm}$ is given by $c\partial_{\mathbf{A}}\tilde{E}_{\pm,\mathbf{A}}$, where $\tilde{E}_{\pm,\mathbf{A}}$ is obtained by replacing $\mathbf{k}$ by $\mathbf{k} + \frac{e}{c}\mathbf{A}$ in ε, $\tilde{\mu}$ and Δ everywhere in Eq. (9) except for the term $\left(\frac{x_{pc}}{x}\Delta\right)^2$, which is kept independent of **A**. Near the node, Δ is negligible so that the current is very close to $e\partial_{\mathbf{k}}\tilde{E} \simeq e\partial_{\mathbf{k}}E^N$ (which becomes exactly $e\partial_{\mathbf{k}}\varepsilon$ along the diagonal), thus reproducing Eq. (1). We have checked numerically that even away from the node in the region of the "FS segment", the current is remarkably close to $e\partial_{\mathbf{k}}E^N$, which can be quite different from the BCS value $e\partial_{\mathbf{k}}\varepsilon$ near the edge of the FS segment.

From Eq. (4), the temperature dependence of the London penetration depth gives a direct measurement of $\alpha^2 \frac{v_F}{v_2}$. Density of states measurements using the T^2 coefficient of the specific heat yields $v_F v_2$. The Fermi velocity can be estimated from transport measurements or high resolution photoemission experiment. Thus in principle the quantities α, v_F and v_2 can be measured. It is of course of great interest to establish how close α is to 1, or whether v_2 is reduced with respect to that extrapolated from the energy gap at $(0, \pi)$ measured by photoemission or tunneling. Crude estimates made in Ref. [11] suggest that α is consistent with 1 but a more precise measurement is clearly called for.

Finally we comment on finite temperature behaviors. In addition to the reduction of superfluid density due to thermal excitation of qp, [11] we expect x_{pc} to decrease with increasing T, leading to a reduction of v_2: $v_2(T) = \frac{x_{pc}(T)}{x_{pc}(0)} v_2(0)$. As T reaches T_c, $x_{pc} = v_2 = 0$ and the nodes of $E^{(sc)}$ become the "FS segment" while the spin gap near $(0, \pi)$ remain finite. We see that x_{pc} plays the role of the order parameter of the transition, so that we may expect the temperature dependence of x_{pc} to be described by a Ginzburg-Landau theory with X-Y symmetry near the transition.

In summary, we believe the $SU(2)$ slave boson theory captures the basic physics of the underdoped cuprates, which is spin-charge separation in the normal state and spin-charge recombination in the SC state. The quasiparticle spectrum in the SC state is remarkably similar to BCS theory, even though the microscopic mechanism is different in the sense that these are not formed out of pairing of normal state qp via exchange of some effective interaction.

References

[1] P.W. Anderson, *Science* **235**, 1196 (1987).

[2] G. Kotliar and J. Liu, *Phys. Rev. B* **38**, 5142 (1988).

[3] Y. Suzumura *et al.*, *J. Phys. Soc. Jpn.* **57**, 2768 (1988); H. Fukuyama, *Prog. Theo. Phys. Suppl.* **108**, 287 (1992).

[4] G. Baskaran and P.W. Anderson, *Phys. Rev. B* **37**, 580 (1988); L. Ioffe and A. Larkin, *Phys. Rev. B* **39**, 8988 (1989); P.A. Lee and N. Nagaosa, *Phys. Rev. B* **45**, 966 (1992).

[5] X.-G. Wen and P.A. Lee, *Phys. Rev. Lett.* **76**, 503 (1996).

[6] P.A. Lee, N. Nagaosa, T.K. Ng, and X.G. Wen, *Phys. Rev. B* **57**, 6003 (1998).

[7] I. Affleck, Z. Zou, T. Hsu, and P.W. Anderson, *Phys. Rev. B* **38**, 745 (1988).

[8] A.G. Loeser *et al.*, *Science* **273**, 325 (1996).

[9] H. Ding *et al.*, *Nature* **382**, 51 (1996).

[10] D.H. Kim, P.A. Lee, and X.G. Wen, *Phys. Rev. Lett.* **79** 2109 (1997).

[11] P.A. Lee and X.G. Wen, *Phys. Rev. Lett.* **78**, 4111 (1997).

[12] Y. Uemura *et al.*, *Phys. Rev. Lett.* **62**, 2317 (1989); **66**, 2665 (1991).

[13] D.A. Bonn *et al.*, *Czech. J. Phys.* **46** S6, 3195 (1996).

[14] A.J. Millis, S.M. Girvin, L.B. Ioffe, and A.I. Larkin, *J. Phys. Chem. Solids* (1998).

[15] X.G. Wen and P.A. Lee, *Phys. Rev. Lett.* **80**, 2193 (1998).

Gauge Field, Confinement, and Superconductivity in Underdoped High-T_c Cuprates

Naoto Nagaosa

Department of Applied Physics, University of Tokyo, Bunkyo-ku, Tokyo 113, Japan

Abstract

The issue of the confinement of the gauge field is studied for the underdoped cuprates in terms of the newly developed SU(2) formulation. The physical conductivity $\sigma(q,\omega)$ controls the confinement-deconfinement (C-DC) transition. In the non-superconducting state, the crossover occurs at the inter-hole distance $\sim x^{-1/2}$ and the holon localization length ξ. The superconducting transition is identified with the confinement, i.e., minopole condensation, triggered by the holon pairing. Then the strong coupling nature of the gauge field has two aspects, i.e., the large inelastic scattering above T_c (classical flucutations) and the confinement in the superconducting state. This explains also the drastic change of the quasi-particles across T_c.

It is now established that the parent material of high-Tc cuprates are well described in terms of the Heisenberg model with $S = 1/2$, and shows the antiferromagnetic long range ordering (AFLRO). With the hole doping this AFLRO is rapidly destroyed, a spin-glass region with localized holes at $0.02 < x < 0.05$, and eventually the superconductivity appears at $0.05 < x$ [1]. A theoretical formulation describing these doped holes is the slave boson method where an electron is regarded as a composite particle of a fermion (spinon) and a boson (holon), which are coupled to the gauge field [2,3]. This gauge field expresses the constraint, and has no dynamics at the starting, i.e., in its strong coupling limit. Then one expect that the confinement occurs, and only the gauge singlets appear in the physical spectra. This is true in the Heisenberg model where the spinon is bound into the spin operator $\vec{S}_i = \frac{1}{2} f^\dagger_{i\alpha} \vec{\sigma}_{\alpha\beta} f_{i\beta}$. Note that the spinon f has spin $S = 1/2$ (doublet) while $\vec{S}$ represents the triplet excitation. Heisenberg model with $S = 1/2$ can be formulated in terms of an SU(2) lattice gauge theory with the staggered fermions in its strong coupling limit [4,5]. As expected the numerical simulations suggest the confinement of the gauge field, and the chiral symmetry breaking of the staggered fermions, which is the natural consequence of the confinement, corresponds to the AFLRO.

One of the exception is the one-dimensional Heisenberg chain where the solitons with $S = 1/2$ are the fundamental objects, which offered the physical intuition to the spinon for higher dimensions. In the field theoretical language in terms of the CP^1 formulation, the topological Berry term with the θ-angle π corresponds to the charge $\pm e/2$ at the end of the sample, which allows the creations of soliton-antisoliton pairs without confinement [6]. The two dimensional analogue of this scenario is the Chern-Simons term which appears in the chiral spin state with broken T (time reversal) and P (parity) symmetry [7-9]. However the experiments are negative for the broken T,P-symmetry.

Starting from the fact that the confinement occurs in the undoped case, the deconfinement, if it exists, is solely due to the doped holes with finite concentration and probably the finite temperature. Detailed analysis of the numerical calcuations on the Green's function and conductivity concludes the binding of the spinons and holons with the linearly increasing potential [10]. This is exactly what is expected in the confining phase of the gauge field. Based on this, Laughlin further proposes that the spin-charge separation occurs only at the quantum critical point, which controls the quantum critical phenomena over a wide range of the temperature [11]. Experimentally many transport properties, e.g., the Hall constant scaling as $1/x$ [12], the large saturation value of the resistivity, and the residual resistivity due to impurities [13-15], strongly suggests that the holons are nearly free quasi-particles in the normal state of high-Tc cuprates. However recent analysis of the magnetic penetration depth well below T_c concluded that the Dirac electrons near the nodal points of $d_{x^2-y^2}$ paired state are the only quansi-particles without large fermi liquid corrections by F_{1s} [16]. Then there should be a drastic change of the excitaion spectra other than the BCS type pairing across the superconducting T_c.

On the other hand the issue of the confinement has been studied for the effective U(1) gauge model after integrating over spinons and holons [17,18]. There it was argued that the dissipative term for the gauge field, which comes form the particle-hole excitation across the large spinon fermi surface and is represented by the spinon conductivity, is very effective to suppress the confinement. This is because the dissipation suppresses the quantum coherency of the gauge flux motion. In the conventional U(1) gauge theoretical formulation, however, the above features of the undoped case are not taken into account properly and an SU(2) theory has been developed to describe the underdoped cuprates [19].

In this paper we study the confinement problem of the gauge field for underdoped cuprates in terms of the SU(2) theory. This includes the identification of the AFLRO state, spin-glass state, and the superconductivity in terms of the confining gauge field but with different length scales.

In the SU(2) formulation, the doublets $\psi_{1i} = \begin{pmatrix} f_{1i} \\ f_{2i}^\dagger \end{pmatrix}$, $\psi_{2i} = \begin{pmatrix} f_{2i} \\ -f_{1i}^\dagger \end{pmatrix}$ and $h_i = \begin{pmatrix} b_{1i} \\ b_{2i} \end{pmatrix}$ are introduced for spinon and holon, respectively [4,19]. The electron operator can be written as an $SU(2)$ singlet, i.e.

$c_{1i} = \frac{1}{\sqrt{2}} h_i^\dagger \psi_{1i} = \frac{1}{\sqrt{2}} \left(b_{1i}^\dagger f_{1i} + b_{2i}^\dagger f_{2i}^\dagger \right)$, $c_{2i} = \frac{1}{\sqrt{2}} h_i^\dagger \psi_{2i} = \frac{1}{\sqrt{2}} \left(b_{1i}^\dagger f_{2i} - b_{2i}^\dagger f_{1i}^\dagger \right)$. Then the link variable is the 2×2 matrix U_{ij} which includes the spinon pairing order parameter Δ_{ij} and the hopping order parameter χ_{ij} as the matrix elements.

$$U_{ij} = \begin{bmatrix} -\chi_{ij}^* & \Delta_{ij} \\ \Delta_{ij}^* & \chi_{ij} \end{bmatrix}. \tag{1}$$

This SU(2) theory has been studied at the mean field level [19]. The spin-gap state in the underdoped region corresponds to the staggered-Flux state with $U_{ii+x}^{(0)} = -\chi\tau_3 - i(-1)^{i_x+i_y}\Delta$, $U_{ii+y}^{(0)} = -\chi\tau_3 + i(-1)^{i_x+i_y}\Delta$. In this staggered fulx state, the effective action describing the holons $h_i = [b_{1i}, b_{2i}]$ in the underdoped spin gap region is given by

$$L = \int dr h^\dagger(r,\tau)[\partial_\tau + ia_0^3\tau_3 + iA_0 + \frac{1}{2m}(-i\nabla + \vec{a}^3\tau_3 + \vec{A})^2 - \mu]h(r,\tau)$$
$$+ \sum_{q,\omega} a_\mu^3(q,\omega)\Pi_{S\mu\nu}(q,\omega)a_\nu^3(-q,-\omega) \tag{2}$$

where the spinons have been integrated over to give the polarization function Π_S in the effective acton of the gauge field. Because the gauge symmetry is broken from SU(2) to U(1) in the staggered flux state, only a^3 gauge field remains massless. Note also that the Ioffe-Larkin composition rule [17] no longer applies because the external vector potential A_μ

is coupled to h_i with the unit matrix and not with τ_3. Then the conductivity is determined primarily by that of the holon system. A fermionization approach has been proposed where the strong gauge field fluctuation and the repulsive interaction between bosons can be approximately treated by transforming the holons into fermions in terms of the Chern-Simons gauge field [15]. This Chern-Simons gauge field is coupled to b_1 and b_2 with opposite charges, and partially screens the a^3 gauge field fluctuation.

Let us first consider the normal state. The leading order contributions of the holons to the effective action for the gauge fields A_μ and a_μ^3 are given as

$$\sum_{q,\omega} \sigma(q,\omega)|\omega|(a_\perp^3(q,\omega)a_\perp^3(-q,-\omega) + A_\perp(q,\omega)A_\perp(-q,-\omega)) \tag{3}$$

where $a_\perp^3$ and $A_\perp$ are the transverse components. The contribution from the spinons Π_S, on the other hand, is small compared with eq.(3) in the limit of small q, ω because of the d-wave type gap in the spinon spectrum. Also we already know that the spinons can not prevent the confinement in the undoped ($x = 0$) limit. Then the strength of the dissipation for the gauge field is determined by the holon's and hence the physically observed conductivity $\sigma(q,\omega)$ of the system. Especially C-DC transition is characterized by the force between the static charges, the static conductivity $\sigma(q) = \sigma(q,\omega = 0)$ is relevant. The schematic behavior of $\sigma(q)$ is given as follows when we employ the fermionization approach. (We take the unit where $e^2/h = 1$).

$$\sigma(q) = 0 \qquad\qquad 2k_F < q$$
$$\sigma(q) = x/(v_F q) \qquad\qquad \ell^{-1} < q < 2k_F$$
$$\sigma(q) = k_F \ell \cong x^{1/2}\ell \qquad\qquad \xi_l^{-1} < q < \ell^{-1}$$
$$\sigma(q) = 0 \qquad\qquad q < \xi_l^{-1} \tag{4}$$

where ℓ is the mean free path, and ξ_l is the localization length of the holons. Here both the Fermi wavenumber k_F and the Fermi velocity v_F scale with $\xi_x^{-1} \cong x^{1/2}$ where ξ_x is the inter-hole distance. We have assumed $\ell >> \xi_x$, which is valid except for the small x. (See the discussion below.) This structure of $\sigma(q)$ gives rise to the crossovers of the gauge field dynamics. We consider the strong coupling limit of the gauge field, and generalize the model eq.(3) to take into account the discreteness of the lattice and also the 2π periodicity of a_{ij}. This has been discussed in [18], and we modify the argument there to include the q-dependence of $\sigma(q)$. The winding number $k(r)$ for the transverse part of the gauge field

a_{ij} is the relevant variable to the C-DC tranistion, and the effective model for $k(r)$ is given at zero temperature by

$$S = \frac{1}{2}\sum_{\mu=x,y}\sum_{ij}\sigma_{i-j}(|\Delta_\mu k(i)| + |\Delta_\mu k(j)| - |\Delta_\mu k(i) - \Delta_\mu k(j)|) \tag{5}$$

where σ_{i-j} is the Fourier transform of $\sigma(q)$. This is a absolute solid-on-solid (SOS) model, and we replace this by an easier problem as

$$S = \frac{1}{4}\sum_{\mu=x,y}\sum_{ij}\sigma_{i-j}((\Delta_\mu k(i))^2 + (\Delta_\mu k(j))^2 - (\Delta_\mu k(i) - \Delta_\mu k(j))^2)$$

$$= \frac{1}{2}\sum_{\mu=x,y}\sum_{ij}\sigma_{i-j}\Delta_\mu k(i)\Delta_\mu k(j)$$

$$= \sum_{\mu=x,y}\sum_{q}\sigma(q)(1-\cos q_\mu)k(q)k(-q) \tag{6}$$

By using the Poisson formula the partition function can be written as

$$Z = \int_{-\infty}^{\infty}\Pi d\phi(r)\sum_{m(r)=-\infty}^{\infty}\exp\left[-\frac{1}{2}\sum_{\mu=x,y}\sum_{ij}\sigma_{i-j}\Delta_\mu\phi(i)\Delta_\mu\phi(j) + 2\pi i\sum_i m(i)\phi(i)\right] \tag{7}$$

where the integer $m(r)$ represents the "vortex" excitation, and its fugacity y is of the order of $y \cong e^{-\sigma_{i-j=0}}$. We estimate $\sigma_{i-j=0} = \sum_q \sigma(q) = 2x$. Assuming the small fugacity for simplicity and restricting the sum $m(r) = \pm 1$, one obtain from eq.(7) the sine-Gordon model

$$Z = \int_{-\infty}^{\infty}\Pi_r d\phi(r)\exp\left[-\sum_{\mu=x,y}\sum_q\sigma(q)(1-\cos q_\mu)\phi(q)\phi(-q) + 2y\sum_r\cos(2\pi\phi(r))\right] \tag{8}$$

Then the renormalization group procedure is divided into three regions. Let Λ denote the wavenumber cutt-off, and the original $\Lambda = \Lambda_0$ is of the order of the inverse of the lattice constant $a^{-1} = 1$. Then the lowest order scaling equation for $\tilde{y} = y\Lambda^{-2}$ is

$$\frac{d\ln\tilde{y}(\Lambda)}{d\ln\Lambda} = 2 - \frac{\pi}{2\sigma(\Lambda)} \tag{9}$$

For $\xi_x^{-1} < \Lambda < \Lambda_0$, there occurs no contribution from the holons, and the scaling trajectory is determined by the spinon contribution which is not included in eq.(6). There the physics should be the same as the undoped case, because it is rare to find the holes within the length scale less than the inter-hole distance $\xi_x \cong x^{-1/2}$. For $\xi_l^{-1} < q < \xi_x^{-1}$ on the other hand, $\sigma(q)$ is always larger than unity, i.e., $1 < \sigma(q) < x^{1/2}\ell$. Then $\tilde{y}$ scales to smaller value corresponding to the deconfinement. However for $\Lambda < \xi_l^{-1}$ the scaling equation again predicts the confinement. Therefore we have three regions for the length scale r, i.e.,

(i) $r < \xi_x$: the similar behavior as the undoped case is expected where the spinons are bound into localized spins,

(ii) $\xi_x < r < \xi_l$: the gauge field is deconfining and the spin-charge separation is realized, and

(iii) $\xi_l < r$: the gauge field is confining and the spinons and holons are bound into electrons, and electrons are localized. In Fig. 1 shown a schematic phase diagram of high-Tc cuprates. The small hole concentration region with AFLRO ($x < 0.02$) is identified with the case $\xi_l < \xi_x$ and only the case (i) is relevant, where the system is described by the depleted Heisenberg model. A recent study have shown that the Ionlized vacancies in the non-linear sigma model gives the topological disorder (random Berry phase term), which enhances the classical nature of the staggered magnetization and leads to the AFLRO [20].

For $0.02 < x < 0.05$, we have the crossover as a function of temperature. At finite temperature T there occurs a relevant length scale $\xi_T^{\text{spin,charge}}$. For the spin correlation $\xi_T^{\text{spin}} \sim e^{(cJ/T)}/T$ (c: constant of order unity) in the renormalized classical region and $\xi_T^{\text{spin}} \sim J/T$ in the quantum critical region. Here we recover J as the unit of the energy. The correlation length for charge is given by $\xi_T^{\text{charge}} \cong v_F/T \sim Jx^{1/2}/T$. It is then concluded that $\xi_T^{\text{charge}} << \xi_T^{\text{spin}}$. When $\xi_T^{\text{spin}} \cong \xi_x$, i.e., $T \sim J/|\ln x|$ the crossover from the localized spin to the spin liquid occurs. From the saturation of the magnetic correlation length [1] and the temperature dependence of T_1^{-1} of NMR [21], we regard this crossover temperature around 500K (the horizontal dotted line in Fig. 1). When $\xi_x < \xi_T^{\text{charge}} < \xi_l$, the system shows metallic behavior and the spin-charge separation occurs, while at even lower temperature $T < v_F/\xi$ the electron localization and the spin-glass behavior occur, where the gauge field is again confining.

Now we turn to the superconductivity. As we discussed below eq.(2) there are two species of holons with opposite gauge charges, which attract with each other. In a fermionization scheme [15], the pairing of the holon is the natural scenario for the superconductivity with the order parameter $\Delta_{ij}^h = <b_{1i}b_{2j}>$ having the real charge 2e. Even without the fermionization, it is easy to see that the single holon condensation is forbidden at finite temperature when one consider the vortex excitations. The screening by the a^3 gauge field makes the vortex energy finite, and hence the superfluidity of the b_1 or b_2 boson alone does not occur at finite temperature. The only possibility is the superfluidity due to the paring of b_1 and b_2 bosons where the order parameter $\Delta_{ij}^h = <b_{1i}b_{2j}>$ has no gauge charge. Therefore the

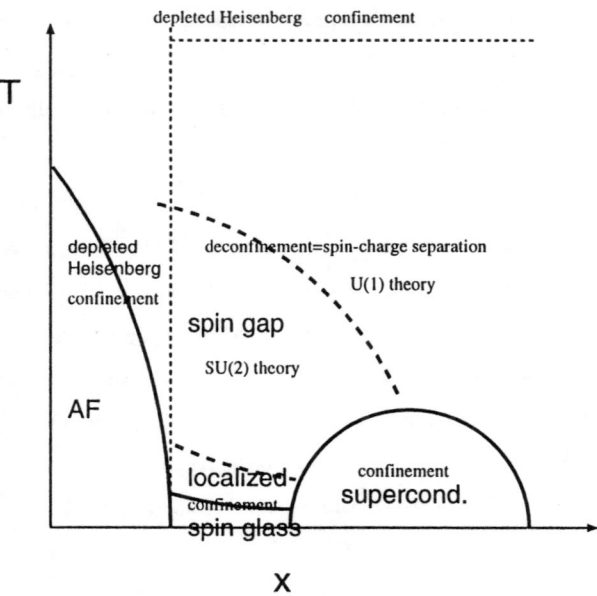

Figure 1. Schematic phase diagram of high-Tc cuprates in the plane of hole concentration x and temperature T.

superconductivity is not accompanied with the Higgs phenomena for the internal gauge field, which resolves the problem of the large energy cost of the gauge field. However the effective action for the gauge field changes. We employ again the fermionization scheme. For the a^3 gauge field the paired state of b_1 and b_2 is the excitonic insulator because b_1 and b_2 have opposite charges with the gap Δ of the order of x. This suppresses $\sigma(q,\omega)$ in eq.(3) for $v_F q, \omega < x$. Note that now $\sigma(q,\omega)$ is different from the physically observed one in the superconducting state because the coherence factor is different. According to eq.(8) this results in the confinement, and the onset of the superconductivity is identified with the confinement transition of the gauge field.

Some remarks are in order about the relation between superconductivity and confinement. It has been discussed in the literature that these two phenomena are dual to each other [22]. The static magnetic monopoles is confined by the string tension of the magnetic flux in the superconductor while the electric charges are confined by the electric string. Considering the conjugate relation of the electric and magnetic field, these two phenomena might appear to compete with each other. However this is not the case when we consider the Higgs boson coupled to the gauge field. It has been discussed by Fradkin and Shenker [23] that the confining phase and the Higgs phase are continued to each other if the gauge charge of the Higgs boson b is fundamental. At finite temperature it is true that the strongly fluctuating magnetic field induces the vortex and hence destroy the superconductivity. However at zero temperature the vortex loop in (2+1)D is the relevant object to destroy the superconductivity, i.e., the appearance of the vortex loop with infinite size or the condensation of the vortex boson in the duality picture corresponds to the insulating ground state. The quantum fluctuation of the gauge field is represented by the instantons in (2+1)D, which are the source or drain of the gauge flux [17,18]. The confinement corresponds to the screening phase of the instantons, and these free instantons cut the vortex loop into finite segments and prevents the appearance of the vortex loop of infinite size [24]. Therefore the Higgs phenomenon is supported by the confinement through the duality between the Higgs field and the gauge field. However the nonzero expectation value of $<b>$ in the confinement-Higgs phase needs the gauge fixing, i.e., the unitary gauge, which actualy does not mean that the gauge field fluctuation is suppressed. Then the two aspects of the strong gauge field fluctuations, i.e., inelastic pair breaking effect above T_c and the confinement-superocnductivity below T_c, are understood as the classical-quantum crossover.

If the gauge charge of the Higgs boson is not fundamental, the Higgs phase and the confining phase are different. In the holon-pairing scenario, this second case applies because $\Delta_{ij}^h = <b_{1i}b_{2j}>$ has no gauge charge and only the spinon order parameter with the adjoint (vector) representation is coupled to the gauge field. Therefore the system appear to be in the Higgs phase with the $U(1)$ symmetry unbroken. However as discussed above the gauge field becomes confining if the superconductivity sets in, and all the gauge charges are confined into gauge singlet. This means that spinons and holons are confined into electrons, which are the only quasi-particle in the superconducting state. Thus the superconducting transition is identified with that of the confinement, and we believe that it is not appropriate to describe the superconducting state by any kind of mean field theory using spinon-holon language. This explains the sudden appearance of the quasi-particle peak in angle-resolved photoemission spectra (ARPES) [25], rapid increase of the lifetime of the quasi-particle [26], and is also consistent with the argument for the temperature dependence of the penetration depth [16]. Another conclusion from this scenario is about the normal core of the vortex under external magnetic field. An estimate of the vortex core size ξ_v has been done by evaluating the superfluidity density $\rho_s(R) = \rho_s(R=\infty) - \rho_n(R)$ with R being the distance from the center of the vortex. By setting $\rho_s(\xi_v) = 0$, Lee and Wen obtained $\xi_v \sim 1/x$ and $H_{c2} \propto x^2$ [27]. Therefore $\xi_v \gg \xi_x$ and hence the spin-charge separated state will appear in the normal core of the vortex. Increasing further the magnetic field above H_{c2} this non-Fermi liquid state with spin-charge separation will continue down to the zero temperature if the system remains metallic.

In summary we have studied the confinement of the gauge field in the SU(2) theory for underdoped cuprates. The conductivity $\sigma(q,\omega)$ is the key quantity to control the confinement. Both the localization of the electron and the on-set of the superconductivity trigger the confinement, and the two length scales , i.e., the inter-hole distance $\xi_x \cong x^{-1/2}$ and the localization length ξ_l are relevant. If there is a quantum critical point at $x = x_{cr}$, where the superconductivity sets in at zero temperature, the critical behavior of C-DC transition would be an interesting issue which needs further studies [28].

The authors acknowledges H.Fukuyama, S.Uchida, N.P.Ong, P.A.Lee, R.B.Laughlin, S.C.Zhang, E.Fradkin, and Z.X.Shen for fruitful discussions. This work is supported by Priority Areas Grants and Grant-in-Aid for COE research from the Ministry of Eduction, Science and Culture of Japan,

REFERENCES

[1] B. Keimewr et al., Phys. Rev. B**46**, 14034 (1992).

[2] P. W. Anderson, Science **235**, 1196 (1987)

[3] G.Baskaran and P.W.Anderson, Phys. Rev. B**37**, 580 (1988); G.Baskaran, Phys. Scr. T**27**, 53 (1989).

[4] I.Affleck, Z.Zou, T.Hsu, and P.W.Anderson, Phys. Rev. B**38**, 745 (1988).

[5] E.Dagotto, E.Fradkin, and A.Moreo, Phys. Rev. B**38**, 2926 (1988).

[6] I.Affleck, in " Strings, Fields and Critical Phenomena" eds. E.Brezin and J. Zin-Justin (North-Holland, 1990) p565.

[7] V.Kalmeyer and R.B.Laughlin, Phys. Rev. Lett. **59**, 2095 (1988).

[8] X.G.Wen, F.Wilczek, and A.Zee, Phys. Rev. B**39**, 11413 (1989).

[9] For the Chern-Simons gauge field see for e.g. S. C. Zhang, Int. J. Mod. Phys. B**6**, 25 (1992).

[10] R. B. Laughlin, in Proceedings of the Inauguration Conference of the Asia-Pacific Center for Theoretical Physics, Seoul National University, Korea 4-10 June 1996, ed. by Y. M. Cho, J. B. Hong, and C. N. Yang (World Sci., Singapore,1998)

[11] R. B. Laughlin, cond-mat/9709195.

[12] S. Uchida et al., in Strong Correlation and Superconductivity (Springer-Verlag, 1989) p194.

[13] T. R. Chien, Z. Z. Wang, and N. P. Ong, Phys. Rev. Lett. **67**, 2088 (1991).

[14] K. Mizuhashi et al., Phys. Rev. B**52**, R3884 (1995); Y. Fukuzumi et al., Phys. Rev. Lett. **76**, 684 (1996).

[15] N. Nagaosa and P. A. Lee, Phys. Rev. Lett. **79**, 3755 (1997).

[16] X. G. Wen and P. A. Lee, cond-mat/9709108.

[17] L.B.Ioffe and A.I. Larkin, Phys. Rev. B**39**, 8988 (1989).

[18] N.Nagaosa, Phys. Rev. Lett. **71**, 4210 (1993).

[19] X. G. Wen and P. A. Lee, Phys. Rev. Lett. **76**, 505; P. A. Lee, N. Nagaosa, T. K. Ng, X. G. Wen, Phys. Rev. B**57(10)**, 6003-6021 (1998).

[20] N. Nagaosa, A. Furusaki, M. Sigrist, and H. Fukuyama, J. Phys. Soc. Jpn. **65(12)**, 3724-3727 (1996).

[21] T. Imai et al., Phys. Rev. Lett. **70**, 1002 (1993).

[22] T. D. Lee, Particle Physics adn Introduction to Field Theory (Harwood academic publihsers, 1981)Chap. 17.

[23] E. Fradkin and S. Shenker, Phys. Rev. D**19**, 3682 (1979).

[24] Y. Nambu, Phys. Rev. D**10**, 262 (1974).

[25] Z. X. Shen and J. R. Schrieffer, Phys. Rev. Lett. **78**, 1771 (1997).

[26] N. P. Ong, this volume.

[27] P. A. Lee, this volume.

[28] H. Fukuyama, Rev. High Pressure Sci. Technol. **7**, 465 (1998).

A Progress Report on the SO(5) Theory of High T_c Superconductivity

Shou-Cheng Zhang

Department of Physics, Stanford University, Stanford CA 94305

Abstract. In this talk I give a brief update on the recent progress in the $SO(5)$ theory of high T_c superconductivity[1]. Reviewed topics include $SO(5)$ ladders, the unification of BCS and SDW quasi-particles in the $SO(5)$ theory and the microscopic origin of the condensation energy.

First of all I would like to thank the Taniguchi foundation and the organizers of the Grand Finale Taniguchi Symposium for inviting me to this wonderful conference. This Symposium is appropriately entitled "The Physics and Chemistry of Transition Metal Oxides", and it covers a vast and broad range of topics whose intimate relations remain to be discovered. In order to achieve a coherent grand synthesis in this subject, we must first overcome the language barrier which separates workers in different sub-fields. At the welcome party of the Symposium, Professor George Swatzsky and I were casually chatting about the $SO(5)$ theory of high T_c superconductivity. A distinguished chemistry professor looked more and more puzzled as he overheard our conversation. Finally, he couldn't hold his curiosity and asked "SO_5? I didn't know that sulfer-pentoxide is a superconductor!"

In a series of conference proceedings, I tried to give a on-going update about the status of the $SO(5)$ theory of HTSC[2, 3]. This is the third one in this series. In the mean time, Auerbach wrote a pedagogical review[4] explaining the $SO(5)$ theory in terms of more familiar concepts in $SO(3)$ quantum magnetism, and Hanke et al wrote a extensive review[5] on the numerical calculations within the $SO(5)$ theory. Currently, the $SO(5)$ approach to HTSC is actively being investigated by many groups, focusing both on the microscopic origins and phenomenological consequences. Much progress has been made in understanding the logical structure and examining the internal consistency of the theory. In this report, I would first like to summarize recent developments in understanding the microscopic realization of the $SO(5)$ symmetry by using ladder systems as a theoretical laboratory[6, 7, 8, 9, 10, 11, 12, 13, 14, 15] and address the unification of BCS and SDW quasi-particles in the $SO(5)$ theory[16]. These developments reveal a fascinatingly rich internal structure of the $SO(5)$ theory and could ultimately lead to a microscopic foundation of the theory. Although the $SO(5)$ theory appears to be a natural framework to understand many experiments of HTSC in a unified fashion, no experiment has directly tested the fundamental validity of the theory. Some more tests have been recently proposed in [17, 18]. However, I would like to focus on some recent works[19, 20] concerning the microscopic origin of the condensation energy in HTSC, which could lead to a direct and quantitative understanding of the microscopic mechanism of HTSC.

Shortly after the $SO(5)$ proposal, various groups have constructed microscopic models with exact $SO(5)$ symmtry[21, 22, 23]. However, all this models involve long range interactions which are not familiar and natural. It appears that as long as there is only one orbital per unit cell, this problem is unavoidable. For this reason the $SO(5)$ symmetry was investigated for the two-legged ladder system, which has two orbitals per unit cell, so that long ranged interaction can be avoided[6]. Another reason for investigating $SO(5)$ symmetry in the ladder system is because it is a example of a Mott insulator without any long range order at half-filling. $SO(5)$ symmetry was original proposed as a theory to unify antiferromagnetism (AF) with superconductivity (SC). It would be a interesting question to see how it applies to Mott insulators without any (quasi-) AF long range order. The third reason for investigating the $SO(5)$ symmetry in the ladder system is to address the question of how this symmetry could emerge at long wave length without being present at the microscopic level[7, 8, 9, 15]. Because of quasi-one-dimensionality, well controlled weak coupling RG calculations can be performed to address this issue.

The fundamental quantity in the microscopic $SO(5)$ models is the concept of a $SO(5)$ spinor[21], which has four components. On a two-legged ladder, one could naturally combine the two sites on a rung to form such a spinor, and $SO(5)$ invariant models can be easily constructed by writing down the most general invariant interactions[6]. The parameter space for $SO(5)$ models is surprisingly large. Among the usual five local parameters t, $t_\perp$, U, V and J, only one condition is required to satisfy the $SO(5)$ symmetry ($U + V = J/4$). The Mott insulating state at half-filling is not only a total spin singlet, but also a $SO(5)$ singlet. The lowest energy excitation on top of this singlet ground state is a five fold degenerate manifold of triplet magnons and Cooper pairs. A uniform magnetic field or chemical potential can lower the energy for one of these bosons, leading to a condensate with AF or SC quasi-long-range order. Within this framework, Mott insulator, AF and SC states are intimately related and can be understood in a unified way. Many theoretical ideas about $SO(5)$ symmetry can be tested in the ladders system. The eigenstates of the $SO(5)$ ladder models can all be classified into general irreducible representations of the $SO(5)$ group. These states form a beautiful and revealing pattern and have been identified in numerical calculations by Eder, Dorneich, Zacher, Hanke and the author[11]. The photoemission spectrum of the $SO(5)$ ladder has been studied in details[11, 13]. The single electron Green's functions can be related by exact Ward identities and are shown by direct diagonalization to evolve continuously from the Mott insulator to the superconducting state[11]. The structure of the exact π resonance can also be understood in detail both analytically and numerically[14, 11].

Due to the quasi-long-range order of the superconducting state, the π resonance is not a delta function peak, but a threshold singularity at energy -2μ[6, 14]. Other physical quantities can also be calculated[25].

While the $SO(5)$ models offer a nice testing ground for many interesting theoretical ideas, the parameters of these models are not "realistic". It is therefore desirable to understand whether they share some common features with more realistic ladder models. For example, in the strong coupling limit, the ground state of a generic ladder model is a product state of singlet rungs. Rather surprisingly, this state is not only a total spin singlet, but also a $SO(5)$ singlet, since this product state is annihilated by the π operators. In any $SO(5)$ singlet state, the behavior of the static AF and SC correlation functions are identical. Therefore, one would expect these correlations to scale towards a common behavior in the strong coupling limit of a generic ladder model. Unfortunately, this argument does not apply to the dynamic correlation functions. One can use the exact $SO(5)$ models as a point of departure to systematically vary the symmetry violating perturbations and compare the results with the partial multiplet structures found in the $t-J$ model[24]. Investigation towards this direction has been taken in ref.[10, 11].

Fortunately, in the weak coupling regime, one can perform controlled RG calculations to see how exact symmetries emerge from nonsymmetric interactions[7, 8, 9, 15]. Lin, Balents and Fisher[9] showed that there are in general 9 marginal operators for a two-legged ladder at half-filling, 5 of them preserve $SO(5)$ symmetry, while 4 of them violate it. Surprisingly, the symmetry violating interactions scale to zero. Within the remaining $SO(5)$ manifold, there are four stable fix points, with even higher symmetry, namely $SO(8)$. Arrigoni and Hanke[8] also studied the symmetry violating band structure effects, for example the next-nearst-neighbor hopping t', and demonstrated that after a suitable redefinition, the symmetry violating effects can be completely absorbed. More recently, Schulz[15] extended the RG calculations from half-filling to general filling, and showed that $SO(6)$ and $SO(5)$ symmetries can emerge dynamically. These developments are very exciting, and offer hope that similar symmetry restoration effects can occur in higher dimensional systems near a quantum critical point[1, 26].

The original $SO(5)$ theory was formulated purely in terms of bosonic collective degrees of freedom[1]. It is clear that a complete theory has to include the fermionic sector. The bosonic theory explains how various collective modes connect to each other at the transition between the AF and SC states, it is natural to ask how the fermionic BCS and SDW quasi-particles connect to each other. The answer to this question turns out to be surprisingly rich and beautiful. In elementary quantum mechanics, we learned about a remarkable demonstration of the spinor nature of the electron. If two strong magnetic fields polarize the electron spins in two orthogonal directions, then the electron wave functions corresponding to these two orthogonal fields are not orthogonal to each other. This non-orthogonality of the electron wave functions allows transmission of a electron beam through the orthogonal field regions. It is also the origin of the Berry's phase. Quite similarly, even though a antiferromagnet and a superconductor have "orthogonal" order parameters, the quasi-particles associated with these two states are not orthogonal to each other. This non-orthogonality leads to a novel generalization of Berry's phase to a $SU(2)$ holonomy[16]. The generalization from the $U(1)$ Berry's phase of a $SO(3)$ spinor to the $SU(2)$ holonomy of a $SO(5)$ spinor is a unique generalization in a precise mathematical sense, and involves some of the most beautiful, and yet seeming disconnected mathematical concepts such as Hopf maps, quaternions and the Yang monopole. So far this work is still at a mathematical stage, and the precise physical implications can only be anticipated at this moment. Potential applications include a generalized Bogoliubov-deGennes type of formalism to discuss novel fermionic excitations near topological defects involving twists of the $SO(5)$ superspin vector[27, 28, 29, 30], a novel type of Andreev reflection at the AF/SC boundary, a non-abelian Bohm Aharonov effect associated with regions with non-trivial superspin twists and novel understanding of the single particle properties in the pseudogap regime. The remarkablly rich fermionic structure in the $SO(5)$ model shows that the $SO(5)$ theory is much more than a expanded version of the Landau-Ginzburg theory, it can fully address single particle excitations and their coupling to the collective modes.

The ideas on the $SU(2)$ holonomy can possibly be extended to other physical systems as well, especially transition metal oxide systems with orbital degeneracy. Soon after the discovery of quantum mechanics, Wigner and von Neumann studied the problem of generic level crossings in quantum mechanics. They and later Dyson classified generic level crossings into three categories, now called the orthogonal, unitary and symplectic ensembles. The familiar $U(1)$ Berry's phase occur in the unitary ensemble. In the sympletic ensemble, one deals with time reversal invariant systems with Kramers degeneracy. The level crossing between two Kramers doublets can be described by the four dimensional (since two doublets give four states in total) symplectic group $Sp(4)$, which happens to be isomorphic to $SO(5)$. This type of level crossing phenomenon can not only occur at AF/SC transition, but also in generic problems involving spin-orbit couplings. The ideas on $SU(2)$ holonomy can not only lead to deeper understandings of these systems, but may also help to understand their (formal) relationship to the high T_c problem.

Although the $SO(5)$ theory was originally proposed as a effective theory to understand the interplay between AF and SC, it implicitly points to a microscopic mechanism for HTSC. Within this theory, the microscopic mechanism for SC is basically the "same" as the microscopic mechanism for AF, namely the lowering of the exchange energy $J \sum_{i,j} \mathbf{S}_i \mathbf{S}_j$. Recently, Scalapino and White[19] argued that the lowering of the exchange energy can be quantitatively correlated with the SC condensation energy. This insightful observation allows for quantitative test of various mechanisms of HTSC which should make detailed prediction on how the exchange energy is saved in the SC state. The $SO(5)$ theory predicts a π resonance mode[1, 31, 32] which is identified with the neutron resonance mode observed in the SC state. The theory of the neutron resonance mode is based on a particle particle collective mode near momentum (π, π), which exists both in the normal and SC state. Since a particle particle mode can only make a contribution to the spin correlation function in the SC state, the neutron resonance mode is observed only below T_c. Recently, Demler and I noticed[20] that this argument also provides a concrete microscopic mechanism for HTSC. It is straightforward to see that the coupling to a particle particle collective mode around (π, π) gives a *negative* difference between the exchange energy $J \sum_{i,j} \mathbf{S}_i \mathbf{S}_j$ in the SC and the normal state. Therefore, the SC saves more exchange energy compared to the normal state, and the amount of saving is precisely given by J times the (dimensionless) integrated spectral weight of the π resonance. From the neutron scattering experiments by Fong et al[33], one can see that the change in the dimensionless quantity $\mathbf{S}_i \mathbf{S}_j$ due to the π resonance is on the order of few per cent, which gives a saving of exchange energy of $35K$ per unit cell. On the other hand, the condensation energy of optimally doped $YBCO$ superconductor is about $5K$[34]. Therefore, we see that the emergence of the π resonance could be the dominant mechanism responsible for the superconducting condensation energy. Upon going to the SC state, the kinetic energy usually increases, so that the saving in exchange energy could be balanced by the cost in kinetic energy to give the right condensation energy. It is therefore highly desirable to find direct ways to measure the change in kinetic energy when the system enters the SC state.

This line of reasoning leading to a microscopic mechanism of SC is somewhat unfamiliar, since most people equate the SC mechanism with a form of attractive interaction between electrons. However, I would like to argue that in a strongly correlated system, this new line of thinking is much more fruitful and experimentally accessible. *The central idea here is to identify a energy saving process which is forbidden in the normal state but possible in the SC state.* In our mechanism, the particle particle resonance is just such a process. The only other example I can think of is the interlayer tunneling mechanism[35]. In this case, the energy saving process is the c axis tunneling, which is forbidden in the normal state if the normal state is not a fermi liquid, but is allowed in the SC state. In both examples we see that once such a process is identified, experimentally falsifiable prediction about the condensation energy follow immediately. Following this line of thinking, we can hopefully move the debate about the microscopic mechanism of HTSC to a new level, where direct comparison with experiments becomes possible.

Since this argument seem to strongly rely on the onset of the π resonance at T_c for the optimally doped superconductors, a alert reader may wonder how this argument applies to the underdoped superconductors, where a broadened resonance peak is observed above T_c but below the pseudogap temperature T_{MF}[36, 37]. Let us first see how the $SO(5)$ theory could explain the broadened peak in the pseudogap regime. The basic process in the SC state is given by the following Feymann diagram.

We see that the spin vertex creates a particle hole pair, but the hole can be converted into a particle by the Gorkov F function, and the multiple scattering in the particle particle channel gives rise to a sharp collective mode

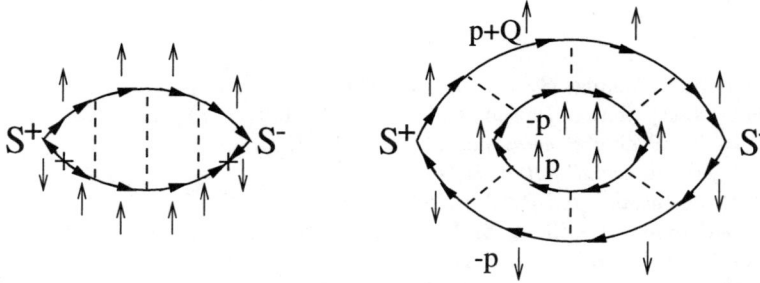

Fig. 1. Feymann diagram for the π resonance in the SC (left panel) and normal (right panel) state. In the SC state, a particle hole pair created by the neutron is converted into a particle particle pair by the SC condensate (marked by two crosses). In the normal state, such a anomalous process is forbidden, but one can open up the crosses and reconnect them to obtain a Cooper pair propagator. In the pseudogap regime, there is no sharp pole due to a single π pair, but a broadened convolution of a π and a Cooper pair.

in the dynamic spin correlation function. In the normal state, the Gorkov F function vanishes identically so that such a process is not possible. However, one could cut two Gorkov F functions and clue them together in the normal state as shown in Fig 1. Such a process does not vanish in the normal state, and represents the preformed Cooper pair fluctuations in the pseudogap regime. In this case, rather than a single resonance peak, we obtain a convolution between the triplet momentum (π,π) particle particle resonance and the singlet zero momentum Cooper pair resonance. If these resonances are weakly dispersive, the convolution spectrum will be reasonably sharp. Since the basic processes below and above T_c are different, we would generally expect some discontinuous behavior at T_c. The experimental plots of the intensity as a function of temperature does appear to consist of two separate curves joining with a discontinuous derivative at T_c. Having understood the SC fluctuation in the normal state as the origin of the neutron intensity above T_c, we would use the general arguments outlined above to correlate the neutron scattering intensity with the condensation energy in the pseudogap regime. Remarkably, the condensation energy obtained from Loram's specific data (Fig. 8 of reference [34]) and π resonance intensity measured in neutron scattering[38] in the underdoped regime have the same qualitative behavior, namely consisting of two separate curves joining with a discontinuous derivative at T_c. The remarkable similarity between these two seemingly different experiments lend strong support to our interpretation, and could not only lead to a quantitative understanding of the microscopic mechanism of HTSC, but also the origin of the pseudogap physics.

I would to thank E. Demler, R. Eder, A. Furusaki, W. Hanke, S. Rabello and D. Scalapino for close collaborations on projects reported above. This work is supported by the NSF under grant numbers DMR-9400372 and DMR-9522915.

References

1. Shou-Cheng Zhang. *Science*, 275:1089, 1997.
2. Shou-Cheng Zhang. *Physica C*, 282-287:265, 1997.
3. Shou-Cheng Zhang. *cond-mat/9709289*, 1997.
4. A. Auerbach. *cond-mat/9801294*, 1998.
5. W. Hanke et al. *cond-mat/9807015*, 1998.
6. D. Scalapino, S.C. Zhang, and W. Hanke. *Phys. Rev. B*, 58:443, 1998.
7. D.G. Shelton and D. Sénéchal. *cond-mat/9710251*, 1997.
8. E. Arrigoni and W. Hanke. *cond-mat/9712143*, 1997.
9. H. Lin, L. Balents, and M. Fisher. *Phys. Rev. B*, 58:1794, 1998.
10. Daniel Duffy, Stephan Haas, and Eugene Kim. *cond-mat/9804221*, 1998.
11. R. Eder et al. *cond-mat/9805120*, 1998.
12. P. Bouwknegt and K. Schoutens. *cond-mat/9805232*, 1998.
13. S-P Hong and S-H. Salk. *cond-mat/9807154*, 1998.
14. A. Furusaki and Shou-Cheng Zhang. *cond-mat/9807375*, 1998.
15. H. J. Schulz. *cond-mat/9808167*, 1998.
16. Eugene Demler and Shou-Cheng Zhang. *cond-mat/9805404*, 1998.
17. H. Bruus et al. *cond-mat/9807167*, 1998.
18. C. Burgess, J. Cline, and A. Lutken. *Phys. Rev. B*, 57:8549, 1998.
19. D.J. Scalapino and S.R. White. *cond-mat/9805075*, 1998.
20. Eugene Demler and Shou-Cheng Zhang. *cond-mat/9806336*, 1998.
21. S. Rabello, H. Kohno, E. Demler, and S.C. Zhang. *Phys. Rev. Lett.*, 80:3586, 1998.
22. C. Henley. *Phys. Rev. Lett.*, 80:3590, 1998.
23. C. Burgess, J. Cline, R. MacKenzie, and R. Ray. *Phys. Rev. B*, 57:8549, 1998.
24. R. Eder, W. Hanke, and S.C. Zhang. *Phys. Rev. B*, 57:13781, 1998.
25. Eugene Pivovarov. *cond-mat/9807291*, 1998.
26. R. B. Laughlin. *cond-mat/9709195*, 1997.
27. D. Arovas, A.J. Berlinsky, C. Kallin, and S.C. Zhang. *Phys. Rev. Lett.*, 79:2871, 1997.
28. Y. Bazaliy, E. Demler, and S.-C. Zhang. *Phys. Rev. Lett.*, 79:1921, 1997.
29. E. Demler, A.J. Berlinsky, C. Kallin, G. Arnold, and M. Beasley. *Phys. Rev. Lett.*, 80:2917, 1998.
30. P.M. Goldbart and D. Sheehy. *Phys. Rev. B*, 58:5731, 1998.
31. E. Demler and Shou-Cheng Zhang. *Phys. Rev. Lett.*, 76:4126, 1995.
32. E. Demler, H. Kohno, and S.-C. Zhang. *cond-mat/9710139*, 1997.
33. H.F. Fong et al. *Phys. Rev. B*, 54:6708, 1996.
34. J. Loram et al. *Journal of Superconductivity*, 7:243, 1994.
35. S. Chakravarty et al. *Science*, 261:337, 1993.
36. P. Dai et al. *Phys. Rev. Lett.*, 77:5425, 1996.
37. H.F. Fong et al. *Phys. Rev. Lett.*, 78:713, 1997.
38. Mook et al. *to be published*, 1998.

Nested Spin-Fluctuation Scenario for Anomalous Metallic Phase of High-T_c Cuprates

K. Miyake and O. Narikiyo

Department of Physical Science, Division of Materials Physics
Graduate School of Engineering Science, Osaka University
Toyonaka, Osaka 560-8531, Japan

Abstract. A nested spin-fluctuation scenario is presented for anomalous metallic phase of high-T_c cuprates. On the basis of the quasiparticles, which is left after the local correaltion effect has been taken into account, it is discussed that the effect of spin fluctuations gives rise to the so-called "spin-gap behaviors" and "spin-charge separation aspects" in the case the Fermi surface of the *renormalized* quasiparticles is technically nested which is expected in the lightly doped cuprates. In particular, anomalous properties of Hall conductivity and magnetoconductivity, and pseudogap structure of one-particle spectral weight are discussed in detail.

1. Introduction

The properties of the normal state of high-T_c cuprates in the lightly doped regime are anomalous. Namely, they are quite different from those of conventional metals and are characterized by the keywords like "spin-gap behaviors" and "spin-charge separation aspects". These properties does not seem to be understood within the framework of *conventional* Fermi liquid theory. Since they are located in the proximity of the antiferromagnetic (AF) insulator, it may be natural to attribute those anomalies to the effect of spin fluctuations. Indeed, these approaches have been successful in the problem of itinerant magnetism which are in the weak or intermediate coupling regime [1].

However, it may be reasonable to suspect that such an approach cannot be applied in its original form to the present problem which is considered to be related to the phenomenon of strongly correlated electrons of $d_{x^2-y^2}$-symmetry on Cu site. Nevertheless, it is possible to develop the spin-fluctuation theory on the quasiparticles which have been renoramlized by the on-site strong correlation among d electrons, because the residual interaction among them is considered to be in the intermediate coupling regime. Namely, the Fermi liquid can give us a good starting point to attack the residual effects of spin fluctuations which can give rise to the non-Fermi liquid aspects.

In order to understood those anomalous properties of high-T_c cuprates, however, it needs to extend the conventional spin-fluctuation theory so as to take into account the key characteristics of high-T_c cuprates in the lightly doped regime, i.e., the nesting tendency of the Fermi surface of *renormalized* quasiparticles. In this article, we discuss that those anomalous properties can be understood in a unified way by applying the mode-mode coupling theory for spin fluctuations to the renormalized fermions with nearly nested Fermi surface.

2. Quasiparticles of high-T_c cuprates as a Fermi liquid

We first discuss the property of quasiparticles which is obtained by taking into account the strong on-site correlations, irrelavant to AF correlations. On this basis, we will discuss later the effect of spin fluctuations which suppress considerably the characteristic energy scale, below which the Fermi liquid is valid, near the AF phase boundary [1].

The minimal Hamiltonian for high-T_c cuprates is given by the so-called dp-model:

$$H = \epsilon_d \sum_{k\sigma} d^\dagger_{k\sigma} d_{k\sigma} + \epsilon_p \sum_{k\sigma}(p^\dagger_{xk\sigma} p_{xk\sigma} + p^\dagger_{yk\sigma} p_{yk\sigma}) + U_d \sum_i n_{di\uparrow} n_{di\downarrow} +$$
$$+ \sum_{k\sigma}[V_{xk} p^\dagger_{xk\sigma} d_{k\sigma} + (x \to y) + \text{h.c.}] + \sum_{k\sigma}(W_k p^\dagger_{xk\sigma} p_{yk\sigma} + \text{h.c.}), \quad (1)$$

where ϵ_d (ϵ_p) denotes the atomic level of d (p) electron, U_d the on-site Coulomb repulsion among $d_{x^2-y^2}$ electrons, $V_{xk} \equiv 2it_{pd}\sin(k_x a /2)$ the hybridization matrix element between $d_{x^2-y^2}$ and p_x electrons, and $W_k \equiv 4t_{pp}\sin(k_x a/2)\sin(k_y a/2)$ the hybridization between p_x and p_y electron.

The parameters in (1) are estimated as $U_d \sim 8t_{pd}$, $\epsilon_p - \epsilon_d \sim 2t_{pd}$, and $t_{pp} \sim 0.5 \times t_{pd}$ [2]. Since the repulsion U_d is the largest, we first renormalize this effect. For that purpose, we combine the formal theory of the Fermi liquid [3] and the Gutzwiller variational method [4] and estimate z_d, the renomalization amplitude, and $\tilde{\epsilon}_d$, the effective level of d electron [5]. The results are shown in Fig. 1 in which we can see that z_d does not vanish as the doping rate δ decreases, $\delta \to 0$, for the realistic values of parameters quoted above. This implies that the effective mass of the quasiparticles so obtained does not diverge at the half-filling. So, the origin of the insulating state at the half-filling is not entirely due to the *conventional* Mott insulator of Brinkman-Rice (BR) type but should be attributed considerably to the appearance of AF state in the intermediate coupling regime.

In other words, the high-T_c cuprates is not in the extremely strongly correlated regime in the sense of BR, but in the intermediate coupling regime. This can be understood physically as follows: The energy difference between $(d_{x^2-y^2})^2$ state of Cu and top of 2p band is the same order as the hybridization

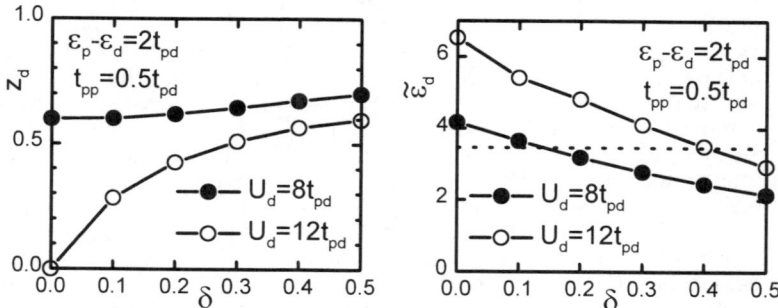

Fig. 1 z_d and $\tilde{\epsilon}_d$ vs δ. The dotted line indicates the top of "p" band for the non-interacting system.

t_{pd} between them; so that doubly occupied state of $d_{x^2-y^2}$ orbital is realized to some extent through the charge transfer process even though $U_d \gg t_{pd}$. Indeed, this can be verified by the fact that n_d^h, the number of d- hole per site which is determined by the NQR method in LSCO [6], is less than 1 by 10~20% as shown in Fig. 2 in which the open triangle ($\triangle$) is for the experimental value, the closed circle ($\bullet$) for the theoretical value calculated by the above method [5], and the closed triangle ($\blacktriangle$) is the result of cluster calculation by Ohta et al. [7]. The theoretical results are in

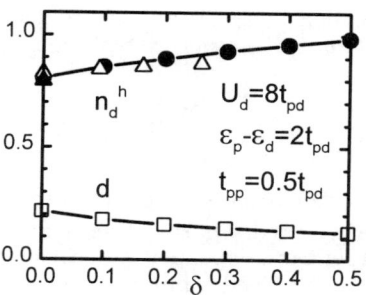

Fig. 2 n_d^h and $d = n_{d\uparrow} n_{d\downarrow}$ vs δ.

good agreement with the experimental one. This implies that $d_{x^2-y^2}$ orbital is doubly occupied by 10~20% as can be seen also from the result for $d = n_{d\uparrow} n_{d\downarrow}$, the rate of doubly occupied state of d orbital, shown by the open square ($\square$).

3. Spin-gap behaviors

In the normal state of lightly doped high-T_c cuprates, the so-called "spin-gap behaviors" are observed. These implies that the low energy spectral weight of spin fluctuations appears to be missing, which is hard to be understood by neither the Fermi liquid theory nor the *conventional* spin-fluctuaion theory [1]. In this section, we review the results due to the nested spin-fluctuation theory [8] on the basis of the quasiparticle concept discussed in the previous section.

While the spin-fluctuation theory for high-T_c has been developed by Moriya et al. applying the SCR theory to the 2D system [9] and also by Pines et al. [10] and Schrieffer et al. [11] on some phenomenological grounds, they do not seem to succeed in explaining the "spin-gap behaviors". In order to derive such behaviors, it is crucial to take into account the fact that the Fermi surface of *renormalized* quasiparticles of lightly doped cuprates should show the nesting tendency because it is located near the AF insulating state in which AF wavevector $\mathbf{Q}_0 \equiv (\pi/a, \pi/a)$, a being the lattice constant, coincides with the reciprocal wavevector connecting between the points on the boundary of the Brillouin zone.

It is also crucial to take into account the dual nature of d-electron on Cu site. Namely, high energy component of d-electron degrees of freedom far way from the Fermi level, i.e., incoherent component, can be described as the local spin degrees of freedom, even in the intemediate coupling regime, while its low energy component is described as the quasiparticles [12].

Following the SCR theory, the dynamical spin susceptibility $\chi(\mathbf{Q}^* + \mathbf{q}, \omega)$ is parameterized as

$$\chi(\mathbf{Q}^* + \mathbf{q}, \omega) = \frac{\tilde{N}_F}{(A_f + A_0)(\kappa^2 + \mathbf{q}^2) - iC\omega}, \qquad (2)$$

where $\mathbf{Q}^*$ ($\simeq \mathbf{Q}_0$) denote the wavevector characterizing incommensurate AF correlation, and $\tilde{N}_F$ is a constant of the order of the DOS of quasiparticles.

The coefficients A_f and C are determined by the expression of $\chi_0(\mathbf{Q}^* + \mathbf{q}, \omega)$, the *non-interacting* susceptibility of quasiparticles, as

$$\frac{\chi_0(\mathbf{Q}^* + \mathbf{q}, \omega + i\delta)}{\chi_0(\mathbf{Q}^*, 0)} \simeq [1 - A_f \mathbf{q}^2 + iC\omega + \cdots]. \tag{3}$$

These coefficients have strong temperature dependences if the Fermi surface is nearly nested. On the other hand, the coefficient A_0 ($\equiv Ja^2 \tilde{N}_F/2$) in (2) has no temperature dependence because it stems from the AF exchange interaction J among localized component of electrons. The inverse correlation length $\kappa \equiv \xi^{-1}$ in (2) is determined self-consistently by the mode-mode coupling approximation of spin fluctuations [1,8,9].

If the Fermi surface of quasiparticles is nearly nested, i.e., the dispersion of quasiparticles $\xi_\mathbf{p}$ satisfies the condition, $\xi_{\mathbf{p}+\mathbf{Q}^*} \simeq -\xi_\mathbf{p}$, then A_f and C have the temperatrue dependence similar to those for superconducting pair fluctuations:

$$A_f = \frac{a_f v_F^2}{T^2 + H^2}, \qquad C = \frac{c_f}{\sqrt{T^2 + H^2}}, \tag{4}$$

where a_f and c_f are numerical constants depending on the band structure and H is a parameter measuring the deviation from perfect nesting ($H = 0$).

In the case nesting rate is high ($H \sim 0$), A_f and C increases rapidly as $T \to 0$. Then, the region of q- and ω-space in which χ, (2), takes large values becomes narrow leading to suppression of summation of spectral weight of spin fluctuations around $q \simeq \mathbf{Q}^*$. Indeed, it is given by

$$\sum_\mathbf{q} \mathrm{Im}\chi(\mathbf{Q}^* + \mathbf{q}, \omega) = \frac{\tilde{N}_F}{4\pi A} \left[\tan^{-1} \frac{A(q_c^2 + \kappa^2)}{C\omega} - \tan^{-1} \frac{A\kappa^2}{C\omega} \right], \tag{5}$$

where $A \equiv A_f + A_0$ and q_c is the cutoff wavenumber of $\mathcal{O}(a^{-1})$. The r.h.s. of (5) decreases as $T \to 0$, because the decreasing tendency of A^{-1} dominates the other factor. Namely, the "spin-gap behaviors" are manifested in the temperature dependence of the physical quantity which is determined by the $\mathbf{q}$-integral of spin fluctuations, even though there exists no spin gap at all in the spin excitation spectrum determined by (2).

The physical quantities exhibiting the "spin-gap behaviors" when the nesting rate is high are expressed in terms of A, C, and κ^2 as follows [8]:

- NMR longitudinal relaxation rates:

$$1/T_1 T \propto \lim_{\omega \to 0} \sum_\mathbf{q} \mathrm{Im}\chi(\mathbf{Q}^* + \mathbf{q}, \omega)/\omega \propto C/A^2 \kappa^2, \tag{6}$$

- Resistivity ρ:

$$\rho \propto \varphi(y+1) - \varphi(y) + \frac{1}{u}[\phi(uy) - \phi(u(y+1))], \tag{7}$$

where $\phi(x) \equiv \ln\Gamma(x) - (x-1/2)\ln x + x$, $\varphi(x) \equiv x[\psi(ux) - \ln(ux)]$, $y = (\kappa/q_c)^2$, and $u \equiv (a_f/2\pi c_f)[1 + \alpha(t^2 + h^2)]/t\sqrt{t^2 + h^2}$. Here, $\alpha \equiv 2J/\pi a_f \epsilon_F$, $t \equiv T/\epsilon_F$ and $h = H/\epsilon_F$, with $\epsilon_F \equiv v_F q_c$ being the effective Fermi energy of quasiparticles.

- Spin-fluctuation contribution to the Sommerfeld coefficient $\gamma_{sf}(T)$:

$$\gamma_{sf} = -\frac{\partial^2 F_{sf}}{\partial T^2}, \quad F_{sf} = -\sum_{|\mathbf{q}|<\tilde{q}_c} \int_0^{\tilde{\omega}_c} d\omega \left(1 + \frac{2}{e^{\omega/T}-1}\right) \tan^{-1}\frac{\omega}{\Gamma_\mathbf{q}}, \quad (8)$$

where F_{sf} is the spin-fluctuation contribution to the free energy [1], $\tilde{\omega}_c \equiv C^{-1}$, $\tilde{q}_c \equiv A^{-1/2}$, and $\Gamma_\mathbf{q} \equiv (A/C)(\kappa^2 + q^2)$.

- Effective carrier number n^* defined as $n^* \equiv (m^*/e^2) \int_0^{\omega_c} d\omega \mathrm{Re}\sigma(\omega)$ where ω_c is the cutoff energy of the order of ϵ_F, and $\sigma(\omega) = (ne^2/m^*)/[\tau^{-1}(\omega) - i\omega]$, with $\mathrm{Re}\tau^{-1}(\omega) \simeq \pi|\omega|$ and $\mathrm{Im}\tau^{-1}(\omega) \simeq \omega/2\tau_0 \cdot \ln y$ [13]:

$$\frac{n^*}{n} \propto \frac{1/\tau_0}{(1/\tau_0)^2 + (1 + 1/\tau_0)^2}, \quad (9)$$

where $1/\tau_0 \equiv h/(1 + \alpha h^2) \propto C/A$ and logarithmic dependence $\ln y$ has been neglected [8].

With the use of κ^2 determined self-consistently by mode-coupling approximation, the temperature dependence of (6)$\sim$(8) are shown in Fig. 3a$\sim$c, where $t^* \equiv T^*/\epsilon_F$, T^* being defined as $A_f(T^*) = A_0$, i.e., $T^* \equiv [(a_f v_F^2/A_0) - H^2]^{1/2}$. The extent in q-space of spin fluctuations is determined by $1/\sqrt{A_f}$ at $T < T^*$, i.e., by the itinerant correlation, while it is determined by $1/\sqrt{A_0}$ at $T > T_*$, i.e., by the correlation of the superexchange type. Namely, the nature of spin fluctuations changes at around $T = T^*$ when T is varied. The parameter y_0 represents the deviation from the AF phase boundary ($y_0 = 0$). We have set the coefficients of (4) as $a_f = 7\zeta(3)/32\pi^2$ and $c_f = \pi/8$, those for 2D circular band, and $J/\epsilon_F = 1/2.4$. γ_{quasi} in Fig. 3c represents the contribution to the Sommerfeld coefficient from the quasiparticles. The relation between the effective carrier number n^*, (9), and the nesting rate h is shown in Fig. 3d. These results show that the better the nesting ($h \to 0$) is the more apparent the "spin-gap behaviors" are. It has been shown that the uniform susceptibility χ has also the spin-fluctuation contribution, $\chi_{sf} = -\partial^2 F_{sf}/\partial B^2$, which exhibits the "spin-gap behaviors" as γ_{sf} when the nesting rate is high enough [8].

The NMR longitudinal relaxation rate $1/T_1T$ exhibits the "spin-gap behavior" at $t < t^*$, where the resistivity ρ bends rather sharply. These are consistent with experimental facts. However, γ_{sf} shows peak at $t \sim t^*/3$ in contrast with experiments. In order to cover this defect, it is necessary to take into account the incoherent component of spin fluctuations which give crucial contribution in the transverse relaxation rate $1/T_{2G}$, which is determined not only by low-energy (coherent) spin fluctuations but by high-energy (incoherent) ones, as discussed in Ref. 14. The incoherent part of spin fluctuations also plays a crucial role to understand the pseudogap structure of one-particle spectral weight as will be discussed in §5, and the so-called marginal Fermi-liquid spectrum of the optical conductivity and the Raman intensity [15].

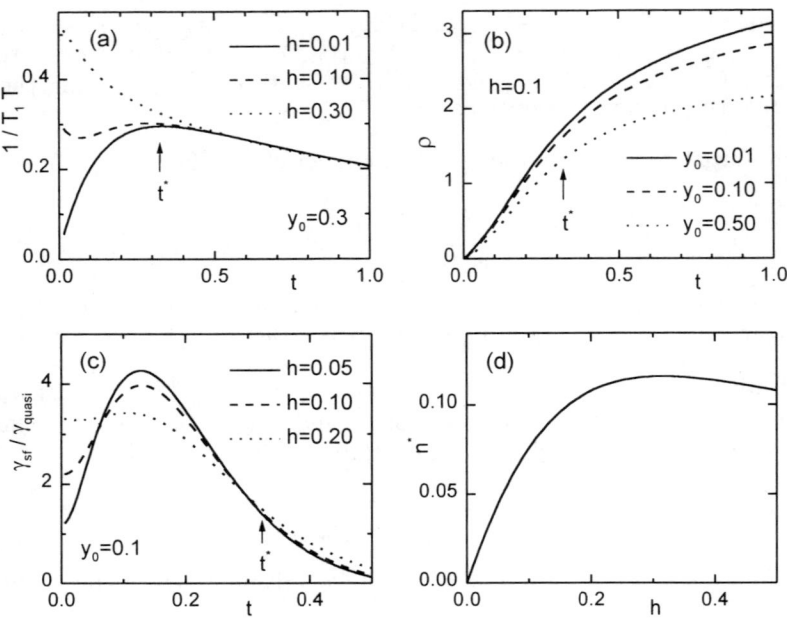

Fig. 3 Physical quantities exhibiting "spin-gap behaviors". Units of ordinates are arbitrary except for (c).

It is noted that the character of *coherent* spin fluctuations can be described by the conventional SCR theory in highly doped region where the effect of nesting is negligible while it is subject to the effect of nesting in the lightly doped region. Indeed, the AF correlation length $\xi = \kappa^{-1}$ determined by T_1 and T_{2G} of NMR exhibits different temperature dependence in two regions. Namely, $\kappa^2 \propto (T + \Theta_1)$ in highly doped region [16] in agreement with SCR theory [9], $\kappa \propto (T + \Theta_2)$ in lightly doped region [17] in agreement with the result of nested spin-fluctuation theory [8]. The latter dependence has also been detected by the neutron scattering in LSCO [18].

4. Spin-charge separation aspects

The so-called "spin-charge separation aspects" observed in lightly doped high-T_c cuprates cannot be understood within the Fermi liquid theory. Indeed, the charge response such as the Hall effect appears to imply that the charge degrees of freedom are not connected with the large Fermi surface near the half-filling but appears to be given by small number of doped holes in the Mott insulator. Moreover, the Hall constant R_H has strong temperature dependence like $\sim 1/T$ as T is decreased from the room temperature. In other words, the so-called Kohler rule is violated [19,20]: $\sigma_{xx} \propto T^{-1}$, $\sigma_{xy} \propto HT^{-3}$ and $\Delta\sigma_{xx} \propto H^2 T^{-5}$. It can be seen that the Kohler rule holds if the magnetotransport coefficients are calculated by means of the Boltzmann equation on the basis of the Fermi liquid theory: $\sigma_{xx} \propto \tau$, $\sigma_{xy} \propto H\tau^2$, and $\Delta\sigma_{xx} \propto H^2\tau^3$ if the momentum dependence of the life time τ of the quasiparticle is weak.

Here we present a theory to explain the violation of the Kohler rule by taking into account the contribution of collective mode beyond the Boltzmann transport [8,21,22]. Indeed, it is well recognized that the pairing fluctuation plays crucial roles for transport phenomena near the superconducting transition: the Aslamazov-Larkin (AL) process gives considerable contribution not only to σ_{xx} [23] but also to σ_{xy} [24] and $\Delta\sigma_{xx}$ [25]. Since the mathematical structure of the theory of nested spin fluctuations is similar to that of superconducting fluctuations, it is expected that the AL-type contribution of nested spin fluctuations can give rise to anomalous contributions leading to violation of the Kohler rule.

The most dominant contributions of the spin fluctuations to the Hall conductivity σ_{xy} and the magneto conductivity $\Delta\sigma_{xx}$ are given by AL-type processes corresponding to the Feynman diagrams shown in Fig. 4a,b where the wavy line represents the spin-fluctuation propagator, the solid line the quasi-particle Green function and the triangle the renormalized current vertex. Differences form the case of pairing fluctuations are that the sign of the fluctuation propagator is opposite, and that the current can couple only with the particle-particle channel so that the vertex correction as shown in Fig. 4c is necessary [21]. The vertex shown in Fig. 4c is evaluated to be $4fC_J q_\mu$ where the factor f is $\mathcal{O}(1)$, and C_J, which is given by A_f in (4), is the coefficient corresponding to that in the case of pairing fluctuations and shows singular temperature dependence $\propto T^{-2}$ if the Fermi surface is perfectly nested. After all, we obtain the fluctuation contributions to σ_{xy} and $\Delta\sigma_{xx}$ as follows [21,22]:

$$\sigma_{xy}^{\rm AL} \approx -H\frac{2|e|^3}{3\pi} \cdot \frac{(fC_J)^3}{A^2} \cdot \frac{T}{\eta^2} \cdot \frac{N_{\rm F}'}{\tilde{U}_d N_{\rm F}^2}. \tag{10}$$

where $N_{\rm F}'$ is the differential DOS of quasiparticles at the Fermi level and $\tilde{U}_d$ the on-site effective interaction among quasiparticles, and

$$\Delta\sigma_{xx}^{\rm AL} \approx -H^2\frac{e^4}{\pi d}CT\frac{(fC_J)^4}{A^2}\frac{1}{\eta^3}, \tag{11}$$

where $\eta \equiv A\kappa^2$ and d is the distance between CuO$_2$ planes.

It is noted that the sign of $\sigma_{xy}^{\rm AL}$ is given by that of $-N_{\rm F}'$. If approximate dispersions of YBCO and LSCO are used, $-N_{\rm F}'$ is positive in lightly doped region while it changes sign when doping rate is increased. This is consistent

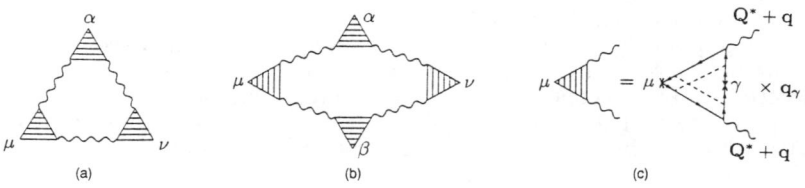

Fig. 4 Feynman diagram giving dominant contribution to (a) σ_{xy}, and (b) $\Delta\sigma_{xx}$, and (c) the current vertex of spin fluctuations.

with the doping dependence of the sign of R_H. Since both (10) and (11) include factors $\chi(\mathbf{Q}^*, 0) \propto \eta^{-1}$ and C_J ($=A_f$), they increase divergently as the AF phase is approached. This is consistent with experiments [19,26]. In the region where $H < T < T^*$ holds, A and $C_J \propto 1/T^2$, and $\chi(\mathbf{Q}^*, 0) \propto 1/T$; so that $\sigma_{xy}^{AL} \propto 1/T^3$. This implies that $R_H = \sigma_{xy}^{AL} \rho^2/H \propto 1/T$ if we use $\rho \propto T$. We can verify that the absolute value of R_H^{AL} is the same order as the observed one in lightly doped region of cuprates. It is also shown that the Hall angle Θ_H obeys approximately the law $\cot \Theta_H \propto T^2$ [8]. In Fig. 5 we plot the temperature dependence of R_H^{AL} [8] and $\Delta\sigma_{xx}^{AL}$ [22] estimated from (10) and (11) using a self-consistent solution for κ [8]. We can see that $\Delta\sigma_{xx}^{AL}$ is proportional to T^{-5} in rather wide temperature range if the system is located near the AF phase boundary, i.e., $h \ll 1$ and $y_0 \ll 1$. This explains experimental results [19,20].

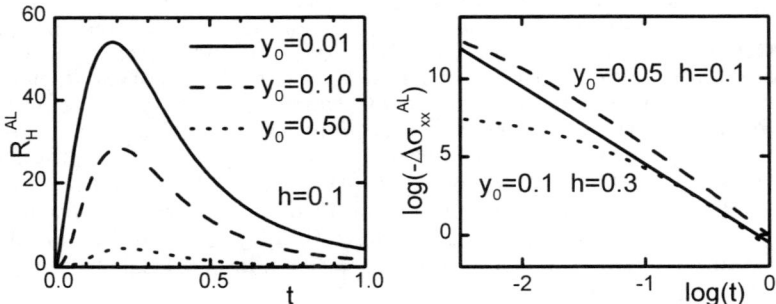

Fig. 5 Temperature dependence of R_H^{AL} [8] and $\Delta\sigma_{xx}$ [22]. Units of ordinates are arbitrary. Solid line in the right figure represents t^{-5} dependence.

5. Pseudogaps of one-particle spectral weight

Recently pseudogaps of one-particle excitation spectrum was found by means of ARPES (angle-resolved photoemission spectroscopy) in the normal state of lightly doped high-T_c cuprates [27-29]. It is remarked that the pseudogaps are characterized by the two energy scales: the higher energy gap of the order of $\sim$100meV which corresponds to the broad peak in ARPES spectrum [27] and the lower energy gap of the order of $\sim$10meV which is measured by the shift of the Fermi edge [28,29]. Both energy gaps appear to have $d_{x^2-y^2}$-like symmetry [27-29]. Here we discuss that such pseudogaps can be understood by the effect of nested spin fluctuations with taking into account the itinerant-localized dual nature of d-electron on Cu sites [30].

The one-particle spectral weight $\rho(k, \varepsilon)$ is approximately decomposed into two contributions: $\rho(k, \varepsilon) = \rho_{\text{coh}}(k, \varepsilon) + \rho_{\text{inc}}(k, \varepsilon)$, where $\rho_{\text{coh}}(k, \varepsilon)$ and $\rho_{\text{inc}}(k, \varepsilon)$ are determined by the contributions from coherent and incoherent spin fluctuations, respectively. The underlying decomposition of spin-fluctuation spectrum is clearly observed by neutron scattering experiments [31,32].

The coherent part is given by

$$\rho_{\text{coh}}(k, \varepsilon) = -\frac{1}{\pi} \frac{\Sigma_{\text{coh}}''(k, \varepsilon + i0^+)}{(\varepsilon/z)^2 + [\Sigma_{\text{coh}}''(k, \varepsilon + i0^+)]^2}, \quad (12)$$

where z being the renormalization amplitude and $\Sigma''_{\rm coh}(k,\varepsilon+i0^+) = -(\lambda^2/2)\sum_q \chi''_{\rm coh}(\mathbf{Q}^*+\mathbf{q},\omega)[\coth(\omega/2T)+\tanh(\epsilon'/2T)]$, which is obtained in the second order perturbation in the coupling λ between electrons and low-energy coherent spin fluctuations $\chi_{\rm coh}(\mathbf{Q}^*+\mathbf{q},\omega)$, given by (2). Here $\omega = \varepsilon - \epsilon'$ and $\epsilon' = \epsilon_{\mathbf{k}-\mathbf{Q}^*-\mathbf{q}}$ with ϵ_k being the renormalized dispersion for quasiparticles measured from the chemical potential. Since we are interested in the imaginary part of the self-energy $\Sigma''_{\rm coh}(k_{\rm F},\varepsilon+i0^+)$, where $k_{\rm F}$ is the Fermi wavevector around $\mathbf{k}=(\pi/a,0)$, we make the flat-band approximation [33], $\epsilon' = \epsilon_{k_{\rm F}} = 0$, assuming that $\mathbf{k}_{\rm F}-\mathbf{Q}^*-\mathbf{q}$ belongs to the other flat band regions: $\mathbf{k}_{\rm F}-\mathbf{Q}^*-\mathbf{q} \sim (0,\pm\pi/a)$, and obtain $\Sigma''_{\rm coh}(k_{\rm F},\varepsilon+i0^+) \simeq -(\lambda^2/2)\coth(\varepsilon/2T)\chi''_{\rm coh}(\varepsilon)$ where $\chi''_{\rm coh}(\varepsilon) = \sum_q \chi''_{\rm coh}(\mathbf{Q}^*+\mathbf{q},\varepsilon)$. With the use of a conventional form of the spin fluctuation propagator (2), we obtain

$$\Sigma''_{\rm coh}(k_{\rm F},\varepsilon+i0^+) \simeq -\frac{\tilde{N}_{\rm F}\lambda^2}{8\pi A}\coth\frac{\varepsilon}{2T}\tan^{-1}\frac{C\varepsilon}{\eta}. \qquad (13)$$

In numerical evaluations we use the following forms of nested spin fluctuations [8]: $\tilde{N}_{\rm F}\lambda^2/8\pi A\epsilon_{\rm F} = r(t^2+h^2)$ with numerical constant r, and $C\varepsilon/\eta = \hat{\varepsilon}/\sqrt{t^2+h^2}(\theta+t)$ where $\hat{\varepsilon} = \varepsilon/\epsilon_{\rm F}$. For simplicity we assume that η approximately takes the Curie-Weiss form, $\eta \propto \theta+t$, where $\theta = \Theta/\epsilon_{\rm F}$ with Θ being the Weiss temperature.

The incoherent part is assumed to be given by $\rho_{\rm inc}(k,\varepsilon) = \rho^0_{\rm inc}\Gamma|\varepsilon|/[\varepsilon^2+\Gamma^2]$ which has broad maxima at $\varepsilon \sim \pm\Gamma$ reflecting the energy spectrum of incoherent spin fluctuations [14] and thus $\Gamma \sim J$.

The numerical results for $\rho(k_{\rm F},\varepsilon)$ are shown in Fig. 6 where we use the following parameters: $h = 0.01$, $r = 0.1$, $w = 0.01$, $z = 0.5$, $\Gamma/\epsilon_{\rm F} = 0.1$ and $\rho^0_{\rm inc} = 50$. It is seen that the spectral weight between $\varepsilon = 0$ and $\varepsilon \sim -\Gamma \sim -J$ decreases as the temperature is lowered. This behavior is qualitatively consistent with ARPES experiments for the lower energy gap [28,29]. On the other hand, the quasiparticle peak at $\varepsilon = 0$ should develop as the temperature is lowered, while it has not been observed because the experimental resolution is not high enough.

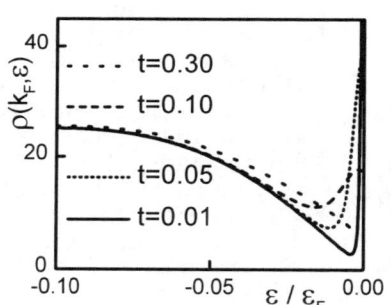

Fig. 6 $\rho(\mathbf{k}_{\rm F},\epsilon)$ for $\mathbf{k}_{\rm F}\simeq(\pi/a,\pi/a)$.

The lower energy pseudogap is signalled by $\Sigma''_{\rm coh}(k_{\rm F},i0^+) \propto (C/A)T\times\chi_{\rm coh}(\mathbf{Q}^*,0)$ whose inverse is proportional to $\rho(k_{\rm F},0)$. Here the factor C/A introduces the characteristic temperature T^* below which the lower-energy pseudogap becomes appreciable. It is noted that the same characteristic temperature T^* appears in the resistivity, $\rho \propto (C/A)T\ln\chi_{\rm coh}(\mathbf{Q}^*,0)$, approximate form of (7), and the NMR relaxation rate, $1/T_1T \propto (C/A)\chi_{\rm coh}(\mathbf{Q}^*,0)$, (6).

For the Fermi surface appropriate to Bi2212 system [34], the pseudogap behavior in ARPES spectrum is expected for electrons on the Fermi surface around $\mathbf{k} = (\pi/a,0)$, since the low-energy state with $\epsilon_k \sim 0$ is scattered by the low-energy spin fluctuation with the momentum $\mathbf{Q}^*$ to another low-energy state with $\epsilon_{k+Q^*} \sim 0$. On the other hand, electrons on the Fermi surface around $\mathbf{k} = (\pi/2a, \pi/2a)$ is transferred to high-energy states by the change $\mathbf{k} \to \mathbf{k} + \mathbf{Q}^*$ and are not scattered by low-energy spin fluctuations, since low-energy spin fluctuations couple only to low-energy electrons [8]. Consequently the pseudogap has $d_{x^2-y^2}$-like symmetry in the momentum space.

References

[1] T. Moriya: *Spin Fluctuations in Itinerant Electron Magnetism* (Springer-Verlag, Berlin, 1985) Chaps. 4.
[2] K. Terakura: Solid State Physics (in Japanese) **25**, 673 (1990).
[3] H. Kohno and K. Yamada: Prog. Theor. Phys. **85**, 13 (1991).
[4] T. M. Rice and K. Ueda: Phys. Rev. Lett. **55**, 995 (1985).
[5] T. Tsuji, O. Narikiyo and K. Miyake: Physica C **244**, 311 (1995).
[6] G.-q. Zheng *et al*: Physica C **208**, 339 (1993).
[7] Y. Ohta *et al*: J. Phys. Soc. Jpn. **61**, 2198 (1992).
[8] K. Miyake and O. Narikiyo, J. Phys. Soc. Jpn. **63**, 3821 (1994)
[9] T. Moriya *et al*: J. Phys. Soc. Jpn. **59**, 2905 (1990).
[10] D. Pines: Physica C **282-287**, 273 (1997).
[11] J. R. Schrieffer and A. P. Kampf: J. Phys. Chem. Solids **56**, 1673 (1995).
[12] Y. Kuramoto and K. Miyake: J. Phys. Soc. Jpn **59**, 2873 (1990);
K. Miyake and K. Kuramoto: Physica B **171**, 20 (1992).
[13] T. Moriya and Y. Takahashi: J. Phys. Soc. Jpn. **60** 776, (1991).
[14] O. Narikiyo and K. Miyake: J. Phys. Soc. Jpn. **63**, 4169 (1994).
[15] O. Narikiyo: Physica C **267**, 204 (1996).
[16] T. Imai *et al*: Phys. Rev. B **47**, 9158 (1993).
[17] Y. Itoh and H. Yasuoka: J. Phys. Soc. Jpn. **63**, 2518 (1994);
R. L. Corey *et al* : Phys. Rev. B **53**, 5907 (1996).
[18] G. Aeppli *et al*: Science **278**, 1432 (1997).
[19] J. M. Harris *et a*: Phys. Rev. Lett. **75**, 1391 (1995).
[20] T. Kimura *et al*: Phys. Rev. B **53**, 8733 (1996).
[21] K. Miyake and O. Narikiyo: Physica C **235-240**, 2237 (1994).
[22] O. Narikiyo and K. Miyake: J. Phys. Soc. Jpn. **66**, 1561 (1997).
[23] L. G. Aslamazov and A. I. Larkin: Sov.-Phys. Solid State **10**, 875 (1968).
[24] H. Fukuyama *et al*: Prog. Theor. Phys. **46**, 1028 (1971).
[25] J. B. Bieri *et al*: Phys. Rev. B **44**, 4709 (1991).
[26] T. Nishikawa *et al*: J. Phys. Soc. Jpn. **62**, 2568 (1993);
H. Y. Hwang *et al*: Phys. Rev. Lett. **72**, 2636 (1994).
[27] D. S. Marshall *et al*: Phys. Rev. Lett. **76**, 4841 (1996).
[28] A. G. Loeser *et al*: Science **273**, 325 (1996).
[29] H. Ding *et al*: Nature **382**, 51 (1996).
[30] O. Narikiyo and K. Miyake: Physica C, (1998) in press.
[31] L. P. Regnault *et al*: Physica B **213-214**, 48 (1995).
[32] S. M. Hayden *et al*: Phys. Rev. Lett. **76**, 1344 (1996).
[33] R. Preuss *et al*: Phys. Rev. Lett. **79**, 1122 (1997).
[34] H. Ding *et al*: Phys. Rev. Lett. **78**, 2631 (1997).

Part V

New Materials: Ladders, Ruthenates and Others

Novel Transition Metal Oxides Prepared at High Pressure and Their Electronic Properties

M. Takano[1], Z. Hiroi[1], M. Azuma[1], S. Kawasaki[1], R. Kanno[2], and T. Takeda[2]

[1] Institute for Chemical Research, Kyoto University, Uji, Kyoto-fu 611-0011, Japan and CREST, Japan Science and Technology Corporation
[2] Faculty of Science, Kobe University, Kobe 657-8501, Japan

Abstract. We have searched for transition metal oxides exhibiting interesting electronic properties by using various preparative techniques. High pressure synthesis at 3GPa and 1200K, typically, has yielded, for example, Fe^{4+}-oxides showing a variety of properties dominated by oxygen-hole character, $SrCu_2O_3$ and $Sr_2Cu_3O_5$ containing 2-leg and 3-leg ladders, respectively, and $Tl_2Ru_2O_7$ exhibiting a sharp metal-insulator transition.

1. Introduction

Recently various types of interplay of the electronic spin, charge, and orbital degrees of freedom have been newly found or re-examined for transition metal (M) oxides [1]. These electronic properties are composition- and structure-sensitive, and, therefore, it is meaningful to search for substances containing different types of M-O sublattices even by using unconventional synthesizing techiniques. Here we briefly report electronic properties like metal-insulator (MI) transition found in oxides containing Fe^{4+}, Cu^{2+}, Ru^{4+} which were synthesized under typical conditions of 3GPa and 1200K.

Unusual valence states like Fe^{4+} can be realized in a strongly oxidizing or reducing atmosphere generated in a high pressure (HP) cell, and this makes it possible to carry out a systematic study of electronic properties as a function of d electron number and other related parameters. For example, though isoelectronic to each other, the charge transfer energy is largely different between oxides containing Fe^{4+} and Mn^{3+}. $CaFe^{4+}O_3$, $Sr_{2/3}La_{1/3}Fe^{11/3+}O_3$, and related phases exhibit unique properties dominated by oxygen-hole character.

$SrCu_2O_3$ and $Sr_2Cu_3O_5$ are a pair of HP phases that have been regarded as the first known spin-ladder compounds. Thanks to their simple compositions and simple structures, the fancy quantum mechanical ground state and its dependence on the leg number have been studied without ambiguity.

Very recently it has been found that $Tl_2Ru_2O_7$ obtained in a strongly oxidizing atmosphere shows a sharp MI transition at 120K, where magnetization drops and the lattice distorts at the same time, while a 4% oxygen deficient phase remains metallic and cubic. The clarification of the transition mechanism would be very interesting.

2. Cupric Spin Ladder Oxides

In 1991 we reported a homologous series of high pressure phases $Sr_{n-1}Cu_nO_{2n-1}$ with $n = 2, 3, ...$ in which two dimensional (2D) Cu_nO_{2n-1} sheets are contained [2]. As can be seen in Fig. 1 these sheets consist of tightly connected n-leg ladders. Both the legs and the rungs being made of linear Cu-O-Cu bond, the nearest-neighbor (NN) interaction within a ladder, J, would be antiferromagnetic (AF) and strong, i.e. $J \approx 10^3$K. Neighboring ladders are phase-shifted from each other. The interladder interaction through the bond bent to $\approx 90°$, J', must be much weaker than J by nature and it is further weakened in effect because of interladder spin frustration caused by the phase shift. The Cu_nO_{2n-1} sheets may thus be regarded as consisting of chemically bound but magnetically separated (quasi-1D) n-leg ladders. Another 2-leg ladder compound $LaCuO_{2.5}$ was found in the La-Cu-O system treated at 6GPa and 1173K [3].

Several theoretical groups studied ladders with different leg numbers for the purpose of studying "the transition from the quasi-long range order in a chain of $S = 1/2$ coupled antiferromagnetically and the true long range order that occurs in a plane" [4]. As a result the dimensional crossover has been found to be far from being smooth. "Ladders with an even number of legs have purely short range magnetic order and a finite energy gap to all magnetic excitations", while all odd-leg ladders are gapless as a single chain is. The magnitude of the spin gap (Δ_{SG}) is equal to $J/2$ for an isotropic 2-leg ladder, but it decreases rather quickly with increasing n toward the gapless nature of the 2D sheet [5]. The isotropy means the equality of the NN interaction in the leg, $J_{//}$, and that in the rung, $J_\perp$. It was further predicted that holes doped in even-leg ladders would pair in approximately $d_{x^2-y^2}$ symmetry and possibly superconduct or be ordered in a charge density wave state. More recently, however, a possibility of superconductivity has been raised for the 3-leg ladder also [6].

2.1 $SrCu_2O_3$ and $Sr_2Cu_3O_5$, Pure and Zn-Substituted

All the experimental data collected by measurements of DC susceptibility [7], NMR [8], μSR [9], and inelastic neutron scattering [10] showed quite consistently that the 2-leg ladder system, $SrCu_2O_3$, assumes a singlet spin liquid state or the so-called short-range Resonance Valence Bond (RVB) state, in which, according to a theoretical prediction, a Cu ion finds a counter Cu ion to be coupled into the singlet state with on the same rung mainly. The estimation of Δ_{SG} ranged depending upon the experimental technique as 420K [7], 410K [10], and 680K [8].

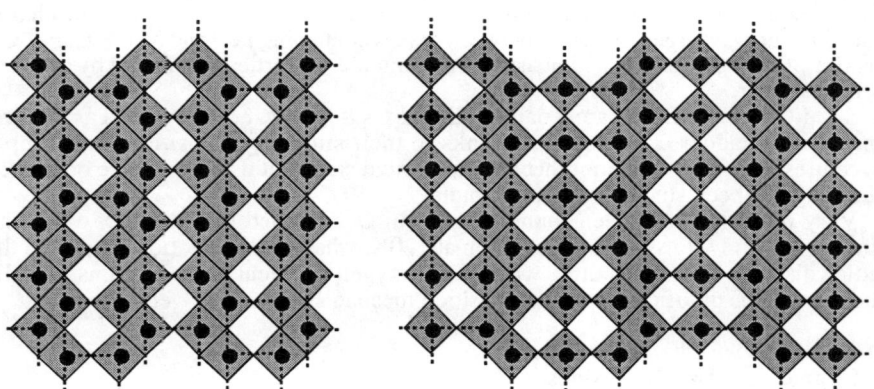

Fig. 1. Schematic representation of a Cu_2O_3 sheet made of 2-leg ladders (left) and a Cu_3O_5 sheet made of 3-leg ladders (right).

By the way, according to Jonston's analysis of the temperature dependence of susceptibility, the ladder is not isotropic but anisotropic as $J_{//} \approx 2000K$ and $J_{\perp}/J_{//} \approx 0.5$ [11]. The experimental values of Δ_{SG} mentioned above correspond to $J_{\perp}/2$ approximately. And according to a theoretical estimation by Mizuno et al., the interladder interaction through the Cu-O-Cu bond angle bent to 92.5° is ferromagnetic, $J' \approx -150K$ [12]. Though not of a negligible magnitude, this is canceled out in effect because of the interladder spin frustration as stated above.

In marked contrast with the 2-leg ladder system that retains the unique coherent singlet ground state state down to at least 20mK [9], the 3-leg ladder compound becomes magnetically ordered at a rather high temperature. The ordering is, however, not a simple Neel order but is such that moments become frozen randomly in a temperature range of 47-57K [9].

We thus revealed successfully their own, leg number-dependent properties of $SrCu_2O_3$ and $Sr_2Cu_3O_5$ but could not test the possibility of high-T_c superconductivity, because the solubility of aliovalent cations at the Sr site like Na$^+$ for hole-doping and La^{3+} for electron-doping is low. Now in progress are the growth of single crystals under pressure and also the carrier doping in $SrCu_2O_3$ in the form of thin film.

It sounds quite reasonable that a nonmagnetic impurity situated in the 2-leg ladder creates a free spin 1/2 localized nearby, without affecting the gapped nature of the host because of its short spin correlation length of only a few lattice spacings. But we have found that a very low concentration of Zn seriously affects the ladder as a whole and leads to an AF ordering at low temperatures [13].

Plotted in Fig. 2 are the magnetic susceptibility data for $Sr(Cu_{1-x}Zn_x)_2O_3$ with $x = 0$ to 0.08. The increase of susceptibility with increasing x indicates the existence of induced paramagnetic moments, and the data between 15 and 30 K could be fitted to the Curie-Weiss law : $\chi(T) = C/(T - \theta) + \chi_0$ with $\chi_0 \approx 10^{-5}$ emu/Cu+Zn mole and $\theta \approx$ -2K. The Curie constant, C, was approximately 0.7 times as small as, but not very far from, the value expected from the picture of a single spin 1/2 per Zn.

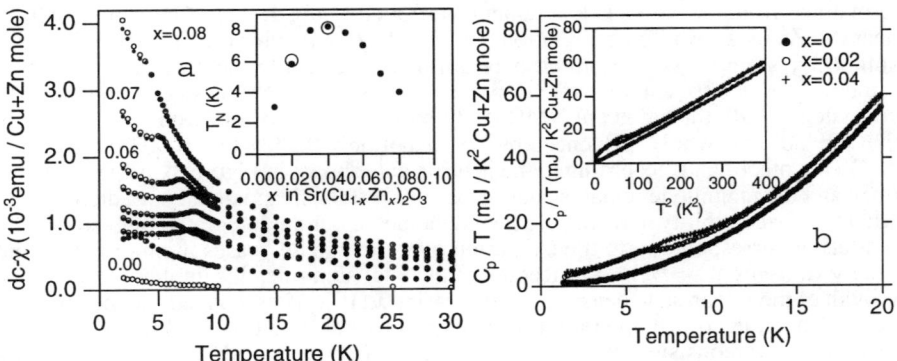

Fig. 2. (a) Temperature dependence of the DC magnetic susceptibility of $Sr(Cu_{1-x}Zn_x)_2O_3$ measured in an external field of 100 Oe on heating (closed circles) and cooling (open circles). Plotted against x in the inset are the temperatures of the magnetic anomaly (closed circles) and the specific heat anomaly (open circles). (b) Total specific heat divided by temperature, C_p/T, plotted against T and against T^2 (inset).

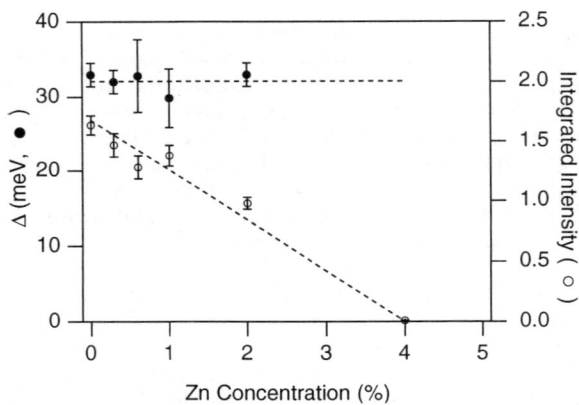

Fig. 3. Magnitude of the spin gap of estimated from inelastic neutron scattering data (•) and the integrated intensity of the magnetic scattering observed between 25 and 50 meV (o) for $Sr(Cu_{1-x}Zn_x)_2O_3$. The incident energy used is 250meV.

Below 10K, however, we found an unexpected cusp-like anomaly suggesting AF ordering even for $x = 0.01$. The cusp temperature increases with Zn content up to $x = 0.04$ and then decreases as shown in the inset to Fig. 2(a). In parallel to this, a λ-type anomaly corresponding to an AF ordering, not a spin glass freezing, was found in specific heat around the same temperature (see Fig. 2(b)). The intrinsic nature of the magnetic ordering has been further confirmed by NQR and NMR measurements very recently [14].

As can be seen in Fig. 2(b) the specific heat contains a finite amount of T - linear term that reveals the presence of a finite density of state at $E = 0$ in the spin excitation spectrum, and the coefficient γ remains almost constant at 3.5mJ/K^2Cu+Zn mole for $x = 0.02$ and 0.04. This independence may be interpreted as indicating that the T - linear term arises from the whole ladder rather than from the spin degree of freedom localized near the Zn ions. Applying a relation $\gamma = 2Nk_B2/3J$ known for the 1D Heisenberg AF spin 1/2 system to the present case assuming N is equal to the number of Cu ions, rather than to that of Zn ions, a reasonable value of $J \approx 1600K$ has been obtained. In consistency with this, a recent NMR study by Fujiwara et al. showed very definitely that the ladder is wholy influenced by Zn ions of only 0.25 % ($x = 0.0025$) [15].

On the other hand, according to the neutron data shown in Figure 3, the gap remains open and the magnitude remains constant up to $x = 0.04$, in apparent contradiction with what the susceptibility and specific heat data mean, but the integrated intensity of the scattering corresponding to the excitation over the gap decreases monotonically and finally vanishes at $x = 0.04$. On account of the total-sum rule, this implies a continuous growth of the in-gap state density with increasing x [10]. Thus, concerning the presence of the Zn-induced in-gap state, all the experimental data obtained using different techniques are consistent with each other. Specifically, the NMR measurement has revealed that it is already present at such a low content of $x = 0.0025$. However, there opens a pseudogap above which the triplet excitation band still exists.

Enhancement of the correlation length and the generation of local staggered moments, the magnitude of which decreasing with increasing distance from the nonmagnetic ion, have been concluded theoretically [16]. The long range ordering sets in at low temperatures because of three dimensional interladder interactions. The interladder interactions would be enhanced near the impurity sites around which relatively large localized moments are generated within each ladder and, at the same time, interladder spin frustration might be lifted. The magnitude of the ordered moment has been found to be very small, ranging from 4×10^{-2} to $0.4\mu_B$ depending upon the distance from the Zn ion for $x = 0.01$ at 1.7K, by the NQR measurement [14].

We note here that the T_N of the Zn-substituted 3-leg system $Sr_2(Cu_{1-x}Zn_x)_3O_5$ monotonically decreased with increasing x, which seems, at least qualitatively, like a simple dilution effect.

2.2 $LaCuO_{2.5}$

The structure of $LaCuO_{2.5}$ is illustrated in Fig. 4 [3,17]. The legs are morphologically almost the same as those in $SrCu_2O_3$, but the arrangement is considerably different. The ladders in $LaCuO_{2.5}$ can be divided into two groups, I and II, which make an angle of 68.5° to each other, and every oxygen array edging the ladders of group I (II) coordinates, as an array of apical oxygen atoms, a neighboring ladder of group II (I). The Cu-O distances in the interladder Cu-O(edge, apex)-Cu bond are 1.941Å and 2.285Å, respectively, and the bond angle is 152.2°. According to a theoretical estimation by Mizokawa et al, $J' \approx -100K$ (ferromagnetic) [18]. This is comparable in magnitude with that for $SrCu_2O_3$, while it is not canceled out as in $SrCu_2O_3$. Thus, $LaCuO_{2.5}$ may be considered to represent a system where J' is apparently important in determining the nature of the ground state.

The temperature dependence of uniform susceptibility suggested a spin gap of Δ_{SG} = 500K [3,17], but NMR [19], μSR [20], and elastic neutron diffraction experiments revealed an AF ordering at about 120K. It has been theoretically pointed out that the singlet ground state should become unstable with increasing J' and be transferred to an ordered state beyond a quantum critical point (QCP) of $|J'|/J \sim 0.11$ [21,22]. $LaCuO_{2.5}$ with an uncanceled interladder interaction seems to lie close to the QCP in the schematical phase diagram illustrated in Fig. 5.

Carrier doping into $LaCuO_{2.5}$ has been done by substituting Sr^{2+} for La^{3+} ions as for $(La,Sr)_2CuO_4$ [3]. An insulator to metal transition was observed, but superconductivity did not emerge even at high pressures up to 8 GPa. By the way, Takagi et al. succeeded in preparing large single crystals and found an unusual correlation effect in their transport measurements [23].

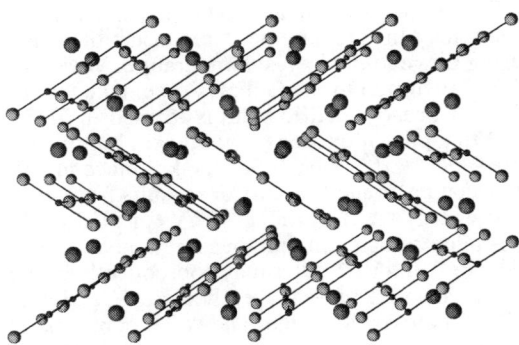

Fig. 4. Structure of $LaCuO_{2.5}$. La: the largest, O: middle, and Cu: the smallest.

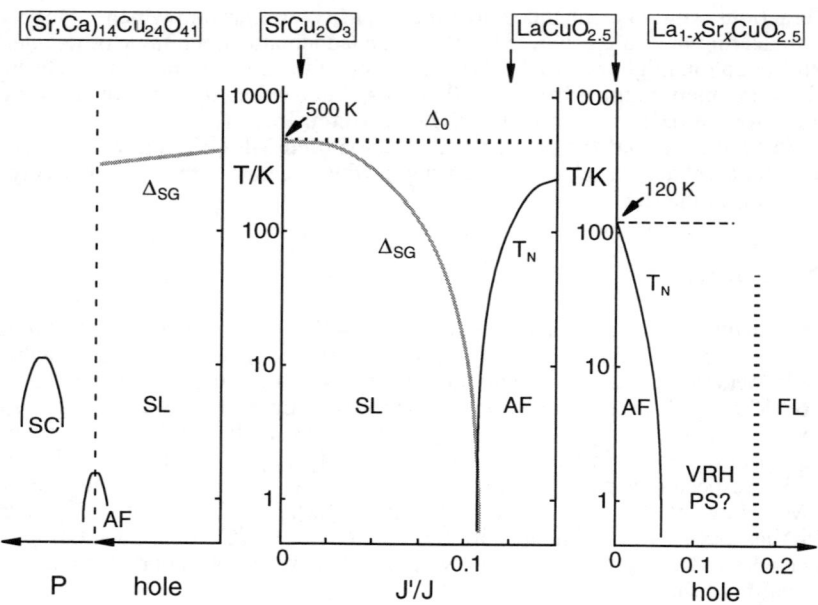

Fig. 5. Schematic phase diagram for spin ladders. $SrCu_2O_3$ and $LaCuO_{2.5}$ represent weakly and strongly coupled ladder systems, respectively, and the corresponding doped ladders are seen in $(Sr,Ca)_{14}Cu_{24}O_{41}$ [24] and $La_{1-x}Sr_xCuO_{2.5}$. The tentative phase diagram for $(Sr,Ca)_{14}Cu_{24}O_{41}$ recently obtained by Akimitsu and Kitaoka's groups [25] are also shown for comparison.

3. Phase Transitions in Fe^{4+} ($3d^4$)-Perovskite Oxides Dominated by Oxygen-Hole Character

There have been known perovskite oxides containing Fe ions in an unusually high valence of Fe^{4+} ($3d^4$). Though stabilized in an orbital-degenerate high-spin state t^3e^1, these ions, unlike isoelectronic Mn^{3+}, remain free from instabilities like the Jahn-Teller effect and orbital ordering down to the lowest temperature. $SrFeO_3$ (SFO), which is cubic, and $CaFeO_3$ (CFO), distorted to the $GdFeO_3$-type, take two different ways to avoid these instabilities. A broad metallic band is formed in SFO, while a charge disproportionation (CD) to a pair of exchange-stabilized, highly symmetric orbital-singlet states, i.e. $2Fe^{4+} \rightarrow Fe^{3+}$ ($3d^5$) + Fe^{5+} ($3d^3$), takes place in CFO [26].

Such a difference between Fe^{4+}- and Mn^{3+}-oxides results from the difference in the effective charge-transfer energy (Δ_{eff}), which is -3 eV for SFO [27] but +1.8 eV for $LaMnO_3$ [28]. Oxygen-hole character should thus be more dominant in Fe^{4+}-oxides than in Mn^{3+}-oxides. The Fe^{4+} and Fe^{5+} states may consequently be considered as Fe^{3+} ions accompanied by a single and double oxygen holes, $d^5\underline{L}$ and $d^5\underline{L}^2$, respectively. In this picture "charge" refers to oxygen holes and "disproportionation" refers to the triplet pairing of oxygen holes. The CD is a narrow oxygen-hole band phenomenon.

The CD is, however, known to occur in varying degrees: The relevant hyperfine parameters measured for CFO by Mössbauer spectroscopy shift continuously with temperature as if $Fe^{(4-\delta)+}$ and $Fe^{(4+\delta)+}$ coexist with δ varying continuously between 0 and 1. It is tempting here to assume that the CD is mediated by the breathing lattice mode, an alternate expansion and contraction of the FeO_6 octahedra. If the amplitude of oxygen displacement varies continuously with temperature, the resulting difference in electronic state between neighboring FeO_6 octahedra would be smooth. A careful neutron diffraction study done recently supports this idea as will be reported elsewhere.

It is interesting to study how to control chemically the delicate oxygen-hole state of CFO. Here we show the melting of CD, metallization, and also a subsequent switching to ferromagnetism in $CaFe_{1-x}Co_xO_3$ (CFCO) [29]. This study was initiated from an expectation that the oxygen-hole character might be finely tuned at low substitution-levels because a $(CoO_6)^{8-}$ octahedron is also dominated by oxygen-hole character [30] but, naturally, to a degree different from that for $(FeO_6)^{8-}$.

Magnetic measurements have revealed that the Co-substitution first lowers T_N but next induces ferromagnetism with a boundary composition of $x \approx 0.2$. The transition temperature is much higher for the ferromagnetic (FM) phase, *i.e.* $T_C \gg T_N$. A typical M-H curve obtained at 5 K is shown in Fig. 6(a) together with resistivity data. $CaFeO_3$ shows a gradual MI transition at 290 K, where the gradual CD also sets in, while the $x = 0.1$ composition remains metallic essentially. The FM substances with $x > 0.2$ are better metallic conductors.

By Mössbauer spectroscopy the CD was found to melt at $x \approx 0.1$, prior to the switching to ferromagnetism. As seen in Fig. 6(b) the distance between the two disproportionated components diminishes gradually with increasing x and then suddenly collapses

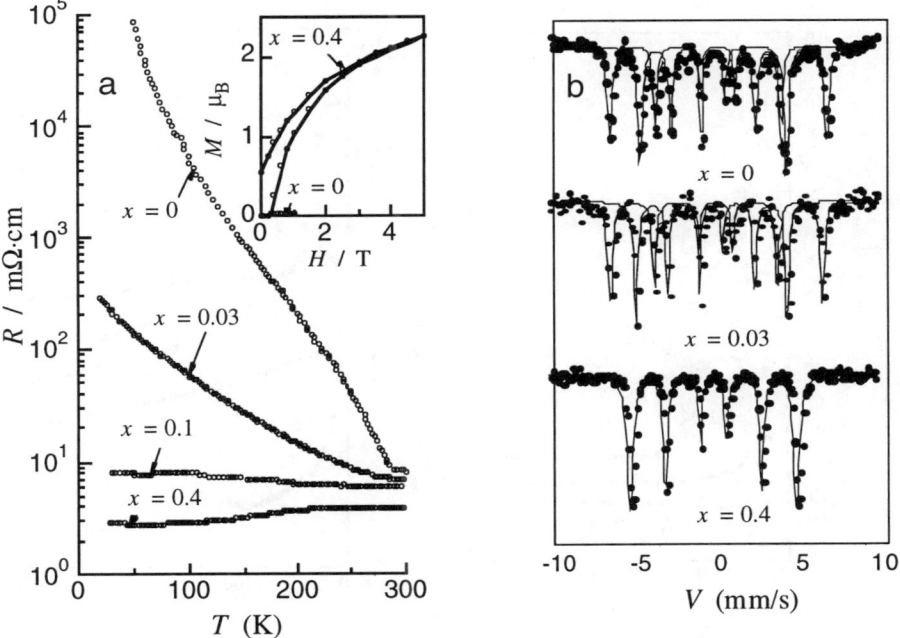

Fig. 6. (a) Temperature dependence of resisitivity (R) and typical magnetization-field (M-H) curve at 5 K (inset) and (b) typical Mössbauer spectra at 4 K for $CaFe_{1-x}Co_xO_3$.

at $0.06 < x < 0.1$. The metallization and the melting of CD may thus be considered to take place at the same Co content of $x \approx 0.1$.

It is now clear that the partial Co-substitution transforms CFO from the charge-disproportionated, AF, insulating phase (Phase I, $x < 0.1$) to the CD-melted, AF, metallic one (Phase II, $0.1 < x < 0.2$), and then to the FM and metallic one (Phase III, $0.2 < x = 0.5$). Considering the fact that the substitution diminishes the cell volume very slightly but does not alter structural symmetry within experimental error, we ascribe the collapse of the CD gap, metallicity, and ferromagnetism to a continuous broadening of the oxygen-hole band width with increasing Co content. As will be reported elsewhere, CFCO has a proper screw type with a propagation vector along the body diagonal, and the dependence of the wavelength upon Co content suggests an enhancement of oxygen hole-mediated ferromagnetic interactions with increasing x.

4. Metal-Insulator Transition in $Tl_2Ru_2O_7$

Ru oxides crystallizing in the pyrochlore structure, $A_2Ru_2O_7$, show interesting electrical properties depending upon the counter cation A and oxygen content. To be reported here briefly is the occurrence of a well-defined MI transition in $Tl_2Ru_2O_7$ which is accompanied by a large drop of magnetization and also by a change in structural symmetry [31,32].

Samples were prepared at 3GPa and 1173K changing the atmosphere in the sample cell. As shown in Fig. 7(a) $Tl_2Ru_2O_7$ shows a sharp MI transition at 120K, but an oxygen-deficiciency of just 1% lowers the transition temperature to about 50K and

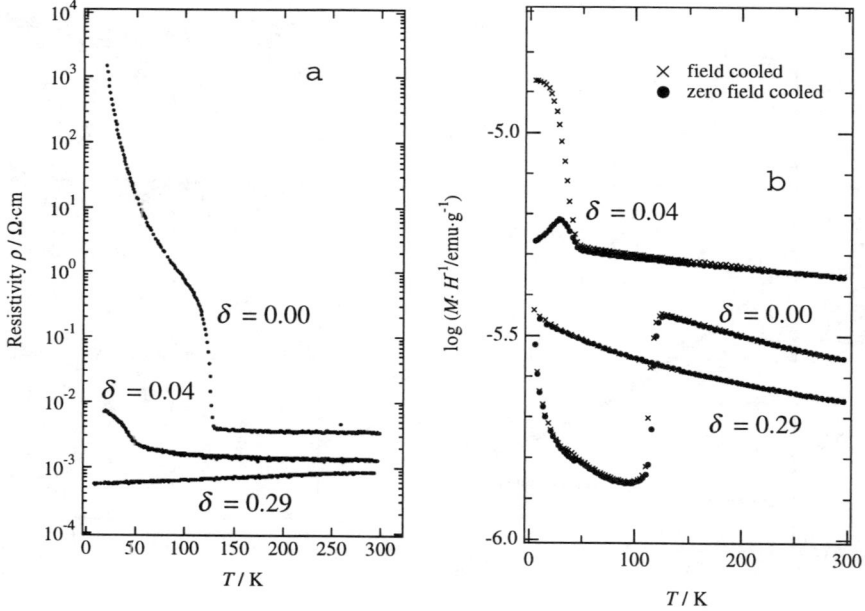

Fig. 7. Temperature- and oxygen content-dependent resistivity (a) and magnetization divided field (b) for $Tl_2Ru_2O_{7-\delta}$.

makes the transition dull. And the 4% deficient sample, $Tl_2Ru_2O_{6.71}$, remains metallic down to the lowest temperature. Magnetization of $Tl_2Ru_2O_7$ drops by 60% at 120K, while $Tl_2Ru_2O_{6.96}$ shows a hysteretic, spin glass-like or canted antiferromagnet-like behavior below 50K (Fig. 7(b)). On the other hand, there is no such anomaly in the 4% deficient sample that remains metallic. The stoichiometric phase changes its structure from cubic ($Fd\bar{3}m$) to orthorhombic ($Pnma$) at the transition temperature, while the other two samples remain cubic.

Clarification of the driving force of this MI transition is a future problem. The coupling of conductivity, magnetism and lattice might prove to result from a specific type of interplay of the spin, charge, and orbital degrees of freedom: The electronic configuration of the trigonally distorted but nearly regular $(RuO_6)^{8-}$ octahedron found in the present structure would not be far from the $^3T_{1g}$ state for a regular octahedron. There might occur, for example, an orbital ordering leading to the formation of Ru-tetramers in the spin singlet state.

By the way, concerning the magnetism in the pyrochlore lattice, we should note here that the terahedral arrangement of antiferromagnetically interacting magnetic atoms causes a spin frustration in this structure, and a theoretical possibility of a 3D quantum spin liquid state has been proposed very recently for the case of the $S=1/2$ Heisenberg antiferromagnet [33].

Another possibility of the instability of the Tl-O sublattice should also be taken into consideration. We mention this possibility because a valence bond calculation using the structure data of the orthorhomic phase has suggested a tendency of CD to $Tl^{(3\pm\delta)+}$.

Acknowledgements

The authors express their hearty thanks for collaboration and discussion to many colleagues including Drs. Y. Kitaoka, K. Ishida, T. Takagi, R.S. Eccleston, N. Nagaosa, T.M. Rice, E. Dagotto, and D.J. Scalapino. This study was supported by a Grant-in-Aid for Scientific Research on Priority Areas given by the Ministry of Education, Science and Culture, Japan and by Core Research for Evolutional Science and Technology of Japan Science and Technology Corporation.

References

[1] M. Imada, A. Fujimori, and Y. Tokura, Rev. Mod. Phys., in press.
[2] Z. Hiroi, M. Azuma, M. Takano and Y. Bando, J. Solid State Chem. **95**, 230 (1991).
[3] Z. Hiroi and M. Takano, Nature **377**, 41 (1995).
[4] E. Dagotto and T.M. Rice, Science **271**, 618 (1996) and references therein.
[5] B. Ammon, B. Frischmuth, and M. Troyer, Phys. Rev. B **54**, R3714 (1996).
[6] T. Kimura, K. Kuroki, and H. Aoki, J. Phys. Soc. Jpn. **67**, 1377 (1998).
[7] M. Azuma, Z. hiroi, M. Takano, K. Ishida, and Y. Kitaoka, Phys. Rev. Lett. **73**, 3463 (1994).
[8] K. Ishida, Y. Kitaoka, Y. Tokunaga, S. Matsumoto, K. Asayama, M. Azuma, Z. Hiroi, and M. Takano, Phys Rev. B **53**, 1 (1996). A theoretical analysis of the temperature dependence of the nuclear spin relaxation time, T_1, indicated Δ_{SG} = 440K, which is consistent with the values reported in refs. 7, 10. See J. Kishine and H. Fukuyama, J. Phys. Soc. Jpn. **66**, 26 (1997).

[9] K. Kojima, A. Keren, G.M. Luke, B. Nachumi, W.D. Wu, Y.J. Uemura, M. Azuma, Z. Hiroi, and M. Takano, Phys. Rev. Lett. **74**, 2812 (1995).
[10] M. Azuma, M. Takano, and R.S. Eccleston, J. Phys. Soc. Jpn. **67**, 740 (1998).
[11] D.C. Johnston, Phys. Rev. B **54**, 13009 (1996).
[12] Y. Mizuno, T. Tohyama, S. Maekawa, T. Osafune, N. Motoyama, H. Eisaki, and S. Uchida, Phys. Rev. B, in press.
[13] M. Azuma, Y. Fujishiro, M. Takano, M. Nohara, and H. Takagi, Phys. Rev. B **55**, R8658 (1997).
[14] S. Ohsugi, Y. Kitaoka, Y. Tokunaga, K. Ishida, K. Asayama, M. Azuma, Y. Fujishiro, and M. Takano, Physica B, in press.
[15] N. Fujiwara, H. Yasuoka, Y. Fujishiro, M. Azuma, and M. Takano, Phys. Rev. Lett. **80**, 604 (1998).
[16] G.B. Martins, M. Laukamp, J. Riera, and E. Dagotto, Phys. Rev. Lett **78**, 3563 (1997) and references therein.
[17] Z. Hiroi, J Solid State Chem. **123**, 223 (1996).
[18] T. Mizokawa, K. Oomoto, T. Konishi, A. Fujimori, Z. Hiroi, N. Kobayashi, and M. Takano, Phys. Rev. B **55**, R13373 (1997).
[19] S. Matsumoto, Y. Kitaoka, K. Ishida, K. Asayama, Z. Hiroi, N. Kobayashi, and M. Takano, Phys. Rev. B **53**, R11942 (1996).
[20] R. Kadono, H. Okajima, A. Yamashita, K. Ishii, T. Yokoo, J. Akimitsu, N. Kobayashi, Z. Hiroi, M. Takano, and K. Nagamine, Phys. Rev. B **54**, R9628 (1996).
[21] B. Normand and T.M. Rice, Phys. Rev. B **54**, 7180 (1996).
[22] M. Troyer, M. E. Zhitomirski, and K. Ueda, Phys. Rev. B **55**, R6117 (1997).
[23] H. Takagi, private communication.
[24] M. Uehara, T. Nagata, J. Akimitsu, H. Takahashi, N. Mori and K. Kinoshita, J. Phys. Soc. Jpn. **65**, 2764 (1996).
[25] J. Akimitsu and Y. Kitaoka, unpublished.
[26] M. Takano, N. Nakanishi, Y. Takeda, S. Naka, and T. Takada, Mat. Res. Bull. **12**, 923 (1977) and references therein.
[27] A.E. Bocquet, A. Fujimori, T. Mizokawa, T. Saitoh, H. Nagatame, S. Suga, N. Kimizuka, Y. Takeda, and M. Takano, Phys. Rev. B. **45**, 1561 (1992).
[28] T. Saitoh, A.E. Bouquet, T. Mizokawa, H. Nagatame, A. Fujimori, M. Abbate, Y. Takeda, and M. Takeda, Phys. Rev. B **51**, 13942 (1995).
[29] S. Kawasaki, M. Takano, R. Kanno, T. Takeda, and A. Fujimori, J. Phys. Soc. Jpn. **67**, 1529 (1998).
[30] R. H. Potze, G. A. Sawatzly and M. Abbate, Phys. Rev. B **51**, 11501 (1995).
[31] T. Takeda, M. Nagata, H. Kobayashi, R. Kanno, Y. Kawamoto, M. Takano, T. Kamiyama, F. Izumi, and A.W. Sleight, J. Solid State Chem., in press.
[32] T. Takeda, R. Kanno, Y. Kawamoto, M. Takano, F. Izumi, A.W. Sleight, and A. W. Hewatt, J. Mat. Chem., in press.
[33] B. Canals and C. Lacroix, Phys. Rev. Lett. **80**, 2933 (1998).

Spin and Charge Dynamics in the Hole-Doped Two-Leg Ladder Compound $Sr_{14-x}Ca_xCu_{24}O_{41}$

J. Akimitsu[1], T. Nagata[1], H. Fujino[1], N. Motoyama[2], H. Eisaki[2], S. Uchida[2], H. Takahashi[3], T. Nakanishi[3], N. Môri[4], M. Nishi[5], K. Kakurai[5], S. Katano[6], M. Hiroi[7], M. Sera[7], and N. Kobayashi[7]

[1] Department of Physics, Aoyama-Gakuin University, Chitosedai, Setagaya-ku, Tokyo 157-8572, Japan
[2] Department of Superconductivity, The University of Tokyo, Hongo, Bunkyo-ku, Tokyo 113-8656, Japan
[3] College of Humanities and Sciences, Nihon University, Sakurajosui, Setagaya-ku, Tokyo 156-0045, Japan
[4] ISSP, The University of Tokyo, Roppongi, Minato-ku, Tokyo 106-0032, Japan
[5] Neutron Scattering Laboratory, ISSP, The University of Tokyo, Tokai, Ibaraki 319-1106, Japan
[6] Neutron Scattering Group, Advanced Science Research Center, Japan Atomic Energy Research Institute, Tokai, Ibaraki 319-1195, Japan
[7] Institute for Materials Research, Tohoku University, Katahira, Aoba-ku, Sendai 980-0812, Japan

Abstract. Our experimental results obtained with the doped ladder material $Sr_{2.5}Ca_{11.5}Cu_{24}O_{41}$ have been reviewed. Our main conclusions are 1) anisotropic resistivity under high-pressure has been measured on the single crystal. The crossover behavior of the charge dynamics from 1D to 2D in this system has been emphasized. 2) The spin gap values have been obtained for the different Ca concentrations with the neutron scattering experiments. Their values are about 32 meV, being independent of Ca concentrations and pressure up to 2.1 GPa. 3) The antiferromagnetism in the $Sr_{2.5}Ca_{11.5}Cu_{24}O_{41}$ at ambient pressure has been found by the specific heat and neutron scattering measurements. The relationship between antiferromagnetism and spin gap state was discussed.

1 Introduction

There has been increasing interest in the $S = 1/2$ two-leg ladder system. Of particular interest are theoretical predictions that the two-leg ladders have a spin gap and hole-doping into the spin ladders would be the bound state of the doped holes, leading to the formations of bipolarons or superconductivity [1]. Such predictions have been extensively studied and spin gap has been observed in the real material such as $SrCu_2O_3$ [2]. Furthermore, the superconductivity

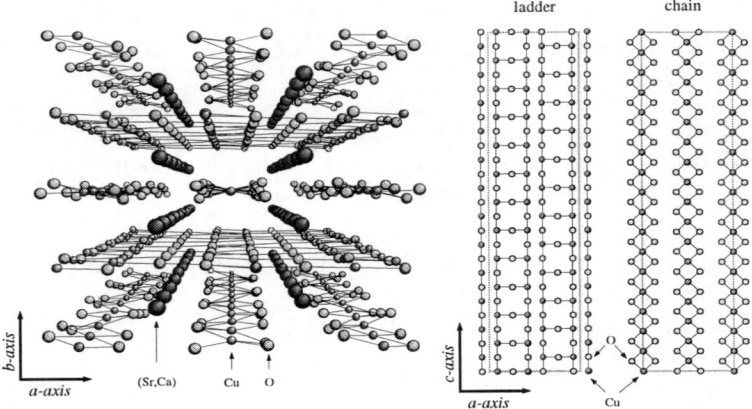

Figure 1: (left figure): Crystal structure of $Sr_{14-x}Ca_xCu_{24}O_{41}$ viewed in perspective along the c-axis. (right figure): Crystal structure of ladder and chain viewed in perspective along the b-axis

has been found in $Sr_{0.4}Ca_{13.6}Cu_{24}O_{41.84}$ at $T_c = 12$ K under pressure [3]. At present, the $Sr_{14-x}Ca_xCu_{24}O_{41}$ system is the only known superconducting cuprate without a two-dimensional (2D) CuO_2 plane.

$Sr_{14-x}Ca_xCu_{24}O_{41}$ is composed of alternating stacks of a plane consisting of an edge-sharing CuO_2 chain, a Sr/Ca layer, and a plane consisting of a two-leg Cu_2O_3 ladder (Fig. 1) [4]. The nominal valence of Cu is +2.25, so the system is inherently hole-doped. From optical conductivity measurements, we know that holes are transferred from the CuO_2 chains to Cu_2O_3 ladders by substituting Sr with Ca. And consequently, a relatively high Cu valence on the ladder site is realized at large x, i.e., +2.2 at $x = 11$, [5]. We report here the present experimental situation on the transport and magnetic properties of the single crystalline $Sr_{14-x}Ca_xCu_{24}O_{41}$.

2 Charge Dynamics – Anisotropic Resistivity and Superconductivity

In this section, we summarize the anisotropic electrical resistivity under pressure and superconductivity was observed between 3.5 and 8 GPa.

For $Sr_3Ca_{11}Cu_{24}O_{41}$ at ambient pressure, ρ_c, the resistivity along the ladder (c-axis), is characterized by metallic, nearly T-linear dependence over a wide temperature region. In contrast, ρ_a, the resistivity between the ladder (a-axis) within the ladder plane, is characterized by a negative temperature coefficient whose absolute value is about two orders of magnitude larger. Recently, Mo-

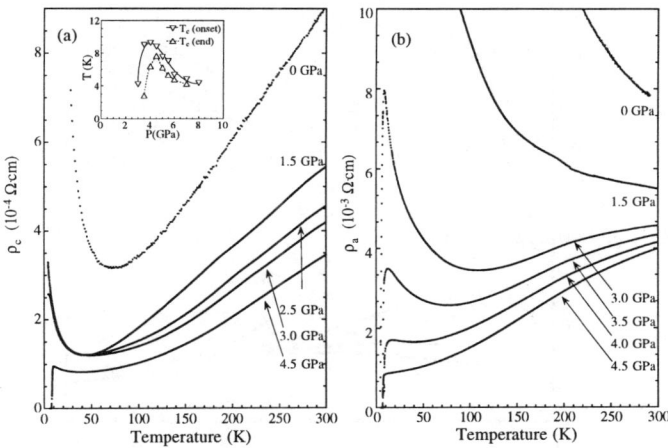

Figure 2: Effect of pressure on temperature dependence of the resistivity (a) along the ladder direction (ρ_c) and (b) across (perpendicular to) the ladder in the ladder plane (ρ_a) of single crystal $Sr_{2.5}Ca_{11.5}Cu_{24}O_{41}$ at indicated pressures. Inset shows the pressure dependence of T_c

toyama et al. systematically studied the Ca concentration dependence of the resistivity ratio ρ_a/ρ_c [6]. They found that anisotropy ratio was increased with increasing the Ca concentration, indicating that the role of Ca substitution is essentially increasing the carrier density in the ladders. Such findings indicate that, at ambient pressure, charge dynamics is essentially one-dimensional (1D) and carriers are confined within each ladder.

Towards a better understanding of the nature of superconductivity in this system, charge dynamics at pressures high enough to bring about superconductivity should be clarified. As a conjecture, one might expect two possible scenarios. In the first, 1D charge dynamics is preserved under high pressure so that superconductivity is produced by the development of the pairing correlation within each ladder such that the interladder coupling plays a minor role. This scenario corresponds to original theoretical arguments which neglect interladder coupling. As for second scenario, application of high pressure increases interladder coupling such that a dimensional crossover from one to two is eventually induced.

To determine which of these scenarios is indeed true, we measured the temperature dependence of the resistivity of a $Sr_{2.5}Ca_{11.5}Cu_{24}O_{41}$ single crystal under high pressure up to 8 GPa.

Figures 2(a) and 2(b) show temperature dependent resistivity ρ_c and ρ_a of $Sr_{2.5}Ca_{11.5}Cu_{24}O_{41}$ at various pressures up to 4.5 GPa. At ambient pressure, ρ_c shows nearly T-linear dependence above 130 K, while it shows the sharp

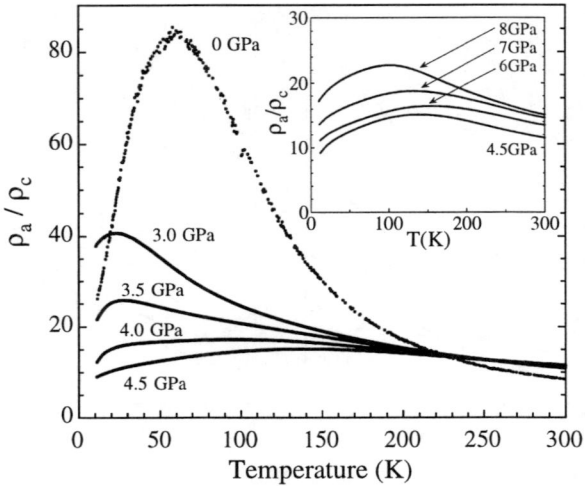

Figure 3: Effect of pressure on temperature dependence of the anisotropic ratio ρ_a/ρ_c of single crystal $Sr_{2.5}Ca_{11.5}Cu_{24}O_{41}$ at indicated pressures. Inset shows the anisotropic ratio ρ_a/ρ_c above 4.5 GPa

upturn below 80 K, which indicate the carrier localization at low temperature. ρ_a, on the other hand, is characterized by a semiconducting behavior indicating the incoherent charge dynamics between ladder. With increasing pressure, both ρ_c and ρ_a rapidly decrease.

Note that at 4.5 GPa, ρ_a shows metallic behavior down to T_c, which indicates that increasing pressure induces not only superconductivity but also coherent charge dynamics perpendicular to the ladder.

Figure 3 shows the temperature-dependent anisotropy ratio ρ_a/ρ_c at various pressures. The ratio at RT is about 10 and almost independent of pressure, while at ambient pressure it gets larger with decreasing temperature, reaching 85 at 50 K. With increasing pressure, however, the enhancement in ρ_a/ρ_c is less pronounced, and the ρ_a/ρ_c vs. T curve is flattened out at about 4.0 GPa where superconductivity appears. This result suggests that the charge dynamics in the superconducting phase is 2D rather than 1D in the a-axis (interladder) hopping. It is noted that the effect of increasing pressure is distinct from that of Ca substitution. Although both ρ_a and ρ_c also decrease with Ca substitution, Ca works opposite for the evolution of ρ_a/ρ_c, enhancing anisotropy as previously mentioned [6]. The Ca substitution is regarded as a chemical pressure because of its smaller ionic radius, leading to the lattice contraction and its major effect on the ladders is to increase the hole density [5]. The application of pressure also has an effect of increasing carrier density in the ladders, however, the contrasting evolution of ρ_a/ρ_c appears to point toward another more relevant effect of pressure, possibly an enhancement of interladder hopping, which would be more crucial for the superconductivity in this system.

3 Spin Dynamics – Coexistence Between Spin Gap State and Antiferromagnetism

3.1 Spin Gap State

In this section we describe the magnetic properties of the two-leg ladder system. Of particular emphasis is the coexistence between spin gap state and antiferromagnetism at ambient pressure.

For the non-doped material $Sr_{14}Cu_{24}O_{41}$, Eccleston et al. [7] have performed inelastic neutron scattering using an array of single crystals to study the spin dynamics of the Cu_2O_3 spin ladder layers and CuO_2 chains. Conclusions on the ladders are summarized as follows.

Their experimental data taken at ISIS are best fitted with a dispersion with a spin gap of 32.5 ± 0.1 meV and a maximum of 193.5 ± 2.4 meV, consistent with a coupling along the ladders, $J_{//} = 130$ meV and a rung coupling $J_\perp = 72$ meV [7].

In the present paper, we describe the inelastic neutron scattering data on the $x = 11.5$ Ca substituted samples in which superconductivity has been observed under high pressure.

Figure 4 shows the inelastic neutron scatteirng intensity at 7K, the difference between the intensity measured at around the antiferromagnetic zone center (0.5 0 1.5) and the background intensity measured at the reciprocal lattice point (0.5 0 1.36) in the constant Q method. A broad but clear peak in the intensity was observed at around 32 meV at the zone center (0.5 0 1.5). The inset present the scattering intensities with 31 meV along the c^* direction. The intensities were peaked at (0.5 0 1.5) and also at (0.5 0 2.5), which shows that the spin gap observed here has the periodicity of the ladders; and thus they are certainly originated from the spin ladders.

Scans for the spin excitations along the c^* direction are also shown in the figure for different positions. With departing from the zone center, the excitation energy is considerably increased, which indicates a dispersive character. For the excitation energy along the rung direction, on the other hand, the peak energy remains at around 32 meV and shows no dispersion. This could be understood from the structure of this system; that is, the interladder exchange energy is quite small because of the frustrations along a^* direction.

The lines shown for these scans are the results convoluted with the instrumental resolution. The parameters are the gap energy and the band maximum in the dispersion relation. Here, following the analysis employed by Eccleston et al. [7], the dispersion relation was assumed as
$$E^2(Q) = \Delta^2 + A^2 \sin^2(2\pi Q),$$
where Δ is the spin gap and A is the band maximum. (See the Eccleston et al.

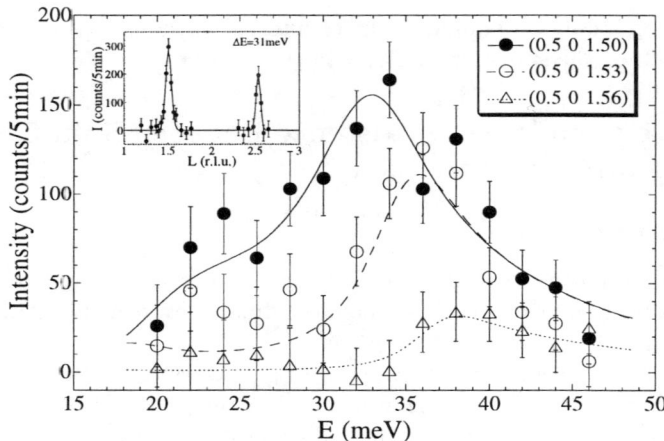

Figure 4: Inelastic neutron scattering intensities after the subtraction of the background intensity measured in the constant Q scans. The data show the spin excitation at the antiferromagnetic zone center and at the points apart from this zone center. The lines are the results of the convolution with the instrumental resolution. The fits give the spin gap energy of 32.1 ± 0.3 meV and the band maximum of the dispersion of 157.8 ± 17.1 meV

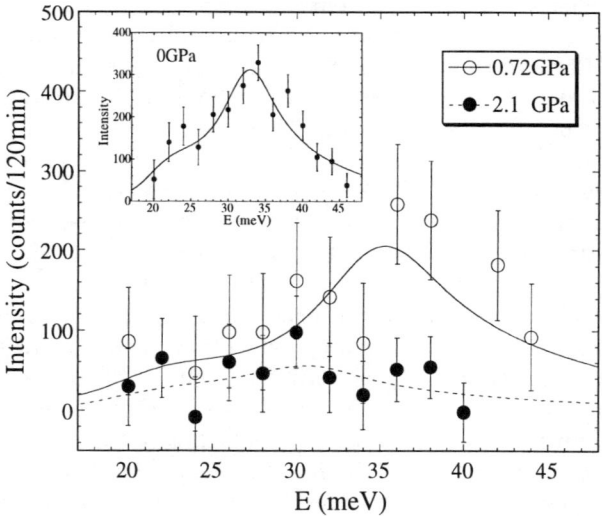

Figure 5: Pressure dependence of the intensities for the spin gap excitation at the antiferromagnetic zone center (0.5 0 1.5). The lines are the results of the convolution with the instrumental resolution

[7].) As shown in the figure, reasonable fits were obtained for $\Delta = 32.1 \pm 0.3$ meV, and A = 157.8 ± 17.1 meV, thus leading to $J_{//} \sim 90 \pm 15$ meV and $J_{\perp} \sim 65 \pm 15$ meV.

Comparing the spin gap value of the parent material $Sr_{14}Cu_{24}O_{41}$ ($\Delta \sim 32.5$ meV), the spin gap of the doped compound does not change even if Ca is substituted up to 11.5. This results is quite different from the NMR data [8, 9].

The spin gap excitations under high pressures are shown in Fig. 5. The data at ambient pressure (in the inset) and at 0.72 GPa were normalized by nuclear peak intensities. As shown in the figure, the errors estimated are rather large because of the high background intensities from the pressure cell. Regarding the spin gap energy, however, the gap energy is obtained to be around 30 meV by the fits; that is, even under pressures the spin gap does not change significantly from that of the parent compound. On the other hand, it is clear that the intensity decreases progressively with increasing pressure. This result implies that the number of the electrons which are forming the spin singlet state is decreased under pressures. It is noteworthy that a recent NMR experiment at the pressure of 3.2 GPa demonstrated that the spin gap is vanished, changing to a pseudo-spin gap in the supperconducting state [10].

3.2 Antiferromagnetic Order

In this section, we report heat capacity and neutron scattering measurements on $Sr_{14-x}Ca_xCu_{24}O_{41}$ single crystal at low temperatures around 2.1 K. Our experiments revealed the new results for the magnetic ground state at high Ca concentration x ($x = 11.5$) under ambient pressure.

From the specific heat results, we observed a magnetic long range order at $x \sim 11.5$ under fairly low temperatures ($T_N \approx 2.1$ K) compared with the spin gap value. Magnetic Bragg peaks below T_N were also observed in the neutron scattering measurements on a $Sr_{2.5}Ca_{11.5}Cu_{24}O_{41}$ single crystal.

We observed a sharp peak at $T_N \approx 2.1$ K in $Sr_{2.5}Ca_{11.5}Cu_{24}O_{41}$ by the heat capacity measurements, and T_N decreases with decreasing Ca contents. In $Sr_{14}Cu_{24}O_{41}$, we cannot see any peak at all temperature.

To elucidate the origin of this phase transition observed by specific heat measurements, elastic neutron scattering was performed for a single crystal $Sr_{2.5}Ca_{11.5}Cu_{24}O_{41}$. From this experiment, we observed a number of Bragg peaks which appeared below 2.1 K. Fig. 6 shows the typical elastic neutron scattering results at (1 0 l), (0 1 l) and (h 0 12), these results were plotted after subtracting the intensity at 5 K from that at 1.4 K. We confirmed that the FWHM of the observed Bragg peaks is nearly the resolution limited.

In order to determine the magnetic structure, we should make clear which layers – ladders or chains – are responsible for the onset of the magnetic long range order. Recent NQR/NMR studies by Ohsugi et al. [11] concluded that (1) the magnetic spins both on ladders and chains are magnetically ordered in

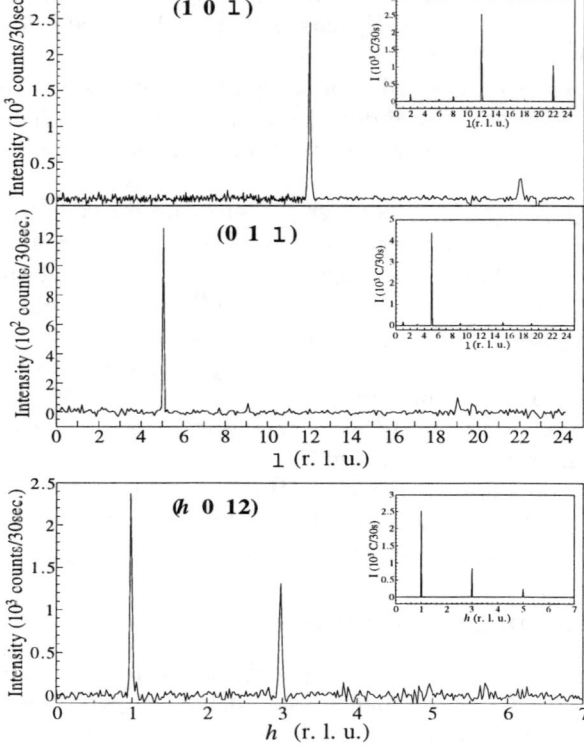

Figure 6: Typical results by elastic neutron scattering at $(1\ 0\ l)$, $(0\ 1\ l)$ and $(h\ 0\ 12)$ of the $Sr_{2.5}Ca_{11.5}Cu_{24}O_{41}$ single crystal. These results were plotted after subtracting the intensity at 5 K from that at 1.4 K. Each insets show the calculated magnetic Bragg peak intensities

this system, (2) spontaneous moments on the ladder sites are rather uniform with small moments less than $\sim 10^{-2}$ μ_B, whereas spontaneous moments on the chain sites are nonuniformaly distributed with rather larger moments (≥ 0.05 μ_B). Therefore, they pointed out that the magnetic Bragg reflection observed in Fig. 6 should mainly come from the magnetic arrangements on the chain site. Actually, assuming that these magnetic Bragg peaks observed in Fig. 6 are originated from the magnetic order at ladder site, the appearance of $(3\ 0\ 12)$ Bragg peaks should lead to the ferromagnetic order in the a-c plane, and this ferromagnetic order aligns antiferromagnetically in the adjacent layers. However, due to the absence of $(0\ k\ 0)$ reflection, the possibility of the ferromagnetic order in the ladder plane will be excluded.

Therefore, we naturally concluded that the magnetic Bragg reflections main-

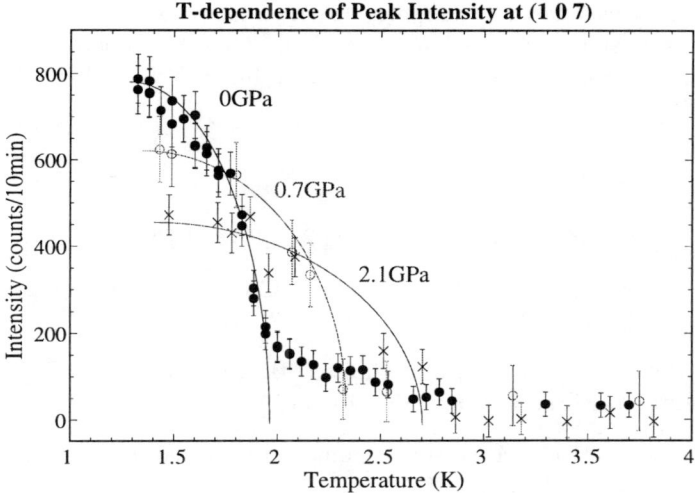

Figure 7: Temperature dependence of the (1 0 12) reflection under the pressure of the $Sr_{2.5}Ca_{11.5}Cu_{24}O_{41}$ single crystal. Solid line is drawn for the guides to the eyes

ly reflect the magnetic order at chain sites and the magnetic order at ladder site can not be detected by the neutron scattering technique due to the small magnetic moments.

We emphasize here the ground state of this system. As described before, the inelastic neutron scattering has indicated that the spin gap energy for the ladders in the $Sr_{2.5}Ca_{11.5}Cu_{24}O_{41}$, the same material in our present experiment, was found to be 32.1 meV. Therefore, it is likely that magnetic order is due to some of the unpaired spins, breaking up spin singlet states by introducing the holes into ladders, and these unpaired spins ordered antiferromagnetically at low temperature. Accordingly the ground state of this system at ambient pressure might be one in which both paired and unpaired spins coexist and the holes and spins float in a \ spin liquid sea " at high temperature. These holes are localized at certain temperature and the spins finally order at $T_N \approx 2.1$ K.

Recently, we have performed the pressure dependence of this antiferromagnetic ordering in connection with its pressure induced superconductivity. Fig. 7 shows the temperature dependence of the intensities from the magnetic reflaction (1 0 12) for the pressure of 0.72 and 2.1 GPa, which was normalized by the nuclear reflecions. As shown in the Fig. 7, the intensity decreases with increasing pressure, however, the Néel temperature T_N increases from 2.3 K to 2.8 K at a rate of 0.3 K/GPa, which can be reasonably explained that the magnetic interaciton in the chains will be increased with increasing the pressure.

This work was partially supported by a Grant-in -Aid for Science Research from the Ministry of Education, Science and Culture, Japan and by a Grant from the CREST, and the Science Fund of Japan Private School Promotion Foundation.

References

[1] E. Dagotto and T.M. Rice, Science **271**, 618 (1996)

[2] M. Azuma, Z. Hiroi, M. Takano, K. Ishida and Y. Kitaoka, Phys. Rev. Lett. **73**, 3463 (1994)

[3] M. Uehara, T. Nagata, J. Akimitsu, H. Takahashi, N. Môri and K. Kinoshita, J. Phys. Soc. Jpn **65**, 2764 (1996)

[4] E.M. MaCarron, M.A. Subramanian, J.C. Calabrese and R.L. Harlow, Mater. Res. Bull. **23**, 1355 (1988)

[5] T. Osafune, N. Motoyama, H. Eisaki and S. Uchida, Phys. Rev. Lett. **78**, 1980 (1997)

[6] N. Motoyama *et al.* (unpublished)

[7] R.S. Eccleston, M. Uehara, J. Akimitsu, H. Eisaki, N. Motoyama and S. Uchida, Phys. Rev. Lett. **81**, 1702 (1998)

[8] K.Kumagai, S. Tsuji, M. Kato and Y. Koike, Phys. Rev. Lett. **78**, 1992 (1997)

[9] K. Magishi, S. Matsumoto, Y. Kitaoka, K. Ishida, K. Asayama, M. Uehara, T. Nagata and J. Akimitsu, Phys. Rev. B **57**, 11533 (1998)

[10] H. Mayaffre, P. Auban-Senzier, M. Nardone, D. Jerome, D. Poilblanc, C. Bourbonnais, U. Ammerahl, G. Dhalenne and A. Revcolevschi, Science **279**, 345 (1998)

[11] S. Ohsugi, K. Magishi, S. Matsumoto, Y. Kitaoka, T. Nagata and J. Akimitsu, (preprint)

NMR Probe of Magnetic Order and Spin Correlation in Hole-Doped Ladder Cuprates

Y. Kitaoka[1*], T. Mito[1], S. Ohsugi[1*], K. Magishi[1*], S. Matsumoto[1*], T. Nagata[2],
J. Akimitsu[2*], N. Motoyama[3], H. Eisaki[3], S. Uchida[3]

[1] Department of Physical Science, Graduate School of Engineering Science, Osaka University,
Toyonaka, Osaka 560, Japan

[2] Department of Physics, Aoyama-Gakuin University, Chitosedai, Setagaya-ku, Tokyo 157, Japan

[3] Department of Superconductivity, University of Tokyo, Bunkyo-ku, Tokyo 113-8656, Japan

*CREST, Japan Science and Technology Corporation (JST), Japan

Abstract

Comprehensive Cu-NQR and -NMR studies of $Sr_{14-x}Ca_xCu_{24}O_{41}$ (denoted as CaX) including CuO_2 chains and Cu_2O_3 ladders have revealed the onset of long-range magnetic order for Ca11.5 below $T_M \sim 2.2-2.5$ K at ambient pressure. Tiny moments with an order of $\sim 10^{-2}\mu_B$ are spontaneously ordered on the ladders, whereas sizable moments larger than $\sim 0.05\mu_B$ are on the chains. This magnetic order which takes place in the charge-localized regime lower than $T_L = 60$ K is suggested to be driven by the formation of pinned CDW-like state on the ladders. In the intermediate T range where resistivity exhibits a quasi-one-dimensional (quasi-1D) metallic behavior, a spin gap persists keeping a quasi-1D feature, even though the pressure increases approaching a superconducting phase boundary. In the high-T metallic regime, the characteristics of spin correlations either in highly-doped compounds or under pressure resemble those in underdoped high-T_c cuprates. In the doped two-leg ladder system with the spin gap, the spin degree of freedom remains alive when the charge is localized, whereas superconductivity seems to occur under the presence of the spin gap when the charge becomes itinerant with analogy to the underdoped high-T_c cuprates.

I. INTRODUCTION

Ladder cuprates attract much interest since the discovery of superconductivity with $T_c \sim 12$ K in polycrystal $Sr_{0.4}Ca_{13.6}Cu_{24}O_{41.84}$ [1] and $T_c \sim 10$ K in single crystal $Sr_{2.5}Ca_{11.5}Cu_{24}O_{41}$ [2], when pressure greater than 3 GPa is applied. Remarkably, these experiments appear to support the theoretical prediction that singlet superconductivity would occur by doping a small amount of holes into even-leg ladders with a spin gap [3–5]. The substitution of Ca for the Sr sites increases the conductivity of $Sr_{14-x}Ca_xCu_{24}O_{41}$ which comprises hole-doped Cu_2O_3 two-leg ladders and CuO_2 chains [6–8]. The crystal structure, the CuO_2 chain and Cu_2O_3 ladder subunit are illustrated in Fig. 1. The optical conductivity experiment clarified that holes are transferred from the chains to the ladders upon the isovalent Ca substitution and hole content n increases progressively as $n \sim 0.14$, 0.2, and 0.22 for $x=6$, 9 and 11, respectively [9].

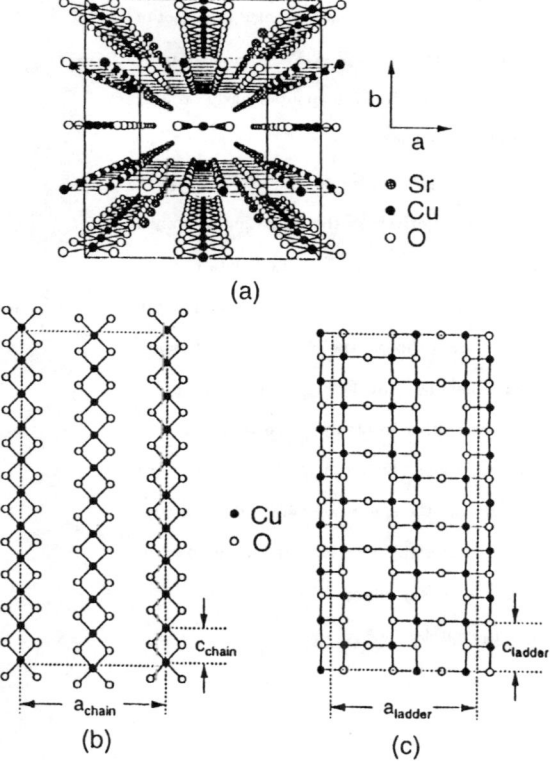

FIG. 1. (a) The crystal structure of $(Sr,Ca)_{14}Cu_{24}O_{41}$. (b) and (c) show the CuO_2 chain and Cu_2O_3 ladder subunit, respectively.

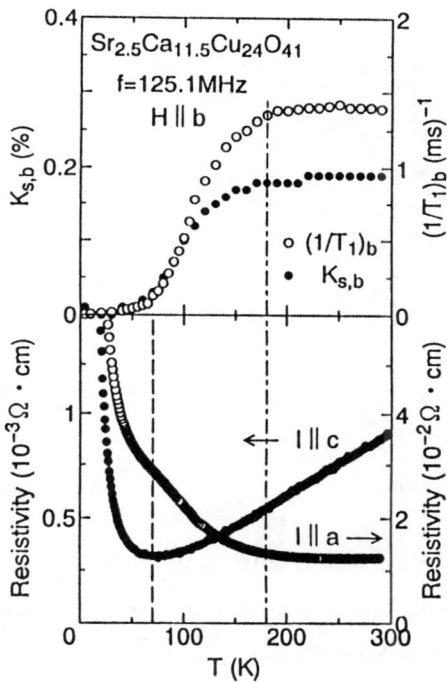

FIG. 2. A comparison between the magnetic and transport properties in Ca11.5 [10]. The upper panel indicates the T dependences of the spin part in the Knight shift $K_{s,b}(T)$ and $(1/T_1)_b$ for $H\|b$ axis. The lower panel is the T dependences of the resistivity along the c axis (scale on the left axis) and the a axis (scale on the right axis).

Comprehensive Cu-NMR investigations on $Sr_{14-x}Ca_xCu_{24}O_{41}$ (hereafter denoted as CaX) clarified the characteristics of magnetic properties in the hole-doped two-leg ladders [10]. The spin gaps obtained from the measurements of the Knight shift and T_1, Δ_K and Δ_{T_1} almost *unchange* with the increasing Ca content from $x=9$ to 11.5. $\Delta_K = 270$ K and $\Delta_{T_1} = 350$ K were estimated in Ca11.5. Systematic T_{2G} measurement demonstrated that spin correlation length, ξ is determined by an average distance of doped holes, which allows us to estimate a hole content n as $\sim$ 0.14, 0.22, and 0.25 per Cu_2O_3 ladder for Ca6, Ca9, and Ca11.5, respectively.

The magnetic and transport properties in Ca11.5 are compared in Fig.2. In the high-T range bounded by the dash-dotted line where the quasi-1D spin correlation is significant, the resistivity ρ_c along the c axis (the ladder leg) exhibits a metallic behavior with a linear-temperature (T) dependence [2,11]. In the intermediate T region where the spin gap opens,

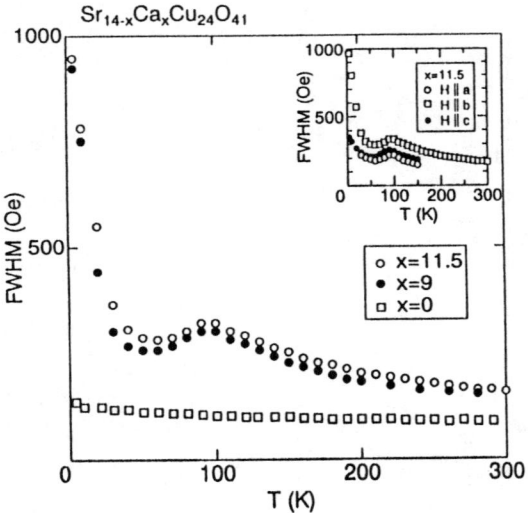

FIG. 3. T dependences of the full width at half maximum (FWHM) in the ladder-Cu NMR spectrum for Ca11.5 (open circles) together with those for Ca9 (closed circles) and Sr14 (open square) [12]. Inset indicates the T dependences of FWHM for $H \| a$ axis (open circles), b axis (open squares) and c axis (closed circles).

$\rho_c(T)$ deviates slightly from the linear-T dependence, but $\rho_a(T)$ parallel to the a axis begins to *increase*. In this regime, hole pairs are formed accompanying the spin gap, but confined on each ladder. In the low-T region below $T_L \sim 60$ K where the resistivity increases following T^{-2}, mobile hole pairs are localized due to some randomness in the quasi-1D conducting channel along the c axis. Since staggered spin fluctuations with low frequencies around $q \sim \pi$ dominate the nuclear relaxation in the ladders for Ca6, Ca9, and Ca11.5, the spin gap was implied to be disrupted in the charge-localized state [10].

With some relevance to this low-T magnetic anomaly, the spin degree of freedom in the charge-ordered state below T_L is evidenced from the T dependence of the full width at half maximum of the NMR spectrum (FWHM) in Ca11.5 [12]. As shown in Fig. 3, the T dependencies of the FWHM in Ca11.5 and Ca9 are found to be scaled to that of measured susceptibility χ_{mes} down to T_L, whereas the FWHM in Sr14 dose not exhibit any significant T dependence over an entire T range [12]. It is considered that the Ca substitution enhances the transferred hyperfine-coupling constant between the ladder-Cu nuclei and the spins on the chains. With the further decreasing T below $T_L \sim 60$ K, the FWHM's in Ca9 and Ca11.5

increase markedly. It is noteworthy that the anisotropy in FWHM, which is indicated in the inset in Fig. 5, coincides with the anisotropy of the hyperfine form factor, since $(FWHM)_b/(FWHM)_{a,c} \sim 2.8$ is compatible to $(A_b-3B)/(A_{a,c}-3B) \sim 2.8$ [10]. A continuous increase of FWHM is hence because field-induced staggered spin polarization develops upon cooling below T_L. Spin degree of freedom manifests on the ladders. Localization of *single hole* is expected to break up the collective spin-singlet state formed by Cu spins, producing unpaired spins near by. Therefore, it should be noted that hole pairs are dissociated into single hole below T_L.

II. MAGNETIC ORDER IN THE CHARGE-LOCALIZED STATE

Recent specific heat $(C(T))$ measurement in Ca11.5 has found a sharp peak around $T_M \sim 2.2$ K well below T_L at zero magnetic field $(H = 0)$ and ambient pressure $(P=1$ bar) [2]. In addition, an elastic neutron diffraction experiment has suggested the development of magnetic Bragg reflections upon cooling below $T_M \sim 2.2$ K, proving that the peak in $C(T)$ is magnetic in origin [13]. It is remarkable that the peak in $C(T)$ becomes broader at $H = 1.8$ T and disappears under $H > 8$ T. These evidences assure the magnetic order below $T_M \sim 2.2$ K. We show that the spin degree of freedom in the ladders plays vital role in an occurrence of the magnetic order.

Figure 4 indicates the T dependence of $^{63}(1/T_1)$ for the Zhang-Rice singlet Cu (ZR) sites in the chains. The NQR spectrum of the ZR sites was observed with a peak at 33.1 MHz regardless of the Ca content. T_1 was measured at $H=0$ and 33.1 MHz by the ^{63}Cu NQR and at $H=1.8$, 4.7 and 8 T by the ^{63}Cu NMR [12]. $1/T_1$ is well fitted in the T range of 20-40 K by the activation form of $1/T_1 \propto \exp(-\Delta/T)$ with $\Delta = 100$ K without any appreciable H dependence as shown by the solid line in Fig. 4. Note that the nuclei at the ZR site is coupled through the transferred hyperfine interaction with the spins on the nearest-neighbor Cu sites (hereafter denoted as dimers for which spin-singlet formation is seemingly developed down to 20 K). The activated behavior of $1/T_1$ implies as if the spins on the dimer would form the spin singlet. The $1/T_1$ at $H = 0$, however, exhibits a sharp peak around 2.5 K close to the temperature where $C(T)$ has a peak. The peak in $1/T_1$ originates from critical spin fluctuations towards a magnetic phase transition and hence corresponds to a magnetic ordering temperature as expected. Furthermore, the peak in $1/T_1$ is suppressed at $H = 1.8$ T and collapses with the further increasing field. It is therefore evident from

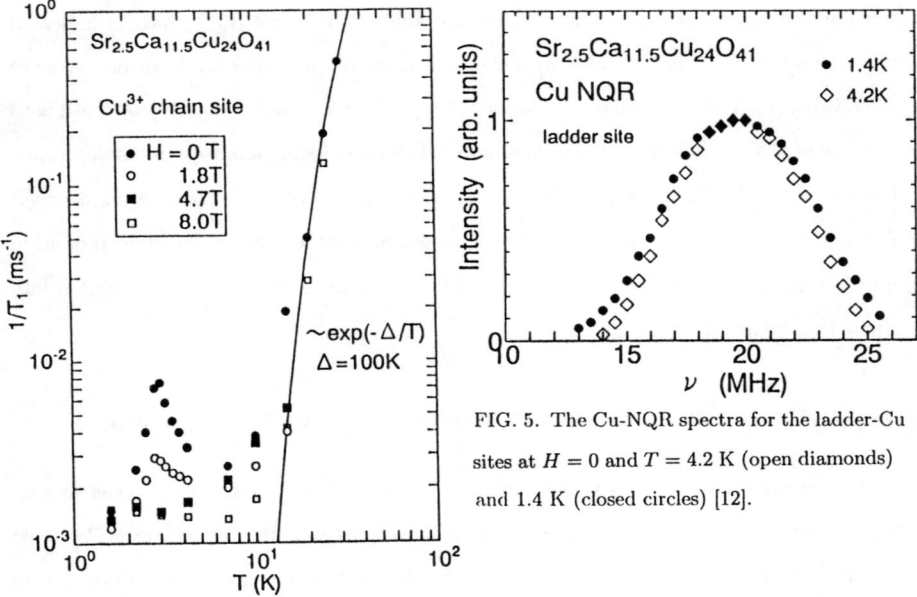

FIG. 5. The Cu-NQR spectra for the ladder-Cu sites at $H = 0$ and $T = 4.2$ K (open diamonds) and 1.4 K (closed circles) [12].

FIG. 4. T dependences of the $^{63}(1/T_1)$ for the ZR sites in Ca11.5 at 33.1 MHz and $H=0$ (closed circles), 1.8 T (open circles), 4.7 T (closed squares) and 8 T (open squares) [12]. Solid line is a fit to the activated form of $1/T_1 \propto \exp(-\Delta/T)$ with $\Delta = 100$ K.

both the measurements of T_1 and $C(T)$ [13] that the magnetic order is suppressed by the field.

From the analysis of the Cu-NQR spectrum, when an internal field $H_n(C)$ is assumed to be parallel to the a, b and c axis, $H_n^{a,b}(C) \sim 0.7$ kOe and $H_n^c(C) \sim 0.6$ kOe are estimated, respectively. Corresponding spontaneous moment $M_{dimer}(C)$ on the dimers is estimated as $M_{dimer}^a(C) \sim 0.05$ μ_B, $M_{dimer}^b(C) \sim 0.04$ μ_B and $M_{dimer}^c(C) \sim 0.04$ μ_B. In Sr14, the dimer was reported to form the spin singlet. In Ca11.5, however, each spin on the dimer is magnetically coupled with spins on "magnetic Cu sites" in the chains on which holes are transferred into the ladders by the Ca substitution. From the fact that $M_{dimer}(C) \sim 0.05\mu_B$ is sizable on the dimer, we therefore expect that a saturation moment on the "magnetic Cu sites" in the chains might be larger than $M_{dimer}(C) \sim 0.05\mu_B$ and the neutron Bragg diffraction probes the periodic array of the magnetic dimers and the "magnetic Cu sites" in the chains.

As for the ladder sites, it has been shown that the ladder-Cu spin degree of freedom which remains alive owing to the charge localization below T_L takes part in the onset of magnetic order [12]. Figure 5 indicates the NQR spectrum broadened over 14 − 22 MHz at $T = 4.2$ K (diamonds) and 1.4 K (closed circles) at $H = 0$. The NQR intensity is normalized by each peak value at 19.8 MHz. The broader spectrum at 1.4 K than that at 4.2 K points to the appearance of the internal field $H_n(L)$ on the ladders. When a spontaneous moment $M_s(L)$ is assumed to be parallel to the a, b and c axis, $H_n(L)$ is estimated as $H_n^a(L) \sim 1.06$ kOe, $H_n^b(L) \sim 1.06$ kOe and $H_n^c(L) \sim 1.9$ kOe and as $M_s^a(L) \sim 2.2 \times 10^{-2}$, $M_s^b(L) \sim 0.9 \times 10^{-2}$ and $M_s^c(L) \sim 4.0 \times 10^{-2} \mu_B$, respectively [12].

Recent optical conductivity measurement has revealed that hole pairs tend to form a CDW-like regular array, although it is not truly three-dimensional (3D) long-range order [14]. From the NMR/NQR results, we have, however, suggested that the CDW-like state is formed by the regular array of *single hole*. This is because that the localization of hole pairs is not expected to break up the spin-singlet state on the ladders. The tiny size of $M_s(L)$ reveals that most of magnetic spectral weight exists in high-energy region comparable to the spin gap of $\Delta_K \sim 270$ K. These features of the magnetic order driven by the localization of holes resemble those in the gapped-1D systems doped with slight impurities such as $CuGe_{1-x}Si_xO_3$ [15], $Cu_{1-x}Zn_xGeO_3$ [16] and $Sr(Cu_{1-x}Zn_x)_2O_3$ [17]. In disordered spin-Peierls $CeCuGe_3$ systems, the long-range AF order and spin-Peierls lattice dimerization was pointed out to coexist with spatially varying order parameter [18,19]. In such the state, the AF moments are largely distributed, whereas it should be emphasized that the tiny spontaneous moments on the ladders in Ca11.5 are rather uniform due to a dense hole content $n \sim 0.25$ [12]. We have proposed that the development of pinned CDW-like periodic array over long-range distance leads to the onset of magnetic order in Ca11.5. It may be due to the disruption of this charge-ordered state that superconductivity was induced by an application of pressure. Next we present the nature of spin gap and spin correlation near the superconducting phase boundary.

III. SPIN GAP AND CORRELATION NEAR THE SUPERCONDUCTING PHASE

As seen in Fig.2, in the T range where the spin gap opens below $T^* \sim 180 - 200$ K in Ca11.5, $\rho_c(T)$ reveals the metallic behavior, whereas $\rho_a(T)$ increases rapidly. The increase

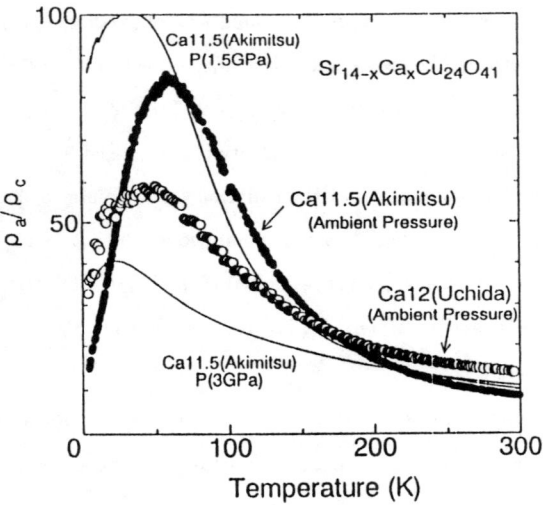

FIG. 6. T dependences of the anisotropy of resistivity $R_A = \rho_a/\rho_c$ at $P=1$ bar in Ca12 (open circles) [20] and at $P=1.5$ GPa (closed circles) and $P=1$ bar and 3 GPa (solid lines) in Ca11.5 [13].

of anisotropy $R_A = \rho_a/\rho_c$ below T^* is indicative of a quasi-1D metallic behavior associated with the confinement of bound hole pairs on each ladder [11,10]. The application of pressure decreases R_A [2,11]. Figure 6 indicates the $R_A(T)$ at $P=1$ bar in Ca12 [20] together with the $R_A(T)$'s at $P=1$ bar, 1.5 and 3 GPa in Ca11.5 reported in the literature [2]. The $R_A(T)$ at $P=1$ bar in Ca12 is smaller than the $R_A(T)$'s at $P=1$ bar and 1.5 GPa in Ca11.5. We expect that the $R_A(T)$ at $P=1.7$ GPa might be close to the $R_A(T)$ at $P=3$ GPa in Ca11.5 where superconductivity sets in. Accordingly, the electronic and magnetic properties of Ca12 at $P = 1.7$ GPa are anticipated to be analogous to those of Ca11.5 at $P = 3$ GPa. Actually, superconductivity has been reported to occur even under pressure smaller than 2 GPa [21].

First we remark from the Cu-Knight shift measurement that a marked evolution of electronic state takes place in the ladders of highly-doled compounds. Measured Knight shift, $K_{mes}(T)$ consists of the T-independent orbital part K_{orb} and the T-dependent spin part $K_s(T)$. In spin-gapped systems where $K_s(T)$ is expected to vanish at low T, K_{orb} is represented by a low-T value of K_{mes}, i.e. , $K_{orb} = K_{mes}(T = 4.2\text{ K})$ in the present case [10]. It should be noted that $K_{orb} = 1.32\%$ is invariant with the increasing Ca content up to Ca11.5, whereas it increases to 1.42% for Ca12. Zheng et al showed in a wide variety

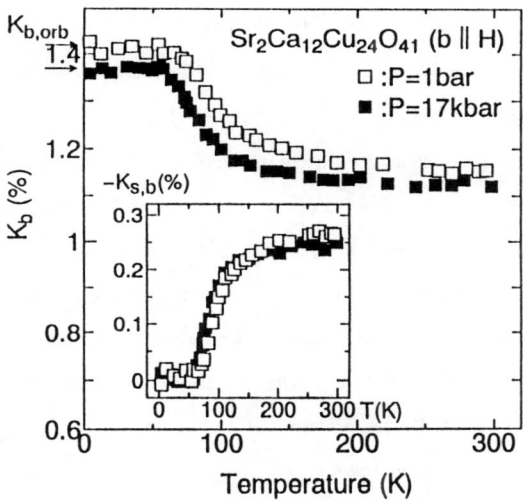

FIG. 7. T dependences of the measured Knight shift K_{mes} for $H\|b$ axis at $P=1$ bar (open squares) and $P=1.7$ GPa (17 kbar) (closed squares) in Ca12 [23]. Inset indicates the T dependences of the spin part $K_{s,b}(T)$ in K_{mes} at $P=1$ bar (open squares) and 1.7 GPa (closed squares) in Ca12.

of high-T_c cuprates that K_{orb} depends on the local hole density in Cu-3d$_{x^2-y^2}$ orbit, $n_{x^2-y^2}$ according to the following equation [22];

$$K_{orb} \propto \chi_{orb} = n_{x^2-y^2} \cdot 2\mu_B^2 \sum_e \frac{|<e|L|x^2-y^2>|^2}{E_e - E_{x^2-y^2}},$$

where χ_{orb} is the orbital susceptibility and L the angular momentum operator and e denote the excited states. Since $K_{orb} = 1.32\%$ is independent of the Ca content up to Ca11.5, holes are mainly transferred from the chains into the O sites on the ladders. Remarkably in Ca12, holes are transferred not only into the O sites but also the Cu sites, resulting in the *increase* of $n_{x^2-y^2}$ in Ca12.

Figure 7 indicates the $K_{mes}(T)$'s at $P=1$ bar (open squares) and $P=1.7$ GPa (17 kbar) (closed squares). It is interesting to note that the K_{mes} at $P=1.7$ GPa *decreases* by $\sim 0.05\%$ than the K_{mes} at $P=1$ bar over the entire T range. This means that the pressure-induced reduction in K_{mes} is nearly independent of the temperature and hence is not magnetic but orbital in origin. We hence obtain $K_{orb}=1.42$ % at $P=1$ bar and 1.37 % at $P=1.7$ GPa [23]. K_{orb} decreases from 1.42 to 1.37 % as the pressure increases from $P=1$ bar to 1.7 GPa. Mayaffre et al reported in the literature [24] that $K_{mes}(T)$ decreases with the increasing pressure from $P=1$ bar to $P=3$ GPa, which is qualitatively analogous to the present result under 1.7 GPa in Ca12. They, however, ascribed this reduction in $K_{mes}(T)$ to the increase of the spin part $|K_s|$, namely, the absence of the spin gap. Since the resistivity measurement at $H=0$ revealed a precursor towards a superconducting transition below 5 K, they concluded that the gapless magnetic excitation in the normal state is relevant to the onset of superconductivity [24]. We have, however, found that the reduction in K_{mes} at $P=1.7$ GPa should be attributed to the *reduction* of K_{orb} not to the *increase* of $|K_s|$. Applying pressure causes a local hole transfer from the Cu to O site, resulting in the *decrease* of K_{orb} or $n_{x^2-y^2}$. This pressure-induced hole redistribution on the ladders increases the mobility of p holes and an inter-ladder interaction. These effects are considered to decrease R_A and suppress the charge order. As a result, the superconductivity appears.

Noting that K_{orb} decreases with the increasing pressure, the $K_s(T)$'s at $P=1$ bar and 1.7 GPa have been found to decrease below $T^* \sim 200$ K due to the opening of the spin gap as shown in the inset in Fig. 7 [23]. An activated fit to the $K_s(T)$ data at low T allows us to estimate a size of the spin gap Δ_K as ~ 300 K, which is independent of the pressure. Persistence of the spin gap has also been assured from the T dependence of $1/T_1$ at $P=1$ bar and 1.7 GPa shown by the open and closed squares in Fig. 8, respectively. A fit of $1/T_1$ at low T to the activation form of $\exp(-\Delta_{T_1}/T)$ yields $\Delta_{T_1}=378$ and 308 K at $P=1$ bar and 1.7 GPa, respectively [23]. The electronic structure undergoes gradual crossover from the quasi-1D to anisotropic 2D one, whereas the size in the spin gap almost *unchanges*, keeping the quasi-1D feature.

In the high-T regime for $T > T^*$, a $[T_{2G}/T_1\sqrt{T}]$=constant behavior was found in Ca6, Ca9 and Ca11.5 [10]. Here $1/T_{2G}$ is the Gaussian spin-echo decay rate dominated by an indirect nuclear-spin coupling through electronic excitations and related to the static wave-number dependent susceptibility, $\chi(q)$. This behavior was consistent with the scaling theory for the $S=1/2$-1D spin chain. The quasi-1D spin correlation was observed in Ca6 Ca9 and Ca11.5 [10], whereas it is not seen in Ca12 at either $P=1$ bar or $P=1.7$ GPa. In Ca12, on the other hand, a scaling of T_{2G}/T_1T stays constant as displayed in Fig. 9 [23] where the same plots are presented as for Ca9 and Ca11.5 [10]. The $[T_{2G}/T_1T]=$ constant behavior was observed in the underdoped cuprate YBa$_2$Cu$_4$O$_8$ (Y124) [25], as drawn by the solid line in Fig. 9. Thus, the nature of spin correlations in Ca12 in the high-T regime no longer reveals the quasi-1D feature as in Ca6, Ca9 and Ca11.5, but is rather close to that in the 2D underdoped cuprates. In this context, the pressure-induced superconductivity in the

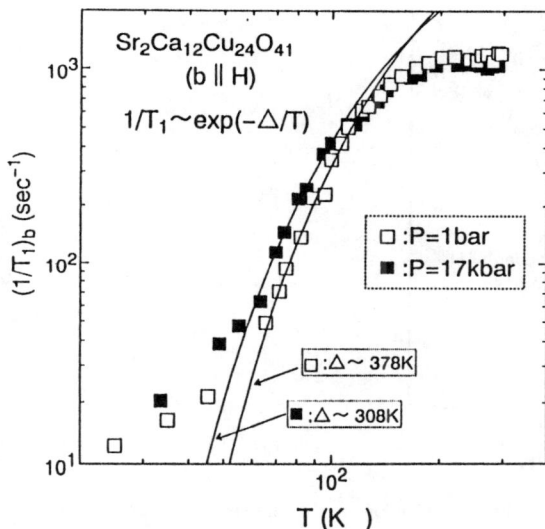

FIG. 8. T dependences of $(1/T_1)_b$ for $H \parallel b$ axis in both logarithmic scales at $P=1$ bar (open squares) and 1.7 GPa (closed squares) in Ca12 [23]. Solid line indicates a fit of the data below 110 K to the activated form of $1/T_1 \propto \exp(-\Delta_{T_1}/T)$ with $\Delta_{T_1} = 378$ and 308 K at $P=1$ bar and 1.7 GPa, respectively.

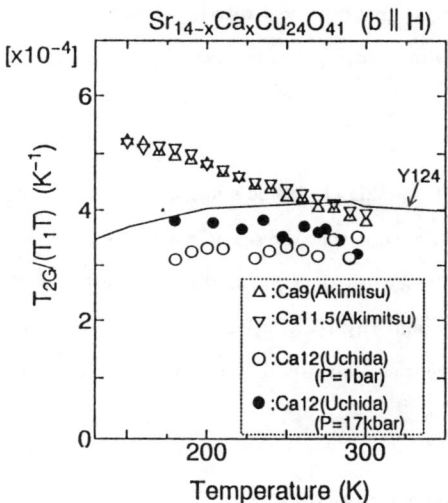

FIG. 9. A comparison with the 2D scaling exhibiting the $[T_{2G}/T_1T]$=constant behavior in the underdoped high-T_c cuprate YBa$_2$Cu$_4$O$_8$ (solid line) [25]. This 2D scaling is valid in Ca12 at $P=1$ (open circles) bar and 1.7 GP (closed circles), whereas the 1D scaling exhibiting the $[T_{2G}/T_1\sqrt{T}]$=constant behavior is valid in Ca9 and Ca11.5 [10]

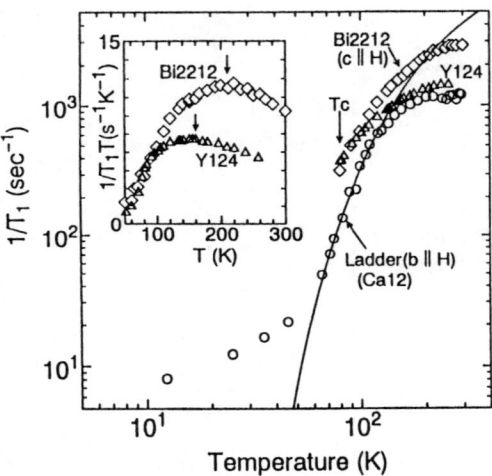

FIG. 10. A comparison with the spin-gap behavior in the underdoped high-T_c cuprates YBa$_2$Cu$_4$O$_8$ (Y124) [25] and Bi$_2$Sr$_2$CaCu$_2$O$_{8+\delta}$ (Bi2212) [26]. The T dependence of $(1/T_1)_b$ for $H\|b$ axis at $P=1$ bar in Ca12 is compared with those of $1/T_1$ in Y124 and Bi2212. Solid line indicates an activated fit of $(1/T_1)$ in Ca12 in the T range of 60-120 K. Inset indicates the T dependence of $(1/T_1 T)$ which decreases below the spin-gap temperature T^* being 150 and 205 K in Y124 and Bi2212, respectively. T^* is marked by the arrow in the inset. Note that the T dependence of $1/T_1 T$ is smooth between T^* and T_c and not consistent with the activated behavior presumably due to the pseudo-2D gap structure with the $d_{x^2-y^2}$-like symmetry.

ladder compound seems to occur in the anisotropic 2D electronic state with the spin gap. On the one hand, as demonstrated in Fig. 10, we stress that the seeming activated decrease of $1/T_1$ at low T in Ca12 is in a marked contrast with their moderate decrease below T^* in the underdoped high-T_c cuprates Y124 [25] and Bi$_2$Sr$_2$CaCu$_2$O$_{8+\delta}$ (Bi2212) [26]. This suggests that the topology on the Fermi surface is significantly different between the ladder and underdoped high-T_c cuprates.

IV. CONCLUDING REMARKS

The holes which are doped into the two-leg ladder with the spin gap tend to form bound pairs, but are dissociated presumably due to the formation of the CDW-like state in the charge-localized regime at low temperatures. When this CDW-like state spatially grows over long-range distance, the 3D magnetic order appears with tiny spontaneous moments with an order of $10^{-2}\mu_B$ on the ladders. As the hole content or the pressure increases, the quasi-1D transport evolves progressively into the anisotropic 2D one, whereas the gap continues to survive near the superconducting phase boundary. When the charge degree of freedom is retained under pressure, the superconductivity appears under the presence of the spin gap. Further systematic study on Ca12 under higher pressure is desired to deepen insight into mechanism for superconductivity in doped low dimensional quantum-spin systems.

V. ACKNOWLEDGMENTS

This work has been supported by CREST (Core Research for Evolutional Science and Technology) of Japan Science and Technology Corporation (JST) and also partly by Grants-in-Aid for the Scientific Research and for the COE Research (10CE2004) from the Ministry of Education, Science, Sports and Culture, Japan.

REFERENCES

[1] M. Uehara et al, J. Phys. Soc. Jpn. **65** (1996) 2764.

[2] T. Nagata et al, Physica C **282-287** (1997) 153.

[3] E. Dagotto, J. Riera, and D.J. Scalapino, Phys. Rev. B **45** (1992) 5744.

[4] T.M. Rice, S. Gopalan, and M. Sigrist, Europhys. Lett. **23** (1993) 445.

[5] M. Sigrist, T.M. Rice, and F.C. Zhang, Phys. Rev. B **49** (1994) 12058.

[6] M. Uehara, M. Ogawa, and J. Akimitsu, Physica C **255** (1996) 193.

[7] M. Kato, K. Shiota, and Y. Koike, Physica C **255** (1996) 284.

[8] S.A. Carter et al., Phys. Rev. Lett. **77** (1996) 1378.

[9] T. Osafune, N. Motoyama, H. Eisaki, and S. Uchida, Phys. Rev. Lett. **78** (1997) 1980.

[10] K. Magishi, Y. Kitaoka, S. Matsumoto, K. Ishida, K. Asayama, M. Uehara, T. Nagata, and J. Akimitsu, Phys. Rev. B **57** (1998) 11533.

[11] N. Motoyama, T. Osafune, T. Kakeshita, H. Eisaki, and S. Uchida, Phys. Rev. B **55** (1997) R3386.

[12] S. Ohsugi et al., submitted.

[13] T. Nagata et al., submitted.

[14] T. Osafune et al., submitted.

[15] L.P. Regnault et al., Europhys. Lett. **32** (1995) 579.

[16] Y. Sasago et al., Phys. Rev. B **54** (1996) R6835.

[17] M. Azuma et al., Phys. Rev. B **55** (1997) R8658.

[18] H. Fukuyama, T. Tanamoto and M. Saito, J. Phys. Soc. Jpn. **35** (1996) 1182.

[19] M. Saito and H. Fukuyama, J. Phys. Soc. Jpn. **66** (1997) 3259.

[20] N. Motoyama et al., unpublished.

[21] N. Motoyama, H. Esisaki and Uchida, private communication.

[22] G.-q. Zheng, Y. Kitaoka, K. Ishida, and K. Asayama, J. Phys. Soc. Jpn. **64** (1995) 2524.

[23] T. Mito et al., in Proc. Int. Conf. on Strongly Correlated Electron System, Paris, 15-19, July (1998); ibid, submitted.

[24] H. Mayaffre et al., Science **279** (1998) 345.

[25] N.J. Curo et al., Phys. Rev. B **53** (1996) 5907.

[26] K. Ishida et al., Phys. Rev. B **57** (1998) No.9 .

Metal-Insulator and Magnetic Transitions in Layered Ruthenates

Y. Maeno,[1,2] S. Nakatsuji,[1] and S. Ikeda[1,3]

[1] *Department of Physics, Kyoto University, Kyoto 606-8502, Japan.*
[2] *CREST, Japan Science and Technology Corporation, Kawaguchi, Saitama 332-0012, Japan.*
[3] *Venture Business Laboratory, Kyoto University, Kyoto 606-8501, Japan.*

Abstract. We present the phenomenological systematics of a series of ruthenates, $(Ca, Sr)_{n+1}Ru_nO_{3n+1}$, which exhibits a rich variety of ground states. We will also discuss in some detail the physical properties of the two-dimensional Mott-transition system, $Ca_{2-x}Sr_xRuO_4$, in which the Mott insulator Ca_2RuO_4 evolves into the unconventional superconductor Sr_2RuO_4 by means of a band-width control.

§1. Introduction

Recently, ruthenium oxides (ruthenates) have shown to be an ideal material system to study the metal-insulator transition by the band-width control, as well as to study unconventional superconductivity near the ferromagnetic instability [1]. In this article, we will first present the systematics of a series of ruthenates $(Ca, Sr)_{n+1}Ru_nO_{3n+1}$, which exhibits a rich variety of ground states depending on the dimensionality and the band width. In Section 3, we will focus on the two-dimensional, band-width control system $(Ca,Sr)_2RuO_4$ and show in some detail how the Mott insulator in the layered perovskite structure evolves into the superconductor. In Section 4, we will summarize the present status of understanding of the superconductivity of Sr_2RuO_4. We will then conclude with a few remarks.

§2. Systematics of Ruthenates

Ruthenates exhibit a rich variety of ground states, ranging from antiferromagnetic insulators to ferromagnetic or paramagnetic metals, as well as the superconductor. It is true that the effects of strong correlations play a key role in determining their physical properties, but the results of band-structure calculations provide a very useful starting point. Let us consider the Ruddelsden-Popper series of ruthenates, $(Ca, Sr)_{n+1}Ru_nO_{3n+1}$, and try to gain some phenomenological systematics based on the results of the band-structure calculations. In these compounds, the electronic states near the Fermi level are antibonding $pd\pi^*$ bands derived from Ru^{4+}-$4d^4$ electrons (low-spin state, $S=1$) hybridized with O^{2-}-$2p^6$ electrons. Their ground-state properties are summarized in Fig. 1, in which respective compounds are plotted in the plane of

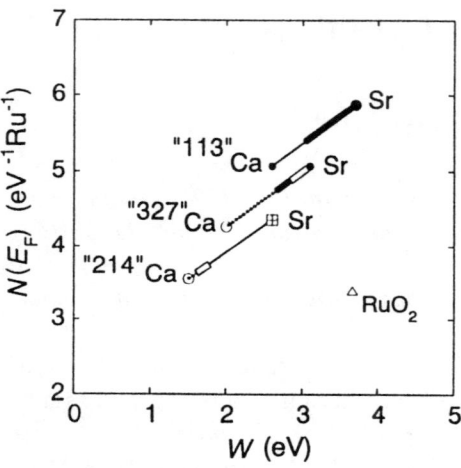

Fig. 1. Systematics of a series of ruthenates in the plane of the band width W vs. the density of states at the Fermi level $N(E_F)$. 113: $(Sr,Ca)RuO_3$, 327: $(Sr,Ca)_3Ru_2O_7$, 214: $(Ca,Sr)_2RuO_4$. Thick solid lines: ferromagnetic metal, open lines: magnetic metal, solid lines: paramagnetic metal, dotted lines: antiferromagnetic insulator, and square: superconductor. For comparison, the location of a typical d-band metal RuO_2 is also shown by a triangle.

the band width W vs. the density of states (DOS) at the Fermi level $N(E_F)$. Here we evaluate W from the dispersion curves as the total width of the $pd\pi^*$ bands which crosses the Fermi level. The coordinates for four of the end members, Sr_2RuO_4, $Sr_3Ru_2O_7$, $SrRuO_3$ and $CaRuO_3$ are known in the literature [2-4]. The open circles for the other end members as well as the straight lines connecting the circles indicate crude expectations. Since the atomic positions have now been determined precisely for Ca_2RuO_4 [5], it is desirable that the band-structure calculations be extended to this Ca end member (open circle). For comparison, the coordinates for RuO_2 are also shown [6]. RuO_2 is an ordinary d-band metal which exhibits Pauli paramagnetism and only a weak effect of electron correlations.

With increasing n, which results in the increasing dimensionality of the system from two to three with stronger hybridization, W increases: $W = 2.6, 3.1$, and 3.7 eV for $n = 1$ (Sr_2RuO_4), 2 ($Sr_3Ru_2O_7$), and ∞ ($SrRuO_3$). Naively, one expects a corresponding *decrease* in $N(E_F)$, since the total number of the states has to be conserved. However, the opposite tendency of *increasing* $N(E_F)$ clearly exists: $N(E_F) = 4.36, 5.07$, and 5.87 in units of $eV^{-1}Ru^{-1}$ for $n = 1, 2$, and ∞, respectively. This is because the orbital degeneracy of the three bands originating from d_{xy}, d_{yz}, and d_{zx}, in which four electrons are accommodated, becomes more complete for less anisotropic systems with larger n and the distribution of the DOS in the immediate vicinity of E_F becomes sharp, resulting in higher $N(E_F)$. As in the system with $n=\infty$, the substitution of Ca for Sr for given n should result in simultaneous decrease in W and $N(E_F)$, since it mainly induces distortion in the RuO_6 networks, resulting in reduced d-p hybridization and at the same time in reduced overlap of the three d-bands.

We expect a ferromagnetic ground state for large $N(E_F)$. In fact, the Stoner enhancement factor $[1-N(E_F)I]^{-1}$ diverges for $SrRuO_3$ [4]. (Here the exchange-correlation integral I is assumed to be 0.187 eV following Ref. 2.) This is consistent with the observed ferromagnetic ground state of $Sr_{1-x}Ca_xRuO_3$ with $0 \leq x \leq 0.7$ [7], shown by the thick solid line in Fig. 1. In contrast, the system tends to be a Mott insulator for small W, since it corresponds to large effective Coulomb energy U/W, in which the on-site Coulomb energy U is estimated to be about 3 eV for these ruthenates [8]. In fact, the ground states of Ca_2RuO_4 and $Ca_3Ru_2O_7$ are antiferromagnetic insulators [9, 10]. This demonstrates the crucial importance of the effect of strong electron correlations, not fully accounted for in the band-structure calculations. Hense the finite $N(E_F)$ showm in Fig. 1 probably does not represent the correct ground states for these compounds. The dotted lines in Fig. 1 indicate the extent of these magnetic-insulator regions. In short, we expect Stoner ferromagnets in the upper right region of the phase diagram and Mott insulators in the lower left region.

An important issue concerning the system with $n=\infty$ has been why $CaRuO_3$ does not exhibit any magnetic ordering, in spite of its large "antiferromagnetic" Weiss temperature in the susceptibility. Recently, Kiyama et al. [11] have shown that the specific heat, as well as the susceptibility, is well characterized by the ferromagnetic spin-fluctuations, not antiferromagnetic ones. Therefore, $CaRuO_3$ is most probably a nearly ferromagnetic metal, and the absence of the magnetic long-range order is attributable to the insufficient magnitude of $N(E_F)$.

In the middle of the diagram are the ferromagnetic metal $Sr_2CaRu_2O_7$ with T_c = 3 K [12], described well as a weak ferromagnet in the framework of the Self-Consistent Renormalization (SCR) theory, and the superconductor Sr_2RuO_4 with T_c = 1.5 K. $Sr_3Ru_2O_7$ appears to be very close to the ferromagnetic instability. Single crystals grown by a floating-zone method exhibit a nearly isotropic peak in the magnetization at about 17 K, as shown in Fig. 2 (a). This behavior is fully consistent with that of the sintered polycrystals [12]. In sharp contrast with this behavior, single crystals grown by a flux method exhibit a clear ferromagnetic ordering at 104 K, with a large saturated magnetic moment of about 1.3 μ_B and the strongly anisotropic magnetization at low temperatures as in Fig. 2 (b) [13]. The c-axis parameter, corresponding to twice the inter-bilayer distance, of the flux-grown crystals of $Sr_3Ru_2O_7$ is shorter by nearly 1 %, and the resulting difference in the lattice distortion seems to be sufficient to drive the ferromagnetic instability.

The lattice parameters of $Sr_{3-x}Ca_xRu_2O_7$ change discontinuously also with x [12]. Therefore, non-monotonic variations in W and $N(E_F)$ are expected in this system, which accounts for the presence of a weak ferromagnet ($0.7<x(Ca)<1.2$)

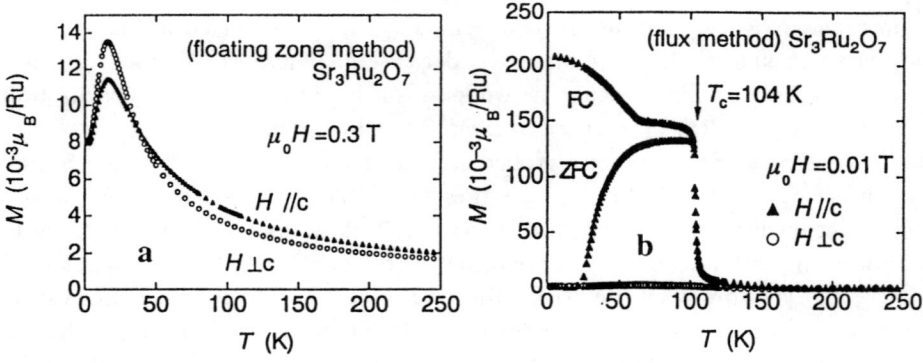

Fig. 2 Contrasting magnetic behavior of $Sr_3Ru_2O_7$ grown (a) by a floating-zone method and (b) by a flux method [13].

between a nearly ferromagnetic metal and an antiferromagnetic insulator. Contrary to the simplified straight line in Fig. 1, therefore, the variations of W and $N(E_F)$ with x for the system with $n=2$ should exhibit discontinuous changes across the structural boundaries at $x(Ca)=0.7$ and 1.2.

§3. Band-Width Control in $Ca_{2-x}Sr_xRuO_4$

A number of Mott transition systems are known in which the metal-insulator transitions are driven by the control of the bandwidth, rather than by the control of the electron filling (carrier doping). Among them, $Ca_{2-x}Sr_xRuO_4$ is unique in the quasi-two-dimensionality of its electronic structure and in the presence of the superconductor as its end member.

Changes in the crystal structure play an important role also in this system. Sr_2RuO_4 ($x=2$) is the copper-free layered perovskite superconductor [14] with $T_c = 1.5K$. Study of phonon dispersion by inelastic neutron scattering revealed that the mode corresponding to the *rotation* of the RuO_6 octahedra about the c axis shows a remarkable softening, whereas the mode corresponding to the *tilting* around an axis in the RuO_2 plane does not [15]. Nevertheless, Sr_2RuO_4 barely retains the tetragonal K_2NiF_4 structure without any static distortion down to the lowest temperatures. We note here that the tilting of the CuO_6 octahedra governs the lattice distortions in La_2CuO_4, whereas the rotation is the main instability in both Gd_2CuO_4 [16] and Sr_2IrO_4 [17].

The other end member Ca_2RuO_4 ($x=0$) is a Mott insulator with canted antiferromagnetic ordering at low temperatures [9]. There are two structural variants of Ca_2RuO_4 (the "S" and "L" phases), which differ in oxygen content. They involve severe distortions: both the rotation of the octahedra about the c axis (by 13 and 12 degrees) and the tilting about an axis in the plane (by 5 and 11 degrees) [5]. On the other hand, the average Ru-O bond length remains nearly unchanged between the Sr- and Ca-end members. Thus, it is mainly by these distortions and the resulting decrease in the 4d-band width that open up the Mott-Hubbard-type gap in Ca_2RuO_4 [1, 5, 9].

We have recently succeeded in synthesizing $Ca_{2-x}Sr_xRuO_4$ in the whole region of x: the complete solution of Ca_2RuO_4 and Sr_2RuO_4. We have also determined the phase diagram, which consists of the following three regions:

I $(0 \leq x < 0.2)$ Metal-insulator (M-I) transition by varying temperature except $x = 0$. Ca_2RuO_4 ($x = 0$) is a Mott insulator at least up to 300K.
II $(0.2 \leq x < 0.5)$ Magnetic metal phase with an almost isotropic peak in the susceptibility. Magnetic unstable point at $x \approx 0.5$.
III $(0.5 \leq x \leq 2.0)$ Paramagnetic metal. The susceptibility gradually changes with x from Curie-like paramagnetic ($x=0.5$), through Curie-Weiss-like paramagnetic $(0.5<x<2.0)$, and to Pauli paramagnetic ($x \approx 2$). Superconductivity at $x=2.0$.

The M-I transition in the region I occurs at temperatures which decreases with x from 250 K at $x=0.05$ to 50 K at $x=0.15$. At nearly the same temperatures, magnetic transition ascribable to canted antiferromagnetism occurs. The transitions are of the first order since they are accompanied by hystereses involving structural change. This transport and magnetic behavior is very similar to that in oxidized Ca_2RuO_4 (O-CRO) [5].

Figure 3 shows the variation of the low-temperature susceptibility $\chi(0)$ with x. The susceptibility $\chi(0)$ represented here is the value of M/H under the field of 1 T at 2 K. It is clear that $\chi(0)$ exhibits a diverging behavior toward $x=0.5$.

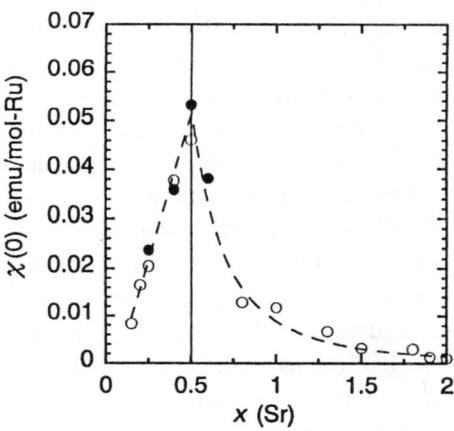

Fig. 3. Magnetic susceptibility of $Ca_{2-x}Sr_xRuO_4$ at 2 K, denoted as $\chi(0)$, as a function of x. It exhibits a diverging tendency toward $x = 0.5$. Open circles: polycrystals, solid circles: isotropic part in single crystals.

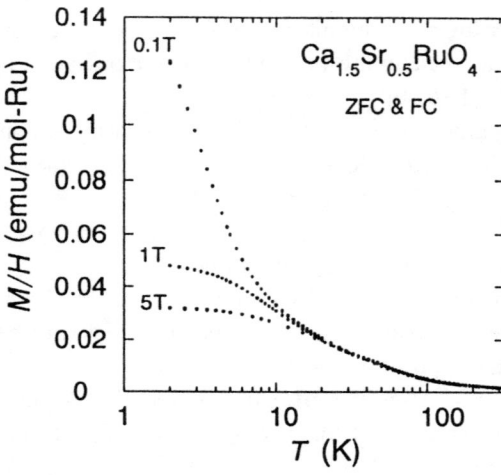

Fig. 4. Temperature dependence of the magnetization M/H of polycrystalline $Ca_{2-x}Sr_xRuO_4$ near its magnetic unstable point at $x = 0.5$ under different fields.

For $0.2 \leq x < 0.5$ (the region II), $M(T)$ shows a peak at a temperature T_M, which decreases with x from about 10 K at $x=0.20$ to zero toward $x=0.5$. At $x=0.5$, the susceptibility shows Curie-like paramagnetic behavior with the Weiss temperature $\theta = 0$ K. With increasing x, it gradually changes to Curie-Weiss behavior with increasing "antiferromagnetic" Weiss temperature, and finally to the Pauli paramagnetic behavior (corresponding formally to infinite $|\theta|$) at $x \approx 2$. It is interesting to note that the effective moment deduced from the Curie constant corresponds to nearly $S=1/2$ for $0.2 < x < 1.5$. It is in sharp contrast with the Curie-Weiss behavior in $(Sr,Ca)_3Ru_2O_7$ and $(Sr,Ca)RuO_3$ with $S=1$, which coincides with the localized spin of a tetravalent Ru ion in the low-spin state.

The anomaly at $x=0.5$ indicates the magnetic unstable point at $T=0$ K. There has been no evidence for any structural transition across $x=0.5$. Figure 4 shows the temperature dependence of the magnetization under different fields. At high fields, the diverging low-temperature behavior is suppressed and the Fermi liquid behavior with T-independent M/H is recovered at low temperatures.

The normal state of $x=2$ end member Sr_2RuO_4 is characterized well as an essentially two-dimensional Fermi liquid [18]. The magnetic susceptibility shown in Fig. 5 is nearly isotropic and shows only very weak temperature dependence. The small anisotropy is attributable to that of the orbital Van Vleck contribution. The magnitude of the Pauli paramagnetic contribution χ_0 is enhanced from the expectation of the band-structure calculation χ_b by a factor of 6.5. Combined with the enhanced electronic coefficient of the specific heat (inset of Fig. 5), we obtain the enhanced Wilson ratio $R_W \equiv (\chi_0/\chi_b)/(\gamma/\gamma_b)=1.7-1.9$. We should also note the details of the weak temperature dependence of the susceptibility. It has a broad maximum at about

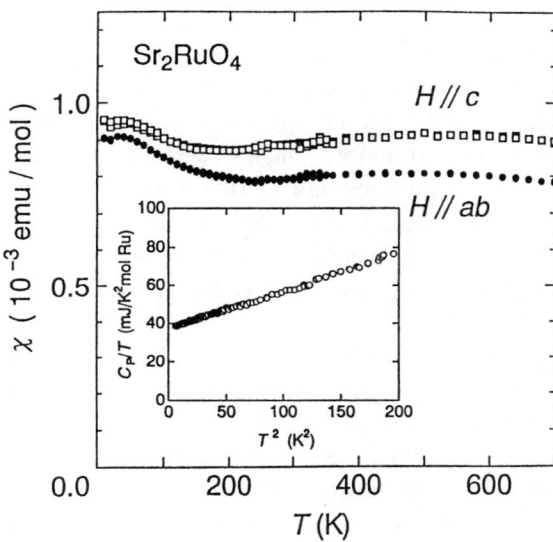

Fig. 5. Magnetic susceptibility of Sr_2RuO_4 up to 700 K, showing enhanced Pauli paramagnetism [18]. A very broad peak at about 500 K, a gradual increase below 200 K, and a broad peak at 30 K are all considered intrinsic to this compound. Inset: specific heat divided by temperature.

500 K. On cooling it starts to increase below about 200 K. Since this increase in the spin susceptibility is also observed from the Knight-shift measurement of ^{17}O NMR [19, 20], we believe that this behavior is intrinsic to Sr_2RuO_4, rather than due to the inclusion of ferromagnetic $SrRuO_3$ as an impurity. The peak at 30 K and the saturated susceptibility at lower temperatures may be related to the formation of the three-dimensional Fermi liquid state, in which the resistivities in all three dimensions exhibit the T-squared dependence [18]. Since the nuclear-spin relaxation rates T_1^{-1} on ^{17}O and Ru exhibit almost the same temperature dependence, Imai et al. have concluded that the ferromagnetic spin fluctuations dominate the relaxation mechanism [20].

§4. Superconductivity in Sr_2RuO_4

The first report of superconductivity in Sr_2RuO_4 [14] was promptly followed by the suggestion of spin-triplet superconductivity by Rice and Sigrist [21] and later by Baskaran [22]. In this section, we will summarize the main experimental advancements concerning its superconductivity within the last year. Preceding developments have already been described in Ref. 1.

(1) Extreme sensitivity of T_c to disorder introduced by non-magnetic impurities has been characterized [23]. Superconductivity is lost when the quasi-particle mean-free-path becomes comparable to the coherence length, while

the Fermi-liquid behavior persists. This result clearly reflects the unconventional nature of superconductivity in Sr_2RuO_4.

(2) The intrinsic T_c as high as 1.5 K was suggested from the impurity effect. Recently, we have indeed attained T_c of 1.50 K in single crystals grown by a floating-zone method. The intrinsic coherence lengths, derived from the anisotropic upper critical fields, are $\xi_{ab}(0)=66$ nm and $\xi_c(0)=3.3$ nm.

(3) The absence of the coherence peak and the presence of the large residual DOS induced by disorder have previously been observed by Ru-NMR and NQR [24]. Very recently, the ^{17}O Knight-shift experiments have also been performed to identify the Cooper-pair spin state [19].

(4) The specific heat measurements indicated that the large residual electronic coefficient γ_0 depends strongly on T_c [25]. With the improvement of the sample quality, γ_0 becomes lower than 25 % of the normal-state electronic coefficient γ_N. Therefore, the nonunitary pairing state in its simplest form, for which $\gamma_0=\gamma_N/2$ is expected, is not compatible with the experiment.

(5) Tunneling experiments with two closely-spaced pads of Pb, evaporated on the surface of a Sr_2RuO_4 crystal, revealed remarkable temperature dependence of the Josephson current [26]. With decreasing temperatures below T_c of Pb (7.2 K), zero-bias current I_c starts to flow between the Pb pads. This is attributable to the induced spin-singlet wave function in Sr_2RuO_4 by the proximity effect. With further decreasing temperature below T_c of Sr_2RuO_4, however, I_c exhibits a truly remarkable, sharp drop, followed by a recovery increase at lowest temperatures. Such anomalous behavior has successfully been explained by the theories assuming the spin-triplet wave function for Sr_2RuO_4 [27-29].

(6) Observation of the spontaneous internal magnetic field below T_c, probed by muon spin rotation (μSR), indicates that the time-reversal symmetry (TRS) is broken in the superconducting state [30]. Examination of the allowed Cooper-pair symmetries in the D_{4h} crystal symmetry leads to the conclusion that the broken TRS most probably originates from the orbital part of the spin-triplet wave function.

(7) Neutron diffraction experiments revealed that the vortices form a square lattice in Sr_2RuO_4 in a wide region in the H-T phase diagram [31]. Comparing this result with the recent theory by Agterberg [32], we have deduced that the γ-sheet of the Fermi surface among the three cylindrical sheets α, β and γ [33] is mainly involved in the superconductivity.

(8) The material origin of the superconducting phase with $T_c = 3$ K has been clarified [34]. The doubling of T_c in Sr_2RuO_4 occurs at the interface of metallic Ru, introduced by eutectic solidification.

§5. Concluding Remarks

The two-dimensional Mott transition system $Ca_{2-x}Sr_xRuO_4$ exhibits a number of remarkable phenomena. It is hoped that nature of spin fluctuations is clarified by the precise characterization of the physical properties, especially near the magnetic unstable point at $x=0.5$.

It is now possible to discuss in detail the relevant spin-triplet wave function for the superconductivity of Sr_2RuO_4. The wave function compatible with all the existing experiments is expressed with the d-vector $d(k)=z(k_x \pm ik_y)$ [21], but the role of the orbital degeneracy [35] needs to be further clarified, especially by thermodynamic measurements at very low temperatures.

Acknowledgements

We wish to thank to Z.Q. Mao, S. NishiZaki, T. Ando, T. Akima, A.P. Mackenzie, M. Sigrist, and many other collaborators for their contributions. We are also grateful to T. Ishiguro and T. Fujita for their support and encouragement.

References

[1] Y. Maeno, Physica C **282-287** (1997) 206.
[2] T. Oguchi, Phys. Rev. B **51** (1995) 1385.
[3] I. Hase and Y. Nishihara, J. Phys. Soc. Jpn. **66** (1997) 3517.
[4] I.I. Mazin and D.J. Singh, Phys. Rev. B **56** (1998) 2556; D.J. Singh, J. Appl. Phys. **79** (1996) 4818.
[5] M. Braden, G. André, S. Nakatsuji, and Y. Maeno, to appear in Phys. Rev. B (1998) July 1 issue.
[6] K.M. Glassford and J.R. Chelikowsky, Phys. Rev. B **49** (1994) 7107.
[7] A. Kanbayashi, J. Phys. Soc. Jpn. **44** (1978) 108.
[8] I. H. Inoue *et al.*, Physica B **223&224** (1996) 516.
[9] S. Nakatsuji, S. Ikeda, and Y. Maeno, J. Phys. Soc. Jpn. **66** (1997) 1868.
[10] G. Cao, S. McCall, J.E. Craw, and R.P. Guertin, Phys. Rev. Lett. **78** (1997) 1751.
[11] T. Kiyama, K. Yoshimura, K. Kosuge, H. Michor, and G. Hilscher, J. Phys. Soc. Jpn. **67** (1998) 307.
[12] S. Ikeda, Y. Maeno, and T. Fujita, Phys. Rev. B **57** (1998) 978.
[13] G. Cao, S. McCall, and J. E. Crow, Phys. Rev. B **55** (1998) R672.
[14] Y. Maeno, H. Hashimoto, K. Yoshida, S. Nishizaki, T. Fujita, J. G. Bednorz, and F. Lichtenberg, Nature **372** (1994) 532.
[15] M. Braden, W. Reichardt, S. Nishizaki, Y. Mori, and Y. Maeno, Phys. Rev. B **57** (1998) 1236.
[16] M. Braden, W. Paulus, A. Cousson, P. Vigoureux, G. Heger, A. Goukassov, P. Bourges, and D. Petitgrand, Europhys. Lett. **25** (1994) 625.
[17] M.K. Crawford, M.A. Subramanian, R.L. Harlow, J.A. Fernandes-Baca, Z.R. Wang, and D.C. Johnston, Phys. Rev. B **49** (1994) 9198.
[18] Y. Maeno, K. Yoshida, H. Hashimoto, S. Nishizaki, S. Ikeda, M. Nohara, T. Fujita, A.P. Mackenzie, N.E. Hussey, J.G. Bednorz, and F. Lichtenberg, J. Phys. Soc. Jpn. **66** (1997) 1405.
[19] K. Ishida, H. Mukuda, Y. Kitaoka, K. Asayama, Z. Q. Mao, Y. Mori, and Y. Maeno, preprint (1998).
[20] T. Imai, A.W. Hunt, K.R. Thurber, and F.C. Chou, preprint (1998).
[21] T. M. Rice and M. Sigrist, J. Phys. Condens. Matt. **7** (1995) L643.
[22] G. Baskaran, Physica B **223&224** (1996) 490.

[23] A. P. Mackenzie, R. K. W. Haselwimmer, A. W. Tyler, Y. Mori, S. Nishizaki, and Y. Maeno, Phys. Rev. Lett. **80** (1998) 161.
[24] K, Ishida, Y. Kitaoka, K. Asayama, S. Ikeda, S. Nishizaki, Y. Maeno, K. Yoshida, and T. Fujita, Phys. Rev. B **56** (1997) R505.
[25] S. NishiZaki, Y. Maeno, S. Farner, S. Ikeda, and T. Fujita, J. Phys. Soc. Jpn. **67** (1998) 560.
[26] R. Jin, Yu. Zadorozhny, D. G. Schlom, Y. Mori, Y. Maeno, and Y. Liu, preprint (1998).
[27] M. Yamashiro, Y. Tanaka, and S. Kashiwaya, preprint (1998).
[28] C. Honerkamp and M. Sigrist, preprint (1998).
[29] M. Sigrist, C. Honerkamp, D. Agterberg, T.M. Rice, M.E. Zhitomirsky, and A. Furusaki, in this volume (1998).
[30] G. M. Luke, Y. Fudamoto, K. M. Kojima, M. I. Larkin, J. Merrin, B. Nachumi, Y. J. Uemura, Y. Maeno, Z. Q. Mao, Y. Mori, H. Nakamura, and M. Sigrist, to appear in Nature (1998) Aug. 6 issue.
[31] T. M. Riseman, P. G. Kealy, E. M. Forgan, A. P. Mackenzie, L. M. Galvin, A. W. Tyler, S. L. Lee, C. Ager, D. McK. Paul, C. M. Aegerter, R. Cubitt, Z. Q. Mao, S. Akima, and Y. Maeno, preprint (1998).
[32] D. F. Agterberg, Phys. Rev. Lett. **80** (1998) 5184.
[33] A. P. Mackenzie, S. Ikeda, Y. Maeno, T. Fujita, S. R. Julian, and G. G. Lonzarich, J. Phys. Soc. Jpn. **67** (1998) 385.
[34] Y. Maeno, T. Ando, Y. Mori, E. Ohmichi, S. Ikeda, S. NishiZaki, and S. Nakatsuji, preprint (1998).
[35] D. F. Agterberg, T. M. Rice, and M. Sigrist, Phys. Rev. Lett. **78** (1997) 3374.

Spin Triplet Superconductivity in Sr_2RuO_4 – A New Territory for Experimentalists and Theorists

M. Sigrist[1], C. Honerkamp[2], D. Agterberg[2], T.M. Rice[2], M.E. Zhitomirsky[3] and A. Furusaki[1]

[1] Yukawa Institute for Theoretical Physics, Kyoto University, Kyoto 606-8502, Japan
[2] Theoretische Physik, ETH-Hönggerberg, 8093 Zürich, Switzerland
[3] Department of Physics, University of Toronto, Canada M5S 1A7

Abstract. The symmetry of the possible superconducting states in Sr_2RuO_4 is analyzed. Based on this discussion we argue that recent μSR-experiments give evidence for a time reversal symmetry breaking odd-parity state. The anomalous temperature dependence of a Pb-Sr_2RuO_4-Pb-device is interpreted based on the assumption of odd-parity pairing. We conclude that the most likely pairing state has the structure, $\mathbf{f}(\mathbf{k}) = \hat{\mathbf{z}}(k_x \pm i k_y)$ with two-dimensional order parameter.

1. Introduction

Since more than a decade a large part of the research on superconductivity has been devoted to the so-called unconventional superconductors where mechanism and pairing symmetry are different from the standard electron-phonon interaction mediated case. These systems include heavy Fermion compounds, high-temperature superconductors and organic superconductors. Four years ago the superconductor Sr_2RuO_4 joined this group of strongly correlated electron systems [1]. A certain analogy to ^3He and the relation to ferromagnetic compounds led to the suggestion that here spin-triplet (odd-parity) superconductivity might be realized [2, 3]. Indeed in recent years a number of experimental studies have revealed the unconventional nature of the superconducting state. Probably the earliest indication came from the absence of a Hebel-Slichter peak in NQR [4]. A strong support for unconventional pairing is also provided by strong sensitivity of T_c on non-magnetic impurities [5].

In the following we will discuss in more detail two very recent experiments which give indirect evidence for odd-parity pairing. One is the observation of magnetic properties by μSR which indicate broken time reversal symmetry $\mathcal{T}$ in the superconducting state [6]. The other is the anomalous temperature dependence of the critical current in a Pb-Sr_2RuO_4-Pb device [7].

2. Symmetry analysis of the superconducting state

We first discuss the symmetry of the possible Cooper pairing channels in Sr_2RuO_4 which has the same layered perovskite crystal structure as La_2CuO_4, a parent compound of high-temperature superconductors. Like in the high-temperature superconductors, the electronic properties are dominated completely by the planes, in our case the RuO_2-planes. We find here three electron bands, crossing the Fermi level, based on the t_{2g}-orbitals of Ru^{4+} ($4d^4$-configuration). The carriers in the t_{2g}-orbitals propagate via π-hybridization with the intermediate O-$2p$-orbitals. Due to the symmetry of the orbitals there is no hybridization

Γ	$f_0(\mathbf{k})$	Γ	$\mathbf{f}(\mathbf{k})$
A_{1g}	1	A_{1u}	$\hat{\mathbf{x}}k_x + \hat{\mathbf{y}}k_y$
A_{2g}	$k_x k_y (k_x^2 - k_y^2)$	A_{2u}	$\hat{\mathbf{x}}k_y - \hat{\mathbf{y}}k_x$
B_{1g}	$k_x^2 - k_y^2$	B_{1u}	$\hat{\mathbf{x}}k_x - \hat{\mathbf{y}}k_y$
B_{2g}	$k_x k_y$	B_{2u}	$\hat{\mathbf{x}}k_y + \hat{\mathbf{y}}k_x$
E_g	-	E_u	$\{\hat{\mathbf{z}}k_x, \hat{\mathbf{z}}k_y\}$

Table 1: The basis pair wave functions of the irreducible representations Γ in the tetragonal point group for the d_{xy}-orbital. $\hat{\mathbf{x}}, \hat{\mathbf{y}}\hat{\mathbf{z}}$ denote the vector component of $\mathbf{f}$ along the corresponding directions, x, y, z respectively. The representations A_1, A_2, B_1, B_2 are one-dimensional and E is two-dimensional.

connecting the d_{xy}-orbital with the other two, d_{yz} and d_{zx}. The d_{xy}-orbital forms a band with tetragonal symmetry and with an electron-like Fermi surface. On the other hand, the d_{yz}- and the d_{zx}-orbital are equivalent and form quasi-one dimensional bands with carriers propagating prefered along the y- or x-direction, respectively. There is, however, hybridization between these two orbitals via next-nearest neighbor hopping which leads to two bands, one creating an electron-like and the other a hole-like Fermi surface as seen in de Haas-van Alphen experiments [5].

We consider now the pairing states associated with these orbitals. Since the RuO$_2$-planes are rather well separated yielding a nearly *two-dimensional* (strongly correlated) Fermi liquid, we will in the following restrict our discussion to intra-plane pairing and discard inter-plane pairing. Furthermore, we assume that spin-orbit coupling is sufficiently large so that spin and orbital degrees of freedom cannot be rotated independently (this is important only for the odd-parity states). The superconducting state is characterized by the mean field $F_{ss',\alpha} = \langle c_{-\mathbf{k}\alpha s} c_{\mathbf{k}\alpha s'} \rangle$ where α is the orbital index. In the even-parity (spin singlet) case this 2×2-matrix can be represented by a scalar function $f_{0\alpha}(\mathbf{k})$ ($\hat{F}_\alpha = i\sigma_y f_{0\alpha}(\mathbf{k})$), and for the odd-parity (spin triplet) case by a vector function $f_{\mu\alpha}(\mathbf{k})$ ($\hat{F}_\alpha(\mathbf{k}) = i\sum_{\mu=x,y,z} f_{\mu\alpha}(\mathbf{k})\sigma_\mu \sigma_y$).

Let us first consider the superconducting states in the d_{xy}-band which has full tetragonal symmetry D_{4h}. The superconducting states can then be classified according to the irreducible representations of D_{4h} (see Tab.I). For the even-parity states there are only candidates for the one-dimensional (non-degenerate) representations, while odd-parity states exist for all representations. The conventional s-wave channel corresponds to A_{1g}. The degenerate channel corresponding to E_u forms a two-dimensional space, $\mathbf{f}(\mathbf{k}) = \hat{\mathbf{z}}(\mathbf{k} \cdot \boldsymbol{\eta})$ where $\boldsymbol{\eta} = (\eta_x, \eta_y)$ is the two-dimensional complex order parameter. Not all combinations of the degenerate pairing states can be stable. Due to the crystal field anisotropy there are only three possible combinations (η_x, η_y) (see Tab.II) which can be obtained by analyzing the generalized Ginzburg-Landau theory, i.e. the free energy expansion in η_x and η_y (see Eq.(8)). Two of the states break the point group symmetry (tetragonal $\to$ orthorhombic) while the third state violates time reversal symmetry $\mathcal{T}$. This last state is the most stable one within a weak coupling

f(k)	broken symmetry
$\hat{z}(k_x \pm k_y)$	D_{4h}
$\hat{z}k_x, \hat{z}k_y$	D_{4h}
$\hat{z}(k_x \pm ik_y)$	$\mathcal{T}$

Table 2: The stable combinations of the pairing states in the two-dimensional representation E_u.

Γ	$f_0(\mathbf{k})$	Γ	$\mathbf{f}(\mathbf{k})$
A_{1g}	$1, k_x^2, k_y^2$	A_{1u}	$\hat{\mathbf{x}}k_x, \hat{\mathbf{y}}k_y$
A_{2g}	–	A_{2u}	$\hat{\mathbf{z}}k_y$
B_{1g}	$k_x k_y$	B_{1u}	$\hat{\mathbf{x}}k_y, \hat{\mathbf{y}}k_x$
B_{2g}	–	B_{2u}	$\hat{\mathbf{z}}k_x$

Table 3: The basis pair wave functions of the irreducible representations Γ in the orthorhombic point group for the d_{yz}-orbital.

approximation and is the two-dimensional analogue of the A-phase of superfluid ^{3}He.

The d_{yz}- and d_{zx}-orbital are related by 90°-rotation around the z-axis, so we concentrate on the d_{yz}-orbital only. This orbital on its own has only orthorhombic symmetry. Thus, the corresponding intra-orbit pairing states should be classified according to the point group D_{2h}. We neglect here the inter-orbit pairing states with the d_{zx}-orbital, because they are unimportant for the further discussion (note that there are no inter-orbit pairing states possible between d_{xy} and $d_{yz,zx}$-orbitals). The basic pairing states are listed in Tab.III according to the exclusively one-dimensional representations of D_{2h}. The analogous set of states is obtained for the d_{zx}-orbital by applying the 90°-rotation ($x \leftrightarrow y$). This implies that each pairing state for d_{yz} has a degenerate partner state in d_{zx}, e.g. the representation B_{2u} is $\hat{z}k_x$ in d_{yz} and $\hat{z}k_y$ in d_{zx}. The hybridization between the two orbitals establishes a certain phase relation between the two degenerate pairing states. Since the hybridized bands have tetragonal symmetry, the combined states have analogous symmetry properties as those of the d_{xy}-orbital (Tab.I,II). Interestingly the hybridization stabilizes here the $\mathcal{T}$-violating state for the example B_{2u} identical to the situation with the E_u-state for the d_{xy}-orbital.

Our discussion shows that Sr$_2$RuO$_4$ decays basically into two subsystems concerning superconductivity. They are only weakly coupled by pair scattering as pointed out by Agterberg et al. [8, 9]. Either the d_{xy}-orbital or the combination of d_{yz}/d_{zx}-orbital are dominating the superconductivity and the subdominant orbital(s) only participates via an induced pairing amplitude, i.e. a kind of "proximity effect in k-space".

3. Time reversal symmetry breaking states

A superconducting state which violates time reversal symmetry $\mathcal{T}$ consists of at least two pairing components. In this section we briefly analyze various possible $\mathcal{T}$-violating states. We start with the *even parity states*. Here $\mathcal{T}$-violating states are exclusively combinations of pairing states belonging to different representations, e.g. $f_0(\mathbf{k}) = a + ib(k_x^2 - k_y^2)$, commonly called "$s + id$"-wave state ($A_{1g} \oplus B_{1g}$). Because the two components are not degenerate the appearance of this combination requires, in general, multiple phase transitions. This means that immediately below the onset of superconductivity only the dominant order parameter occurs, while the subdominant component becomes finite only below a further transition leading then to the violation of time reversal symmetry. Note that this type of consecutive superconducting transitions were observed in the heavy Fermion compounds UPt$_3$ and U$_{1-x}$Th$_x$Be$_{13}$ [10, 11].

Concerning the odd-parity states we have already identified one $\mathcal{T}$-violating state which has the form $\mathbf{f}(\mathbf{k}) = \hat{\mathbf{z}}(k_x \pm i k_y)$. This state consists of two *degenerate* components (E_u) and, therefore, is a pure $\mathcal{T}$-violating state that appears after a single transition at the onset of superconductivity. This state is a so-called *unitary* state, i.e. $\mathbf{f}(\mathbf{k}) \times \mathbf{f}^*(\mathbf{k}) = 0$. *Non-unitary* odd-parity states can not be found in our classification in any of the orbitals. Such states, however, can be obtained here, similar to the $\mathcal{T}$-violating even-parity states, by combining two states belonging to different one-dimensional representations, e.g. $\mathbf{f}(\mathbf{k}) = (\hat{\mathbf{x}} \mp i\hat{\mathbf{y}})(k_x \pm i k_y)$ (from $A_{1u} \oplus A_{2u}$) [12]. Also here the same argument as above applies, that multiple transitions are required to reach this state. This type of state has close similarity with the A_1-phase in ^{3}He which is stabilized under an external field. It is, however, difficult to stabilize this type of state in a superconductor [9, 13].

Let us now turn to the experimental status concerning phase transition and time reversal symmetry. There is no sign of an additional phase transition within the superconducting state. On the other hand, there is clear evidence for $\mathcal{T}$-violation, starting at the onset of superconductivity, observed by zero-field relaxation in μSR measurements [6]. Muon spins are very suitable to detect local internal magnetic fields of a material. In the $\mathcal{T}$-violating superconducting state spontaneous static supercurrents and magnetic field distributions are generated in the vicinity of inhomogeneities of the superconducting order parameter, i.e. at impurities, lattice defects, domain walls, interfaces or surfaces [10]. This field creates a random field distribution for the muon spins and yields a specific form of the relaxation of the spin polarization. The relaxation rate is directly proportional to the width of the field distribution which is in turn roughly proportional to $i(\eta_1^* \eta_2 - \eta_1 \eta_2^*)$ for the two-component order parameter. Thus, the relaxation rate should increase below the onset to a $\mathcal{T}$-violating state. The observation of this behavior at T_c of Sr$_2$RuO$_4$ suggests $\mathbf{f}(\mathbf{k}) = \hat{\mathbf{z}}(k_x \pm i k_y)$ as the strongest candidate for superconductivity in this material. Based on our symmetry analysis this provides additional indirect evidence for spin-triplet odd-parity pairing.

4. Phase sensitive probe

In a recent experiment on a device consisting of two Pb-films, attached on top of a Sr_2RuO_4 single crystal and separated by a tiny gap, a very anomalous temperature dependence of the critical current from one Pb-film to the other via the single crystal was observed [7]. The Pb-films have a transition temperature T_{cPb} considerably higher than Sr_2RuO_4 (T_{cS}). In the temperature range $T_{cPb} > T > T_{cS}$ the critical current increases monotonically because a proximity-induced s-wave component in Sr_2RuO_4 leads to a growing superconducting coupling between the two Pb-films. With the onset of superconductivity in the single crystal the critical current drops abruptly. It recovers, however, after a similarly sharp anomaly at a slightly lower temperature and increases then monotonically.

Yamashiro et al.[14] and Honerkamp et al.[15] gave an interpretation of this phenomenon based on the assumption that Sr_2RuO_4 realizes an odd-parity superconducting state. It was proposed many years ago that an odd parity superconductor sandwiched between two conventional superconductors should have an intrinsic π-phase shift, if used like a Josephson device [16]. It can be shown that the geometry of the present device satisfies the condition that a similar π-phase shift for the Josephson effect between the two s-wave states of Pb and the p-wave state of Sr_2RuO_4 is generated. Within a Ginzburg-Landau treatment we obtain for the two channels the following contributions,

$$I(\phi, T) = I_s(T) \sin\phi - I_p(T) \text{sign}(\sin\frac{\phi}{2}) \cos\frac{\phi}{2} \qquad (1)$$

where ϕ is the order parameter phase difference between the two Pb-films, and $I_s(T) > 0$ for $T < T_{cPb}$ (s-wave component) and $I_p(T) > 0$ for $T < T_{cS}$ (p-wave component), both monotonically growing [15]. The different phase dependences and signs lead to a sharp drop of the maximal current I_{max} immediately below T_{cS}. At a lower temperature the maximizing phase ϕ_{max} ($I_{max} = I(\phi_{max})$) changes discontinuously leading to a further abrupt change in the temperature dependence of I_{max} (see Fig.1). This behavior closely resembles that seen in the experiment [7].

This experiment is basically a phase sensitive probe on the superconducting state of Sr_2RuO_4, although it seems not particularly designed in that way. The behavior seen is the result of a subtle competition between the two channels, coupling the two Pb-films, which incorporate different intrinsic phase shifts. With varying temperature the form of the Josephson current-phase relation is modified and influences critical current of the device in a drastic way. The change of character should also have observable effect in other properties. In particular, we would like to mention two effects which could be used to verify the above scenario.

The first test is based on the ac-Josephson effect for the device. An applied voltage V leads to a time-dependent phase $\phi = \omega t$ with $\omega = 2eV/\hbar$. Decomposing the Josephson current into the basic Fourier components we obtain,

$$I(t, T) = [I_s(T) - \frac{8}{3}I_p(T)]\sin\omega t - \sum_{n \geq 2} \frac{8I_p(T)n}{4n^2 - 1}\sin(n\omega t). \qquad (2)$$

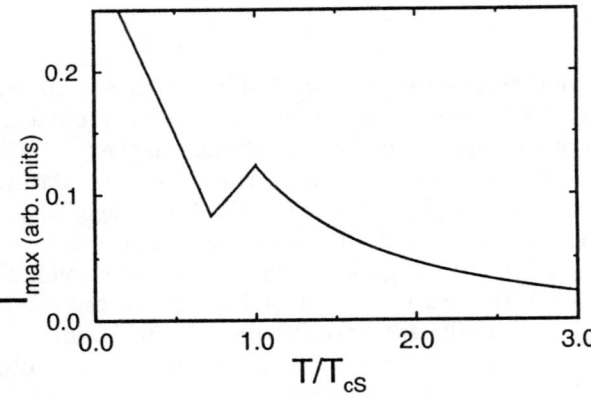

Fig.1: Anomalous temperature dependence of the maximal current calculated by using Eq.(1) [15].

The basic component shows an anomaly at T_{cS} and a sign change so that it vanishes at the temperature where $3I_s(T) = 8I_p(T)$. All the other components have a monotonic temperature dependence.

The second effect is connected with the fact that the phase ϕ_{min}, minimizing the energy of the device

$$E(\phi) = -\frac{\hbar}{2e}\left[I_s(T)\cos\phi - 2I_p(T)\left|\sin\frac{\phi}{2}\right|\right], \qquad (3)$$

turns away from the standard values $\phi = 2\pi n$ continuously below T_{cS},

$$\left|\sin\frac{\phi_{min}}{2}\right| = \frac{I_p(T)}{4I_s(T)}. \qquad (4)$$

This yields a phase difference between the two Pb-films. Hence, if we connect the two films by another conventional superconductor forming a closed loop, then a spontaneous current can flow for large enough inductance of the whole superconducting loop. This is a consequence of the fact that the Pb-Sr$_2$RuO$_4$-Pb device violates time reversal symmetry, since the lowest energy state is given for a phase $\phi_{min} \neq \pi(0, \pm 1, \pm 2, ...)$ with two-fold degeneracy.

5. Phenomenological theory of Sr$_2$RuO$_4$

The realization of unconventional superconductivity in Sr$_2$RuO$_4$ may be considered as an established fact. In particular, we have now a number of experiments which suggest that the superconducting state has odd parity as previously suggested on theoretical grounds [2]. Two of those we have introduced above. The result of our symmetry analysis is that the state which is compatible with all experiments has the form

$$\mathbf{f}(\mathbf{k}) = \hat{\mathbf{z}}(k_x \pm ik_y) \qquad (5)$$

and is $\mathcal{T}$-violating. Note that this is the only state in our classification which profits from the spin fluctuation feedback effect [17].

Further experimental evidence for spin-triplet pairing is coming from recent Knight shift measurements of the spin susceptibility based on ^{17}O-NMR [18]. For the state in Eq.(5) we expect that the spin susceptibility behaves like

$$\chi(T) = \chi_{\text{Pauli}} \begin{cases} Y(T) & \mathbf{H}_{ex} \parallel \hat{z} \\ 1 & \mathbf{H}_{ex} \perp \hat{z} \end{cases} \tag{6}$$

where $Y(T)$ is the corresponding Yosida function [2]. The present NMR data demonstrate that there is no change of $\chi(T)$ in any direction of the field within the basal plane [18]. The field direction parallel to the z-axis is not accessible because of the rather low critical field H_{c2}. Hence, these data are, at least, consistent with the expected behavior in Eq.(6).

The above superconducting state can be described by a two-component order parameter, $\boldsymbol{\eta} = (\eta_x, \eta_y)$,

$$\mathbf{f}(\mathbf{k}) = \hat{z}(\mathbf{k} \cdot \boldsymbol{\eta}) \tag{7}$$

which obeys the Ginzburg-Landau free energy functional,

$$F = \int d^3r \left[a(T - T_c)|\boldsymbol{\eta}|^2 + b_1|\boldsymbol{\eta}|^4 + \frac{b_2}{2}(\eta_x^{*2}\eta_y^2 + \eta_x^2\eta_y^{*2}) + b_3|\eta_x|^2|\eta_y|^2 \right.$$

$$+ K_1(|D_x\eta_x|^2 + |D_y\eta_y|^2) + K_2(|D_x\eta_y|^2 + |D_y\eta_x|^2)$$

$$+ K_3((D_x\eta_x)^*(D_y\eta_y) + c.c.) + K_4((D_x\eta_y)^*(D_y\eta_x) + c.c.)$$

$$\left. + K_5(|D_z\eta_x|^2 + |D_z\eta_y|^2) + \frac{1}{8\pi}(\nabla \times \mathbf{A})^2 \right] \tag{8}$$

where a, b_i and K_i are real coefficient and $\mathbf{D} = \nabla - 2ei\mathbf{A}/\hbar c$ enters into the gauge invariant gradient terms. In order to stabilize the $\mathcal{T}$-violating state (Eq.(5)) we have to choose $b_2 > b_3 > 0$ and $b_2 > 0$.

The fact that two order parameter components appear leads to the possibility of various extraordinary structures in the superconducting state. Note that the superconducting phase is two-fold degenerate. Therefore, in general, there will be domains formed separated by domain walls. The structure of domain walls have been investigated in detail in the past [10, 19, 20]. Other structures are vortices where the core is split into two separate singularities of each order parameter component [10, 21]. It is also possible to generate vortices with fractional flux quanta [20]. We may ask whether further unusual structures are possible for this type of order parameter. Multi-component order parameters also lead to various collective modes which can couple to external probes such as ultrasound or microwave. The above Ginzburg-Landau theory has also been recently successfully applied to the problem of the vortex lattice structure [22]. It was shown that the vortices arrange in a square lattice structure as observed in recent neutron scattering experiments [23].

In conclusion, we find that it is very likely that the superconductivity in Sr_2RuO_4 is due to a two-component order parameter. From a microscopic point of view the role of the different orbitals and their interplay is a very interesting aspect of this system. This was partially discussed in the context of the orbital dependent superconductivity [8]. The study of the vortex lattice suggests that the d_{xy}-orbital is dominating the superconductivity [22, 23]. It remains a challenging problem to understand the mechanisms leading to this behavior and their implications on various physical properties. The fact that we deal here with a (strongly correlated) Fermi liquid is certainly a helpful starting point. Sr_2RuO_4 provides an interesting example where we might understand more about spin triplet superconductivity than in any of the previously known systems of this type.

Acknowledgement: We are grateful to Y. Maeno and his collaborators for many stimulating discussions. This work was financially supported by the Swiss Nationalfonds and Japanese Ministry of Education, Sciene and Culture and the Japanese Society for Promotion of Science.

References

[1] Y. Maeno, Physica C **282 - 287** (1997), 206.

[2] T.M. Rice and M. Sigrist, J. Phys. C: Condens. Matter **7** (1995), L643.

[3] G. Baskaran, Physica B **223 & 224** (1996), 490.

[4] K. Ishida, Y. Kitaoka, K. Asayama, S. Ikdea, S. Nishizaki, Y. Maeno, K. Yoshida and T. Fujita, Phys. Rev. B **56** (1997), R505.

[5] A.P. Mackenzie, R.K.W. Haselwimmer, A.W. Tyler, G.G. Lonzarich, Y. Mori, S. Nishizaki and Y. Maeno, Phys. Rev. Lett. **80** (1998), 161.

[6] G.M. Luke, Y. Fudamoto, K.M. Kojima, M.I. Larkin, J. Merrin, B. Nachumi, Y.J. Uemura, Y. Maeno, Z.Q. Mao, Y. Mori, H. Nakamura and M. Sigrist, to be published in Nature.

[7] R. Jin, Y. Zadorozhny, Y. Liu, Y. Mori, Y. Maeno, D.G. Schlom and F. Lichtenberg, preprint.

[8] D.F. Agterberg, T.M. Rice and M. Sigrist, Phys. Rev. Lett. **78**, 3374 (1997).

[9] I.I. Mazin and D.J. Singh, Phys. Rev. Lett. **79**, 733 (1997).

[10] M. Sigrist and K. Ueda, Rev. Phys. Mod. **63** (1991), 239.

[11] R.H. Heffner and M.R. Normand, Comments on Condensed Matter Physics **17**, 361 (1996).

[12] K. Machida, M. Ozaki and T. Ohmi, J. Phys. Soc. Jpn. **65** (1996), 3720; M. Sigrist and M.E. Zhitomirsky, J. Phys. Soc. Jpn. **65** (1996), 3452.

[13] T. Sugiyama and T. Ohmi, J. Phys. Soc. Jpn. **64**, 2746 (1995).

[14] M. Yamashiro, Y. Tanaka and S. Kashiwaya, preprint.

[15] C. Honerkamp and M. Sigrist, to be published in Prog. Theor. Phys..

[16] V.D. Geshkenbein, A.I. Larkin and A. Barone, Phys. Rev. B **36** (1987), 235.

[17] M. Sigrist and T.M. Rice, *Current Topics in Physics*, Proceedings of the "Inauguration Conference of the Asia-Pacific Center for Theoretical Physics, June 1996", edited by Y.M. Cho, J.B. Hong and C.N. Yang, Seoul, Korea, World Scientific (1998).

[18] K. Ishida, H. Mukuda, Y. Kitaoka,K. Asayama, Z.Q. Mao, Y. Mori and Y. Maeno, preprint.

[19] G.E. Volovik and L.P. Gor'kov, Zh. Eksp.Teor. Fiz. **39**, 550 (1984) [Sov. Phys. JETP **39**, 674 (1984)].

[20] M. Sigrist, T.M. Rice and K. Ueda, Phys. Rev. Lett. **63**, 1727 (1989).

[21] T.A. Tokuyasu, D.W. Hess and J.A. Sauls, Phys. Rev. B**41**, 8891 (1990).

[22] D.F. Agterberg, Phys. Rev. Lett. **80**, 5184 (1998).

[23] T. M. Riseman, P. G. Kealey, E. M. Forgan, A. P. Mackenzie, L. M. Galvin, A. W. Tyler, S. L. Lee, C. Ager, D. McK Paul, C. M. Aegerter, R. Cubitt, Z. Q. Mao, S. Akima, and Y. Maeno, preprint.

Concluding Remarks

It is a great pleasure for me to be attending with you the last of the Taniguchi Symposia on Solid State Physics. It is a great shame that there will be no more such conferences, for I have rarely experienced a week of scientific exchanges as productive as this one. However, since Solid State Physics in Japan is now terrifyingly strong, at least as seen from the other side of the ocean, perhaps it has come to pass that Mr. Taniguchi's work is done. From Professor Akimitsu's discovery of superconductivity in a cuprate chain compound to the discovery of simultaneous superconducting and antiferromagnetic order in LSCO to Professor Nagaosa's work relating the quantum mechanics of these materials to the Strong Interactions, something of great interest to me personally, one sees world leadership in our field being seized by the young physicists steadfastly supported by Mr. Taniguchi and others of his generation. I am honored to be among you.

The subject matter of this conference, highly-correlated electronic materials, is the cutting edge of quantum physics in our time, for we now know that the ordinary laws of quantum mechanics do unpredictable things when the number of particles is large and that these can be discovered only by experimentally studying quantum systems whose behavior is not understood. Solids are the only systems in which such experiments are practical. It is, of course, possible that all possible quantum states of matter there ever could be have already been discovered. But I do not think so. I believe there are more, and that the fractional quantum Hall discovery, which won the Nobel Prize for Physics this year, was not a fluke but rather the tip of the iceberg, a foretaste of many amazing things waiting to be discovered in highly-correlated electronic systems, if only we have the patience to look for them and the eyes to see.

Many of you are aware that at a previous IBM meeting in the Keidanren Guest House on Mount Fuji I brought my Mother along, as she had never been to Japan. On the way home across the Pacific she remarked that she understood why I attended so many conferences. I asked her why, as I did not understand this myself. She said, "Your work is very lonely. You need to remind each other that you are important." So it is. As we leave this beautiful place where pearls grow quietly let us remember Mr. Taniguchi's faith in the scientific process and in us, and take courage from it. He reminded us again and again that what we do is important. Let us honor his memory by doing the same.

Thank you all for making this a wonderful conference.

Stanford University *Robert B. Laughlin*
Department of Physics
Stanford, CA 94305, USA

Good Bye

by R.B. Laughlin

List of Contributors

A

Agterberg, D. 323
Akimitsu, J. 289, 299
Assaad, F.F. 120
Azuma, M. 279

B

Birgeneau, R.J. 182

C

Campuzano, J.C. 152

D

Ding, H. 152

E

Eisaki, H. 111, 163, 289, 299
Endoh, Y. 182

F

Fang, Z. 34
Fiebig, M. 93
Fujimori, A. 111
Fujino, H. 289
Fujita, M. 182
Fukase, T. 182
Fukuda, T. 69
Fukuyama, H. 3, 231
Furukawa, N. 221
Furusaki, A. 323

G

Goodenough, J.B. 9
Greven, M. 182

H

Hanamura, E. 95
Hiroi, M. 289
Hiroi, Z. 279
Hirota, K. 182
Honerkamp, C. 323
Hosoya, S. 182

I

Ichikawa, N. 163
Ikeda, S. 313
Imada, M. 120
Ino, A. 111
Ishihara, S. 84
Ishikawa, T. 55

K

Kakurai, K. 289
Kanamori, J. 3, 19
Kanno, R. 279
Kastner, M.A. 182
Katano, S. 289
Katsufuji, T. 111
Kawasaki, S. 279
Keimer, B. 173
Kim, C. 111
Kim, Y.M. 182
Kimura, H. 69, 182
Kimura, T. 55
Kishio, K. 111
Kitaoka, Y. 219
Kobayashi, N. 289
Kohno, H. 231
Krishana, K. 202
Kurahashi, K. 182

L

Laughlin, R.B.   332
Lee, C.H.   182
Lee, P.A.   241
Lee, S.H.   182
Lee, Y.S.   182

M

Maekawa, S.   84, 136
Maeno, Y.   313
Magishi, K.   299
Matsumoto, S.   299
Matsushita, H.   182
Mito, T.   299
Miyake, K.   267
Mizokawa, T.   111
Môri, N.   289
Motome, Y.   120
Motoyama, N.   289, 299

N

Nagaosa, N.   3, 250
Nagata, T.   289, 299
Nakanishi, T.   289
Nakatsuji, S.   313
Narikiyo, O.   267
Nishi, M.   289
Noda, T.   163
Norman, M.R.   152

O

Ogata, M.   212
Ohsugi, S.   299
Okamoto, N.   69
Ong, N.P.   202

R

Randeria, M.   152
Rice, T.M.   221, 323

S

Sato, M.   192
Sera, M.   289
Shen, Z.-X.   111, 144
Shiba, H.   45
Shiina, R.   45
Shirane, G.   182
Sigrist, M.   323
Solovyev, I.V.   34
Suzuki, T.   182

T

Taguchi, Y.   111
Takahashi, H.   289
Takano, M.   279
Takeda, T.   279
Tanabe, Y.   95
Terakura, K.   34
Tokura, Y.   55, 111
Tsunetsugu, H.   120

U

Uchida, S.   111, 163, 289, 299
Ueki, S.   182

W

Wakimoto, S.   182

X

Xu, Z.A.   202

Y

Yamada, K.   182
Yoshida, T.   111

Z

Zhang, S.-C.   260
Zhang, Y.   202
Zhitomirsky, M.E.   323
Zhou, J.-S.   9

Springer Series in Solid-State Sciences

Editors: M. Cardona P. Fulde K. von Klitzing H.-J. Queisser

1 **Principles of Magnetic Resonance**
   3rd Edition By C. P. Slichter
2 **Introduction to Solid-State Theory**
   By O. Madelung
3 **Dynamical Scattering of X-Rays in Crystals** By Z. G. Pinsker
4 **Inelastic Electron Tunneling Spectroscopy**
   Editor: T. Wolfram
5 **Fundamentals of Crystal Growth I**
   Macroscopic Equilibrium and Transport Concepts
   By F. E. Rosenberger
6 **Magnetic Flux Structures in Superconductors** By R. P. Huebener
7 **Green's Functions in Quantum Physics**
   2nd Edition
   By E. N. Economou
8 **Solitons and Condensed Matter Physics**
   Editors: A. R. Bishop and T. Schneider
9 **Photoferroelectrics** By V. M. Fridkin
10 **Phonon Dispersion Relations in Insulators** By H. Bilz and W. Kress
11 **Electron Transport in Compound Semiconductors** By B. R. Nag
12 **The Physics of Elementary Excitations**
   By S. Nakajima, Y. Toyozawa, and R. Abe
13 **The Physics of Selenium and Tellurium**
   Editors: E. Gerlach and P. Grosse
14 **Magnetic Bubble Technology** 2nd Edition
   By A. H. Eschenfelder
15 **Modern Crystallography I**
   Fundamentals of Crystals
   Symmetry, and Methods of Structural Crystallography
   2nd Edition
   By B. K. Vainshtein
16 **Organic Molecular Crystals**
   Their Electronic States By E. A. Silinsh
17 **The Theory of Magnetism I**
   Statics and Dynamics
   By D. C. Mattis
18 **Relaxation of Elementary Excitations**
   Editors: R. Kubo and E. Hanamura
19 **Solitons** Mathematical Methods for Physicists
   By. G. Eilenberger
20 **Theory of Nonlinear Lattices**
   2nd Edition By M. Toda
21 **Modern Crystallography II**
   Structure of Crystals 2nd Edition
   By B. K. Vainshtein, V. L. Indenbom, and V. M. Fridkin
22 **Point Defects in Semiconductors I**
   Theoretical Aspects
   By M. Lannoo and J. Bourgoin
23 **Physics in One Dimension**
   Editors: J. Bernasconi and T. Schneider
24 **Physics in High Magnetics Fields**
   Editors: S. Chikazumi and N. Miura
25 **Fundamental Physics of Amorphous Semiconductors** Editor: F. Yonezawa
26 **Elastic Media with Microstructure I**
   One-Dimensional Models By I. A. Kunin
27 **Superconductivity of Transition Metals**
   Their Alloys and Compounds
   By S. V. Vonsovsky, Yu. A. Izyumov, and E. Z. Kurmaev
28 **The Structure and Properties of Matter**
   Editor: T. Matsubara
29 **Electron Correlation and Magnetism in Narrow-Band Systems** Editor: T. Moriya
30 **Statistical Physics I** Equilibrium Statistical Mechanics 2nd Edition
   By M. Toda, R. Kubo, N. Saito
31 **Statistical Physics II** Nonequilibrium Statistical Mechanics 2nd Edition
   By R. Kubo, M. Toda, N. Hashitsume
32 **Quantum Theory of Magnetism**
   2nd Edition By R. M. White
33 **Mixed Crystals** By A. I. Kitaigorodsky
34 **Phonons: Theory and Experiments I**
   Lattice Dynamics and Models of Interatomic Forces By P. Brüesch
35 **Point Defects in Semiconductors II**
   Experimental Aspects
   By J. Bourgoin and M. Lannoo
36 **Modern Crystallography III**
   Crystal Growth
   By A. A. Chernov
37 **Modern Chrystallography IV**
   Physical Properties of Crystals
   Editor: L. A. Shuvalov
38 **Physics of Intercalation Compounds**
   Editors: L. Pietronero and E. Tosatti
39 **Anderson Localization**
   Editors: Y. Nagaoka and H. Fukuyama
40 **Semiconductor Physics** An Introduction
   6th Edition By K. Seeger
41 **The LMTO Method**
   Muffin-Tin Orbitals and Electronic Structure
   By H. L. Skriver
42 **Crystal Optics with Spatial Dispersion, and Excitons** 2nd Edition
   By V. M. Agranovich and V. L. Ginzburg
43 **Structure Analysis of Point Defects in Solids**
   An Introduction to Multiple Magnetic Resonance Spectroscopy
   By J.-M. Spaeth, J. R. Niklas, and R. H. Bartram
44 **Elastic Media with Microstructure II**
   Three-Dimensional Models By I. A. Kunin
45 **Electronic Properties of Doped Semiconductors**
   By B. I. Shklovskii and A. L. Efros
46 **Topological Disorder in Condensed Matter**
   Editors: F. Yonezawa and T. Ninomiya

Springer Series in Solid-State Sciences
Editors: M. Cardona P. Fulde K. von Klitzing H.-J. Queisser

47 **Statics and Dynamics of Nonlinear Systems**
Editors: G. Benedek, H. Bilz, and R. Zeyher

48 **Magnetic Phase Transitions**
Editors: M. Ausloos and R. J. Elliott

49 **Organic Molecular Aggregates**
Electronic Excitation and Interaction Processes
Editors: P. Reineker, H. Haken, and H. C. Wolf

50 **Multiple Diffraction of X-Rays in Crystals**
By Shih-Lin Chang

51 **Phonon Scattering in Condensed Matter**
Editors: W. Eisenmenger, K. Laßmann, and S. Döttinger

52 **Superconductivity in Magnetic and Exotic Materials** Editors: T. Matsubara and A. Kotani

53 **Two-Dimensional Systems, Heterostructures, and Superlattices**
Editors: G. Bauer, F. Kuchar, and H. Heinrich

54 **Magnetic Excitations and Fluctuations**
Editors: S. W. Lovesey, U. Balucani, F. Borsa, and V. Tognetti

55 **The Theory of Magnetism II** Thermodynamics and Statistical Mechanics By D. C. Mattis

56 **Spin Fluctuations in Itinerant Electron Magnetism** By T. Moriya

57 **Polycrystalline Semiconductors**
Physical Properties and Applications
Editor: G. Harbeke

58 **The Recursion Method and Its Applications**
Editors: D. G. Pettifor and D. L. Weaire

59 **Dynamical Processes and Ordering on Solid Surfaces** Editors: A. Yoshimori and M. Tsukada

60 **Excitonic Processes in Solids**
By M. Ueta, H. Kanzaki, K. Kobayashi, Y. Toyozawa, and E. Hanamura

61 **Localization, Interaction, and Transport Phenomena** Editors: B. Kramer, G. Bergmann, and Y. Bruynseraede

62 **Theory of Heavy Fermions and Valence Fluctuations** Editors: T. Kasuya and T. Saso

63 **Electronic Properties of Polymers and Related Compounds**
Editors: H. Kuzmany, M. Mehring, and S. Roth

64 **Symmetries in Physics** Group Theory Applied to Physical Problems 2nd Edition
By W. Ludwig and C. Falter

65 **Phonons: Theory and Experiments II**
Experiments and Interpretation of Experimental Results By P. Brüesch

66 **Phonons: Theory and Experiments III**
Phenomena Related to Phonons
By P. Brüesch

67 **Two-Dimensional Systems: Physics and New Devices**
Editors: G. Bauer, F. Kuchar, and H. Heinrich

68 **Phonon Scattering in Condensed Matter V**
Editors: A. C. Anderson and J. P. Wolfe

69 **Nonlinearity in Condensed Matter**
Editors: A. R. Bishop, D. K. Campbell, P. Kumar, and S. E. Trullinger

70 **From Hamiltonians to Phase Diagrams**
The Electronic and Statistical-Mechanical Theory of sp-Bonded Metals and Alloys By J. Hafner

71 **High Magnetic Fields in Semiconductor Physics**
Editor: G. Landwehr

72 **One-Dimensional Conductors**
By S. Kagoshima, H. Nagasawa, and T. Sambongi

73 **Quantum Solid-State Physics**
Editors: S. V. Vonsovsky and M. I. Katsnelson

74 **Quantum Monte Carlo Methods in Equilibrium and Nonequilibrium Systems** Editor: M. Suzuki

75 **Electronic Structure and Optical Properties of Semiconductors** 2nd Edition
By M. L. Cohen and J. R. Chelikowsky

76 **Electronic Properties of Conjugated Polymers**
Editors: H. Kuzmany, M. Mehring, and S. Roth

77 **Fermi Surface Effects**
Editors: J. Kondo and A. Yoshimori

78 **Group Theory and Its Applications in Physics**
2nd Edition
By T. Inui, Y. Tanabe, and Y. Onodera

79 **Elementary Excitations in Quantum Fluids**
Editors: K. Ohbayashi and M. Watabe

80 **Monte Carlo Simulation in Statistical Physics**
An Introduction 3rd Edition
By K. Binder and D. W. Heermann

81 **Core-Level Spectroscopy in Condensed Systems**
Editors: J. Kanamori and A. Kotani

82 **Photoelectron Spectroscopy**
Principle and Applications 2nd Edition
By S. Hüfner

83 **Physics and Technology of Submicron Structures**
Editors: H. Heinrich, G. Bauer, and F. Kuchar

84 **Beyond the Crystalline State** An Emerging Perspective By G. Venkataraman, D. Sahoo, and V. Balakrishnan

85 **The Quantum Hall Effects**
Fractional and Integral 2nd Edition
By T. Chakraborty and P. Pietiläinen

86 **The Quantum Statistics of Dynamic Processes**
By E. Fick and G. Sauermann

87 **High Magnetic Fields in Semiconductor Physics II**
Transport and Optics Editor: G. Landwehr

88 **Organic Superconductors** 2nd Edition
By T. Ishiguro, K. Yamaji, and G. Saito

89 **Strong Correlation and Superconductivity**
Editors: H. Fukuyama, S. Maekawa, and A. P. Malozemoff

Springer Series in Solid-State Sciences

Editors: M. Cardona P. Fulde K. von Klitzing H.-J. Queisser

Managing Editor: H. K. V. Lotsch

90 **Earlier and Recent Aspects of Superconductivity**
Editors: J. G. Bednorz and K. A. Müller

91 **Electronic Properties of Conjugated Polymers III** Basic Models and Applications
Editors: H. Kuzmany, M. Mehring, and S. Roth

92 **Physics and Engineering Applications of Magnetism** Editors: Y. Ishikawa and N. Miura

93 **Quasicrystals** Editors: T. Fujiwara and T. Ogawa

94 **Electronic Conduction in Oxides**
By N. Tsuda, K. Nasu, A. Yanase, and K. Siratori

95 **Electronic Materials**
A New Era in Materials Science
Editors: J. R. Chelikowsky and A. Franciosi

96 **Electron Liquids** 2nd Edition By A. Isihara

97 **Localization and Confinement of Electrons in Semiconductors**
Editors: F. Kuchar, H. Heinrich, and G. Bauer

98 **Magnetism and the Electronic Structure of Crystals** By V. A. Gubanov, A. I. Liechtenstein, and A. V. Postnikov

99 **Electronic Properties of High-T_c Superconductors and Related Compounds**
Editors: H. Kuzmany, M. Mehring, and J. Fink

100 **Electron Correlations in Molecules and Solids** 3rd Edition By P. Fulde

101 **High Magnetic Fields in Semiconductor Physics III** Quantum Hall Effect, Transport and Optics By G. Landwehr

102 **Conjugated Conducting Polymers**
Editor: H. Kiess

103 **Molecular Dynamics Simulations**
Editor: F. Yonezawa

104 **Products of Random Matrices**
in Statistical Physics By A. Crisanti, G. Paladin, and A. Vulpiani

105 **Self-Trapped Excitons**
2nd Edition By K. S. Song and R. T. Williams

106 **Physics of High-Temperature Superconductors**
Editors: S. Maekawa and M. Sato

107 **Electronic Properties of Polymers**
Orientation and Dimensionality of Conjugated Systems Editors: H. Kuzmany, M. Mehring, and S. Roth

108 **Site Symmetry in Crystals**
Theory and Applications 2nd Edition
By R. A. Evarestov and V. P. Smirnov

109 **Transport Phenomena in Mesoscopic Systems** Editors: H. Fukuyama and T. Ando

110 **Superlattices and Other Heterostructures**
Symmetry and Optical Phenomena 2nd Edition
By E. L. Ivchenko and G. E. Pikus

111 **Low-Dimensional Electronic Systems**
New Concepts
Editors: G. Bauer, F. Kuchar, and H. Heinrich

112 **Phonon Scattering in Condensed Matter VII**
Editors: M. Meissner and R. O. Pohl

113 **Electronic Properties of High-T_c Superconductors**
Editors: H. Kuzmany, M. Mehring, and J. Fink

114 **Interatomic Potential and Structural Stability**
Editors: K. Terakura and H. Akai

115 **Ultrafast Spectroscopy of Semiconductors and Semiconductor Nanostructures**
2nd Edition By J. Shah

116 **Electron Spectrum of Gapless Semiconductors**
By J. M. Tsidilkovski

117 **Electronic Properties of Fullerenes**
Editors: H. Kuzmany, J. Fink, M. Mehring, and S. Roth

118 **Correlation Effects in Low-Dimensional Electron Systems**
Editors: A. Okiji and N. Kawakami

119 **Spectroscopy of Mott Insulators and Correlated Metals**
Editors: A. Fujimori and Y. Tokura

120 **Optical Properties of III–V Semiconductors**
The Influence of Multi-Valley Band Structures
By H. Kalt

121 **Elementary Processes in Excitations and Reactions on Solid Surfaces**
Editors: A. Okiji, H. Kasai, and K. Makoshi

122 **Theory of Magnetism**
By K. Yosida

123 **Quantum Kinetics in Transport and Optics of Semiconductors**
By H. Haug and A.-P. Jauho

124 **Relaxations of Excited States and Photo-Induced Structural Phase Transitions**
Editor: K. Nasu

125 **Physics and Chemistry of Transition-Metal Oxides**
Editors: H. Fukuyama and N. Nagaosa

Springer and the environment

At Springer we firmly believe that an international science publisher has a special obligation to the environment, and our corporate policies consistently reflect this conviction.

We also expect our business partners – paper mills, printers, packaging manufacturers, etc. – to commit themselves to using materials and production processes that do not harm the environment. The paper in this book is made from low- or no-chlorine pulp and is acid free, in conformance with international standards for paper permanency.

Printing: Mercedesdruck, Berlin
Binding: Buchbinderei Lüderitz & Bauer, Berlin

OHIO UNIVERSITY LIBRARY

Please return this book as soon as you have finished with it. In order to avoid a fine it must be returned by the last date stamped be-